SCIENTIFIC FRONTIERS IN
DEVELOPMENTAL TOXICOLOGY
AND
RISK ASSESSMENT

Committee on Developmental Toxicology
Board on Environmental Studies and Toxicology
Commission on Life Sciences

National Research Council

NATIONAL ACADEMY PRESS
Washington, DC

NATIONAL ACADEMY PRESS 2101 Constitution Ave., N.W. Washington, DC 20418

NOTICE: The project that is the subject of this report was approved by the Governing Board of the National Research Council, whose members are drawn from the councils of the National Academy of Sciences, the National Academy of Engineering, and the Institute of Medicine. The members of the committee responsible for the report were chosen for their special competences and with regard for appropriate balance.

This project was supported by the U.S. Department of Defense; (DAMD 17-89-C9086 and DAMD 17-99-C9049), the U.S. Department of Health and Human Services; (PR-470644, PR-470645, 273-MH-901198, 273-MH-913119, P-0158-00-6-00, 0009554547, and 263-MD-635973), the U.S. Environmental Protection Agency; (CR-823615, R-827241-01), the U.S. Department of Veterans Affairs; (V101 P-1578), and the American Industrial Health Council. All contracts and awards were between the sponsoring agency or organization and the National Academy of Sciences. Any opinions, findings, conclusions, or recommendations expressed in this publication are those of the author(s) and do not necessarily reflect the view of the organizations or agencies that provided support for this project.

International Standard Book Number 0-309-07086-4
Library of Congress Catalog Card Number 00-106422

Additional copies of this report are available from:

National Academy Press
2101 Constitution Ave., NW
Box 285
Washington, DC 20055

800-624-6242
202-334-3313 (in the Washington metropolitan area)
http://www.nap.edu

Printed in the United States of America

THE NATIONAL ACADEMIES

National Academy of Sciences
National Academy of Engineering
Institute of Medicine
National Research Council

The **National Academy of Sciences** is a private, nonprofit, self-perpetuating society of distinguished scholars engaged in scientific and engineering research, dedicated to the furtherance of science and technology and to their use for the general welfare. Upon the authority of the charter granted to it by the Congress in 1863, the Academy has a mandate that requires it to advise the federal government on scientific and technical matters. Dr. Bruce M. Alberts is president of the National Academy of Sciences.

The **National Academy of Engineering** was established in 1964, under the charter of the National Academy of Sciences, as a parallel organization of outstanding engineers. It is autonomous in its administration and in the selection of its members, sharing with the National Academy of Sciences the responsibility for advising the federal government. The National Academy of Engineering also sponsors engineering programs aimed at meeting national needs, encourages education and research, and recognizes the superior achievements of engineers. Dr. William A. Wulf is president of the National Academy of Engineering.

The **Institute of Medicine** was established in 1970 by the National Academy of Sciences to secure the services of eminent members of appropriate professions in the examination of policy matters pertaining to the health of the public. The Institute acts under the responsibility given to the National Academy of Sciences by its congressional charter to be an adviser to the federal government and, upon its own initiative, to identify issues of medical care, research, and education. Dr. Kenneth I. Shine is president of the Institute of Medicine.

The **National Research Council** was organized by the National Academy of Sciences in 1916 to associate the broad community of science and technology with the Academy's purposes of furthering knowledge and advising the federal government. Functioning in accordance with general policies determined by the Academy, the Council has become the principal operating agency of both the National Academy of Sciences and the National Academy of Engineering in providing services to the government, the public, and the scientific and engineering communities. The Council is administered jointly by both Academies and the Institute of Medicine. Dr. Bruce M. Alberts and Dr. William A. Wulf are chairman and vice chairman, respectively, of the National Research Council.

Committee on Developmental Toxicology

Elaine M. Faustman *(Chair)*, University of Washington, Seattle, Washington
John C. Gerhart *(Vice Chair)*, University of California, Berkeley, California
Nigel A. Brown, St. George's Hospital Medical School, London, United Kingdom
George P. Daston, The Procter & Gamble Company, Cincinnati, Ohio
Mark C. Fishman, Massachusetts General Hospital and Harvard Medical School, Boston, Massachusetts
Joseph F. Holson, WIL Research Laboratories, Inc., Ashland, Ohio
Herman B.W.M. Koëter, Organisation for Economic Cooperation and Development, Paris, France
Anthony P. Mahowald, University of Chicago, Chicago, Illinois
Jeanne M. Manson, University of Pennsylvania, Phildelphia, Pennsylvania
Richard K. Miller, University of Rochester, Rochester, New York
Philip E. Mirkes, University of Washington, Seattle, Washington
Daniel W. Nebert, University of Cincinnati Medical Center, Cincinnati, Ohio
Drew M. Noden, Cornell University, Ithaca, New York
Virginia E. Papaioannou, Columbia University College of Physicians and Surgeons, New York, New York
Gary C. Schoenwolf, University of Utah, Salt Lake City, Utah
Frank Welsch, Chemical Industry Institute of Toxicology, Research Triangle Park, North Carolina
William B. Wood, University of Colorado, Boulder, Colorado

Consultant

Paul W.J. Peters, University of Utrecht, The Netherlands

Staff

Abigail E. Stack, Project Director
Ruth E. Crossgrove, Editor
Mirsada Karalic-Loncarevic, Information Specialist
Leah L. Probst, Senior Project Assistant
Emily L. Smail, Project Assistant

Sponsors

American Industrial Health Council
Centers for Disease Control and Prevention
U.S. Department of Defense
U.S. Environmental Protection Agency
U.S. Department of Veterans Affairs
National Center for Toxicological Research
National Institute of Environmental Health Sciences
National Institute of Child Health and Human Development
National Institute for Occupational Safety and Health

Other Reports of the Board on Environmental Studies and Toxicology

Copper in Drinking Water (2000)
Ecological Indicators for the Nation (2000)
Waste Incineration and Public Health (1999)
Hormonally Active Agents in the Environment (1999)
Research Priorities for Airborne Particulate Matter: I. Immediate Priorities and a Long-Range Research Portfolio (1998); II. Evaluating Research Progress and Updating the Portfolio (1999)
Ozone-Forming Potential of Reformulated Gasoline (1999)
Risk-Based Waste Classification in California (1999)
Arsenic in Drinking Water (1999)
Brucellosis in the Greater Yellowstone Area (1998)
The National Research Council's Committee on Toxicology: The First 50 Years (1997)
Toxicologic Assessment of the Army's Zinc Cadmium Sulfide Dispersion Tests (1997)
Carcinogens and Anticarcinogens in the Human Diet (1996)
Upstream: Salmon and Society in the Pacific Northwest (1996)
Science and the Endangered Species Act (1995)
Wetlands: Characteristics and Boundaries (1995)
Biologic Markers (5 reports, 1989-1995)
Review of EPA's Environmental Monitoring and Assessment Program (3 reports, 1994-1995)
Science and Judgment in Risk Assessment (1994)
Ranking Hazardous Waste Sites for Remedial Action (1994)
Pesticides in the Diets of Infants and Children (1993)
Issues in Risk Assessment (1993)
Setting Priorities for Land Conservation (1993)
Protecting Visibility in National Parks and Wilderness Areas (1993)
Dolphins and the Tuna Industry (1992)
Hazardous Materials on the Public Lands (1992)
Science and the National Parks (1992)
Animals as Sentinels of Environmental Health Hazards (1991)
Assessment of the U.S. Outer Continental Shelf Environmental Studies Program, Volumes I-IV (1991-1993)
Human Exposure Assessment for Airborne Pollutants (1991)
Monitoring Human Tissues for Toxic Substances (1991)
Rethinking the Ozone Problem in Urban and Regional Air Pollution (1991)
Decline of the Sea Turtles (1990)

Copies of these reports may be ordered from
the National Academy Press
(800) 624-6242
(202) 334-3313
www.nap.edu

Preface

Developmental defects are a significant human-health problem. Approximately 3% of human developmental defects are attributed to exposure to toxic chemicals (e.g., lead and mercury) and physical agents (e.g., radiation), including agents found in the environment. Twenty-five percent of developmental defects might be due to a combination of genetic and environmental factors, where those factors are defined broadly to include physical, chemical, and biological agents and conditions, such as infections, nutritional deficiencies and excesses, life-style factors (e.g., alcohol), hyperthermia, ultraviolet radiation, X-rays, and the myriad of manufactured chemicals (e.g., pharmaceuticals, synthetic chemicals, solvents, pesticides, fungicides, herbicides, cosmetics, and food additives) and natural materials (e.g., plant and animal toxins and products). Because of human-health concerns about the developmental toxicity of environmental agents, scientists and regulators have focused efforts on understanding and protecting against the potential hazards of these agents to developing embryos, fetuses, and children.

Recent advances in the fields of developmental biology and genomics provide opportunities to further understand the role of environmental agents in human developmental defects and, therefore, the National Research Council (NRC) undertook a project to explore the opportunities in this area. The first phase of the project consisted of a symposium entitled "New Approaches for Assessing the Etiology and Risks of Developmental Abnormalities from Chemical Exposure." The symposium was held December 11-12, 1995, in Washington, D.C. In the second phase, a multidisciplinary committee with expertise in developmental biology and developmental toxicology was convened by the NRC to prepare this consensus report.

In this report, the Committee on Developmental Toxicology evaluates current approaches used to assess risk for developmental defects and identifies key areas of uncertainty in those approaches. It also evaluates current understanding of the mechanisms of action of chemicals that result in developmental defects. The committee examines recent advances in developmental biology and genomics to highlight how new scientific information can be used to improve risk assessment for developmental toxicants and to elucidate the mechanisms by which toxicants induce developmental defects. Finally, the committee evaluates how the new information and technologies can be integrated into an overall risk-assessment framework.

The number of new discoveries made between late 1995, when the NRC symposium was held, and the beginning of 2000, when this report was completed, is staggering—especially in genomics, human genetics, transgenic mouse studies, and elucidation of signal transduction pathways of central importance to developmental biology and, by extrapolation, to developmental toxicology. The amount of additional information expected in the next 4 years likely will be even more explosive. Hence, it should be emphasized that this report represents a "snapshot in time" during a time of monumental advances in molecular biology and genetic research.

We would like to express our thanks and appreciation to Carole Kimmel, U.S. Environmental Protection Agency, who was instrumental in helping the NRC to initiate this project. The committee was generously assisted by the following people who presented valuable background information during the committee's public sessions: Carole Kimmel; Lewis Holmes, Harvard Medical School; Daniel Krewski, University of Ottawa; Andrew Olshan, University of North Carolina, Chapel Hill; James Ostell, National Center for Biotechnology Information; Allan Spradling, Carnegie Institute of Washington; and Robert Strausberg, National Cancer Institute. We gratefully acknowledge Barbara Abbott, U.S. Environmental Protection Agency, and Patricia Rodier, University of Rochester, who, at the committee's request, contributed information on the mechanism of action of 2,3,7,8-tetrachlorodibenzo-*p*-dioxin and on chemicals that may induce autism, respectively.

We also gratefully acknowledge Paul Peters from the University of Utrecht. Dr. Peters served as a consultant to the committee and provided valuable information on research being done in European countries to improve risk assessment for developmental defects.

The committee wishes to thank the American Industrial Health Council, the Centers for Disease Control and Prevention, the U.S. Department of Defense, the U.S. Environmental Protection Agency, the U.S. Department of Veterans Affairs, the National Center for Toxicological Research, the National Institute of Environmental Health Sciences, the National Institute of Child Health and Human Development, and the National Institute for Occupational Safety and Health for their interest and support of this project.

This report has been reviewed in draft form by individuals chosen for their diverse perspectives and technical expertise in accordance with procedures approved by the NRC's Report Review Committee for reviewing NRC and Institute of Medicine reports. The purpose of this independent review is to provide candid and critical comments that will assist the NRC in making the published report as sound as possible and to ensure that the report meets institutional standards for objectivity, evidence, and responsiveness to the study charge. The review comments and draft manuscripts remain confidential to protect the integrity of the deliberative process. We wish to thank the following individuals, who are neither officials nor employees of the NRC, for their participation in the review of this report: John DeSesso, Mitretek Systems; Barbara Hales, McGill University; Lewis Holmes, Harvard Medical School; John Moore, National Toxicology Program Center for the Evaluation of Risks to Human Reproduction; Gary Shaw, California Birth Defects Monitoring Program; Allan Spradling, Carnegie Institution of Washington; and Patrick Wier, Smithkline Beecham Pharmaceuticals. Donald Mattison, March of Dimes Birth Defects Foundation, served as review coordinator.

The individuals listed above have provided many constructive comments and suggestions. It must be emphasized, however, that responsibility for the final content of this report rests entirely with the authoring committee and the NRC.

We are grateful for the assistance of the NRC staff in the preparation of the report. Staff members who contributed to this effort are Warren Muir, executive director of the Commission on Life Sciences; James Reisa, director of the Board on Environmental Studies and Toxicology; Carol Maczka, director of BEST's Toxicology and Risk Assessment Program; Ruth Crossgrove, editor; Mirsada Karalic-Loncarevic, information specialist; Leah Probst, senior project assistant; and Emily Smail, project assistant. We are especially indebted to Abigail Stack, who served as project director. In this role, Dr. Stack served tirelessly, both with scientific and administrative support, maintaining the integrity of the report yet gently insisting on timely responses. We thank her for her excellent service in this challenging role.

Finally, we would like to thank all the members of the committee for their dedicated efforts throughout the development of this report.

We hope this report reflects the exciting deliberations of the committee that led to its genesis.

Elaine Faustman, Ph.D.
Chair, Committee on Developmental Toxicology

John Gerhart, Ph.D.
Vice Chair, Committee on Developmental Toxicology

Contents

List of Abbreviations

ACE	angiotensin-converting enzyme
ADH	alcohol dehydrogenase
ADI	acceptable daily intake
ADME	absorption, distribution, metabolism, and excretion
AER	apical ectodermal ridge
AHH	aryl hydrocarbon hydroxylase
AHR	aryl hydrocarbon receptor
ALDH	acetylaldehyde dehydrogenase
ARNT	aryl hydrocarbon receptor nuclear translocator
ASD	autism spectrum disorders
ATP	adenosine triphosphate
AUC	area under the curve
BLAST	Basic Local Alignment Search Tool
BMD	benchmark dose
CDC	Centers for Disease Control and Prevention
cDNA	complementary deoxyribonucleic acid
CGAP	Cancer Genome Anatomy Project
CHEST	chick embryotoxicity test
Cmax	peak threshold concentration
CNS	central nervous system
DBCP	1,2-dibromo-3-chloropropane
DDBJ	DNA Data Bank of Japan
DDT	dichlorodiphenyltrichloroethane
DES	diethylstilbesterol
Δ7-DHC	Δ7-dehydrocholesterol

DME	drug-metabolizing enzyme
DNA	deoxyribonucleic acid
DOE	U.S. Department of Energy
DPH	diphenylhydantoin
EBI	European Bioinformatics Institute
EBV	Epstein-Barr virus
ED_{05}	best estimate of a dose at a 5% level of response
EDSP	Endocrine Disruptor Screening Program
EGEE	ethylene glycol monoethylether
EGF	epidermal growth factor
EGME	ethylene glycol monomethyl ether
EGP	Environmental Genome Project
ELSI	ethical, legal, and social implications
EM	extensive metabolizers
EMBL	European Molecular Biology Library
EPA	U.S. Environmental Protection Agency
ER	endoplasmic reticulum
ES cell	embryonic stem cell
EST	expressed sequence tag
EtO	ethylene oxide
F_1	first filial generation
FAE	fetal alcohol effects
FAK	focal adhesion kinases
FAS	fetal alcohol syndrome
FDA	U.S. Food and Drug Administration
FETAX	frog embryo teratogenesis assay-*Xenopus*
FGF	fibroblast growth factor
FOB	functional observational battery
FQPA	Food Quality Protection Act
GFP	green fluorescent protein
GMS	genomic mismatch scanning
GR	glucocorticoid receptor
GRE	glucocorticoid response element
GSDB	Genome Sequence Database
hCMV	human cytomegalovirus
HEPM	human embryonic palatal mesenchymal
HGP	Human Genome Project
HOX	homeobox
Hsp	heat-shock promoter
HSV-1	herpes simplex virus type 1
ICBD	International Clearinghouse for Birth Defects
IPCS	International Programme on Chemical Safety
IRIS	Integrated Risk Information System

JNK	c-Jun terminal kinase
LANL	Los Alamos National Laboratory
LCR-MT	locus control regions of the metallothionein gene
LOAEL	lowest-observed-adverse-effect level
Mb	megabases
MeHg	methylmercury
mm	millimeter
MMTV LTR	mouse mammary tumor virus long-terminal repeat
MOE	margin of exposure
MOT	mouse ovarian tumor
MRI	magnetic resonance imaging
mRNA	messenger RNA
MTD	maximum tolerated dose
NAREP	North American Registry for Epilepsy and Pregnancy
NCBI	National Center for Biotechnology Information
NCGR	National Center for Genome Resources
NCI	National Cancer Institute
NCTR	National Center for Toxicological Research
NHGRI	National Human Genome Research Institute
NIEHS	National Institute of Environmental Health Sciences
NIGMS	National Institute of General Medical Sciences
NIH	National Institutes of Health
NMDA	*n*-methyl-D-aspartate
NOAEL	no-observed-adverse-effect level
NRC	National Research Council
OECD	Organization for Economic Cooperation and Development
ORF	open reading frame
PAH	polycyclic aromatic hydrocarbon
PARP	poly(ADP-ribose)polymerase
PCR	polymerase chain reaction
PUBS	percutaneous umbilical blood sampling
PZ	progress zone
RA	retinoic acid
Raldh2	retinaldehyde dehydrogenase-2
RAR	retinoic acid receptor
RAREs	retinoic acid response elements
RfC	reference concentration
RfD	reference dose
RNA	ribonucleic acid
RNAi	RNA-mediated gene interference
RT	reverse transcription
RT-PCR	reverse transcription polymerase chain reaction
RTECS	Registry of Toxic Effects of Chemical Substances

RTK	receptor tyrosine kinase
SAR	structure-activity relationship
SHH	Sonic Hedgehog
SNP	single-nucleotide polymorphism
SOT	Society of Toxicology
SV40	simian virus 40
2,4,5-T	2,4,5-trichlorophenoxyacetic acid
TCDD	2,3,7,8-tetrachlorodibenzo-*p*-dioxin
TEF	toxicity equivalency factor
TGF	transforming growth factor
Tk	thymidine kinase
TSCA	Toxic Substances Control Act
UM	ultra-metabolizers
VPA	valproic acid
Wnt	Wingless-Int
YPLL	years of potential life lost
ZPA	zone of polarizing activity

SCIENTIFIC FRONTIERS IN

DEVELOPMENTAL TOXICOLOGY

AND

RISK ASSESSMENT

Executive Summary

Of approximately 4 million births per year in the United States, major developmental defects are identified in approximately 120,000 live-born infants. Major defects are defined as ones that are life threatening, require major surgery, or present a significant disability. The most frequently recognized class of developmental defects are the structural abnormalities (e.g., neural tube and heart defects), which represent the majority of the developmental defects identified at birth. Other manifestations of abnormal development include growth retardation (e.g., low birth weight), functional deficits (e.g., mental retardation), and pre- and postnatal death (including early pregnancy losses). Because of differences in definition, detection, and reporting practices, the actual frequency of developmental defects is not known with certainty.

At present, the causes of the majority of developmental defects are not understood. It is known that prenatal exposure to some chemicals (e.g., mercury, lead, and polychlorinated biphenyls) and physical agents (e.g., radiation) found in the environment can cause developmental defects. Scientists generally agree that approximately 3% of all developmental defects are attributable to exposure to toxic chemicals and physical agents, including environmental factors, and that 25% of all developmental defects may be due to a combination of genetic and environmental factors. These environmental factors include infection, nutritional deficiencies and excesses, life-style factors (e.g., alcohol), hyperthermia, ultraviolet radiation, X-rays, and closer to the concerns of this committee, the myriad of manufactured and natural agents encountered by humans.

THE CHARGE TO THE COMMITTEE

The National Research Council (NRC) undertook the study leading to this report to clarify how environmental agents may be impacting human development by using the scientific knowledge gained from major advances in developmental and molecular biology over the past 10-15 years. This study was undertaken with a public health goal of understanding mechanisms of developmental defects to improve our preventive actions. The Committee on Developmental Toxicology was formed to evaluate the current understanding of the mechanisms of action of developmental toxicants and to make recommendations for the improvement of developmental toxicity risk assessment. The specific tasks of the committee were as follows: (1) evaluate the evidence supporting hypothesized mechanisms of developmental toxicity; (2) evaluate the state of the science on testing for mechanisms of developmental effects; (3) evaluate how that information can be used to improve qualitative and quantitative risk assessment for developmental effects; and (4) develop recommendations for future research in developmental toxicology and developmental biology; focus on those areas most likely to assist in assessing risk for developmental defects.

COMMITTEE'S APPROACH TO ITS CHARGE

The project was conducted in two phases. The first phase consisted of a symposium entitled "New Approaches for Assessing the Etiology and Risks of Developmental Abnormalities from Chemical Exposure," which was held December 11-12, 1995, in Washington, D.C. The proceedings from that symposium were published in *Reproductive Toxicology*[1] and were used as background information for the second phase of the project. In the second phase, a multidisciplinary committee with expertise in developmental biology and developmental toxicology was asked to address the tasks described above.

In this report, the committee documents many recent advances in research in the areas of developmental biology and genomics. These extraordinary advances are significant for developmental toxicology and risk assessment because they present opportunities to improve substantially the detection of developmental toxicants and to elucidate the mechanisms by which toxicants induce developmental defects. The committee makes recommendations for incorporating the new scientific information with existing experimental methods to improve the understanding of the role of environmental agents in human developmental disorders.

In approaching its charge, the committee evaluated current methods used to assess risk for developmental defects. Specifically, the committee reviewed the types of data commonly used to evaluate chemicals for potential developmental toxicity and explored the limitations of the risk assessment process. The limitations include the lack of information on the mechanisms of action of chemicals

---

[1] Symposium Proceedings. 1997. New Approaches for Assessing the Etiology and Risks of Developmental Abnormalities from Chemical Exposure. Reprod. Toxicol. 11(2/3):261-463.

and the uncertainties associated with the extrapolation of data among humans due to the variability in their susceptibility to chemicals, and with the extrapolation of data from animals to humans. The committee attempted to determine whether those limitations could be addressed by recent advances in the understanding of normal development, gene-environment interactions, and human susceptibility. In particular, the committee evaluated new developmental biology data from model animals (e.g., fruit fly, roundworm, zebrafish, and mouse), including genetically modified model animals, and from new molecular biology approaches utilizing in vitro and cellular assays. It developed approaches to show how such new information could improve hazard identification and dose-response assessment and clarify the mechanisms of developmental toxicity. The committee also evaluated data on new technologies for assessing human variability in genes involved in developmental processes and the metabolism of chemicals and determined whether the new technologies could improve risk characterization by reducing uncertainty and variability. Finally, the committee evaluated how this information could be integrated into an overall risk-assessment framework. The committee's major conclusions and recommendations, organized in response to each of the committee's tasks, are discussed in the remainder of this summary.

CONCLUSIONS IN RELATION TO THE CHARGE

Charge 1: Evaluate the Evidence Supporting Hypothesized Mechanisms of Developmental Toxicity. There are only a few compounds (e.g., retinoic acid, diethylstilbesterol (DES), and 2,3,7,8-tetrachlorodibenzo-*p*-dioxin (TCDD)) for which the mechanism of developmental toxicity is partially explained and no compound for which it is fully explained. Reasons for this incomplete understanding include the lack of knowledge about normal developmental processes, the complexity of developmental toxicity, the broad spectrum of agents and chemical mixtures present in the environment, and the variety of potential mechanisms by which they might cause toxicity.

Ideally, a full description of the mechanism of action by which a chemical causes developmental toxicity includes the following types of mechanistic information:

(1) the chemical's toxicokinetics (i.e., its absorption, distribution, metabolism, and excretion) within the mother, fetus, and embryo;

(2) the chemical's toxicodynamics (i.e., how the chemical or a metabolite derived from it interacts with specific molecular components of developmental processes in the embryo and fetus or with maternal or extraembryonic components of processes supporting development);

(3) the consequences of those interactions on cellular or developmental processes (also part of toxicodynamics); and

(4) the consequence of the altered process for a developmental outcome, namely, the generation of a defect.

Research has been conducted on the toxicokinetics of some toxicants (e.g., retinoic acid, diphenylhydantoin, methotrexate, and methylmercury). For certain toxicants, routes and rates of exposure to the fetus and embryo have been identified, as have the presence of parent compounds and metabolites in the mother, fetus, and embryo. Additionally, the role of drug-metabolizing enzymes (DMEs) in the metabolism of toxicants has been studied extensively. However, knowledge about critical metabolites and their reactivity with specific target tissues is lacking for most environmental agents.

Some toxicants (e.g., retinoids, DES, and TCDD) are known to act on molecular components that function as signaling proteins and transcriptional regulators (described below). However, the committee found that little information is available on how most chemicals impact those molecular components. Where information is available, it is generally sparse and does not allow for the association of developmental defects with a toxicant's action on specific molecular components of developmental processes.

Charge 2: Evaluate the State of the Science on Testing for Mechanisms of Developmental Effects. Major discoveries have recently been made about the components, mechanisms, and processes of normal development. Developmental processes have been identified at the molecular level in various model animals, including the fruit fly, the roundworm, the zebrafish, the frog, the chick, and the mouse. Molecular components of these processes are substantially conserved (i.e., the structure and function of the components have not changed throughout evolution) among animal phyla, including mammals; they regulate development by signaling specific cells to activate proteins called transcription regulators, which turn specific genes on and off. Seventeen signaling pathways are currently recognized, and probably only a few more remain to be discovered. These conserved pathways are used repeatedly in various combinations at different times and locations in the developing embryo and fetus. Species differences in development involve different times, locations, and combinations of these pathways. Many of the kinds of cell responses to signals also are conserved, including selective gene expression, secretion, cell proliferation, and cell migration.

The sequencing of the human genome and a variety of animal genomes is providing fundamental information about genome organization, genome evolution, and genetic polymorphisms (variations in the deoxyribonucleic acid (DNA) sequence of a particular gene within a population of organisms). Identifying polymorphisms in the human genome might provide opportunities to increase the understanding of genotype-environment interactions and human susceptibility to toxicants. For example, recent insights into the human differences in the activity of various DMEs and the genetic basis for those differences offer a promising direction for research. Components involved in developmental processes such as the signaling pathways that might be important in susceptibility are less well studied.

Analyses of human and model animal gene sequences will increase the understanding of gene function and gene expression. For example, methods are available that can be used to determine the changes in gene expression in embryos following exposure to toxicants and to allow for assessment of the consequences of such changes for development within a species and among various species.

Charge 3: Evaluate How Recent Advances in Developmental Biology and Genomics Can Be Used to Improve Qualitative and Quantitative Risk Assessment for Developmental Effects. The committee concludes that the major recent advances in developmental biology and genomics can be used to improve qualitative and quantitative risk assessments by integrating toxicological and mechanistic data on a variety of model test animals with data on human variability in genes encoding components of developmental processes, genes encoding enzymes involved in the metabolism of chemicals, and genes encoding receptors and transporter proteins that move these chemicals and their metabolites in and out of the cell.

For example, as described in Chapter 7 of this report, chemicals could be evaluated for their potential to alter signaling pathways central to normal development by using nonmammalian model animal systems, such as the fruit fly, roundworm, and zebrafish. Those systems are inexpensive, and the assays can be performed rapidly; therefore, large numbers of chemicals, chemical mixtures, and testing conditions could be evaluated for impacts on these key developmental processes. Such mechanistic information from those systems could be used to improve the identification of potential mammalian hazards, because molecular components and processes of development are well understood in those model animals and because the conservation of signaling pathway components is pervasive and extends to humans. The nonmammalian systems, and the laboratory mouse, can be genetically modified to facilitate the identification of vulnerable developmental pathways, target organs, and times of susceptibility during development.

In addition, the model animals can be assessed to define their differences from humans in toxicokinetic and toxicodynamic properties. They also can be genetically modified to contain human metabolism genes, which will reduce the toxicokinetic differences between experimental animals and humans. Such information could be used to improve extrapolation of toxicological data from model animals to humans.

Individual human susceptibility to toxicants and genotype-environment interactions could be explored using sequence information from genes encoding DMEs and molecular components, such as components of signaling pathways, involved in development. This information would improve the understanding of human variability in metabolism and the identification of genes encoding molecular components that might be particularly susceptible to chemicals during development.

Charge 4: Develop Recommendations for Research in Developmental Toxicology and Developmental Biology; Focus on Those Areas Most Likely to Assist in Risk Assessment for Developmental Effects. The committee recommends a multilevel, multidisciplinary approach to risk assessment that incorporates information from a range of model systems intended to

(1) evaluate chemicals for potential developmental toxicity;
(2) provide mechanistic information on toxicants;
(3) address several key areas of uncertainty about the relevancy of cross species extrapolation of toxicological information from animals to humans; and
(4) further the exploration of the genotype-environment interactions that might underlie a large fraction of developmental defects and could help to explain human variability in response to environmental agents.

This novel approach should provide a guide for obtaining the kinds of data that are needed for a comprehensive cross-species model of exposure and development. Specifically, as described in Chapter 9 of this report, the committee recommends that research be conducted in the following areas.

- *Greater use of model systems for developmental toxicity and risk assessment.* Model systems should be used to assess and understand chemical effects (or absence of effects). This recommendation is based on the conclusion that model-animal research has been highly informative about mammalian development, especially human development, and therefore, is likely to be informative about mammalian developmental toxicity. The model systems that should be considered include in-vitro and cellular assays, nonmammalian (e.g., fruit fly, roundworm, and zebrafish) developmental assays, mammalian (e.g., the mouse) developmental assays, and in-depth mammalian tests of mechanism and susceptibility.
- *Evaluation of chemicals for developmental toxicity.* In-vitro and cellular assays and nonmammalian tests should be used for evaluating chemicals and chemical mixtures so that patterns of toxicity can be more readily recognized. The number of chemicals in commerce is rapidly expanding, and it is a continuing challenge to obtain toxicity data on them. These model systems could be used quickly and are inexpensive, and their use would permit a large number of chemicals and doses to be evaluated for their potential impact on many key developmental processes.
- *Analysis of mechanisms of toxicity.* Mechanistic information is essential to our understanding of how chemicals can perturb development and, thus, is an important component of risk evaluation. To improve the understanding of the mechanisms of action of toxicants, critical molecular targets of components of developmental processes should be identified. Potential critical molecular targets that should be further investigated include (1) evolutionarily conserved pathways of development, such as intercellular signaling pathways (including their

associated transcriptional regulators); (2) conserved molecular-stress and checkpoint pathways; and (3) conserved toxicokinetic components such as those involved in the transport and metabolism of toxicants (e.g., DMEs). It is important to explore how such molecular perturbations can result in altered function and adverse outcomes of development. Model animals such as the fruit fly, roundworm, and zebrafish can be used to study the mechanisms of developmental toxicity. The signaling pathways that operate in the development of the organs of these organisms also operate in the development of mammalian organs; therefore, the effects of chemicals on fundamental processes such as signaling can be detected. Because the same signaling pathways operating in various kinds of organ development in mammals are partially known and will be better known soon, a chemical's toxicological impact on these pathways can be predicted on the basis of the results in nonmammalian organisms and tested in mammals. Molecular-stress and checkpoint pathways are used by cells to counteract damage to basic cellular functions, including functions involved in development, and investigating these pathways is important to understand the broad responses of cells to environmental stimuli. Multiple pathways are used in the development of organs; however, one pathway at a time can be studied for a specific aspect of development (e.g., the development of a particular organ) by using genetically modified (e.g., sensitized) animals.

- *Human variability of response to developmental toxicants.* To define the genetic basis of variability in human response to developmental toxicants, differences in toxicokinetics, signaling pathways, and molecular-stress and checkpoint pathways need to be characterized. Two approaches to studying variability are recommended: (1) a human epidemiological approach making use of genome information, and (2) a model-animal approach making use of molecular biological techniques and insights. Research should be conducted to assess differences in the genes encoding molecular components among various species, including humans, and among human individuals. As human gene polymorphisms are identified, they should be introduced into the mouse, and molecular biological techniques should be used to assess the organisms' sensitivity or resistance to various chemicals.
- *Assaying across the entire developmental period.* All periods of development are susceptible to the actions of toxicants. For example, early fetal loss in human development occurs in 20-30% of initial pregnancies and, although many of these losses are due to chromosomal aberrations, exposure to a toxicant during early times in development can lead to loss of the embryo or fetus as well as specific structural defects and functional deficits. Use of genetically modified model systems could provide mechanistic information to improve the understanding of early fetal loss as well as morphological alterations and later functional deficits by providing sensitized systems for evaluating developmental defects.
- *Extrapolation from animals to humans.* Differences in toxicokinetics and toxicodynamics of experimental animals and humans should be better character-

ized to improve extrapolations from animals to humans. For example, the study of differences in DMEs between humans and experimental animals will improve the ability to extrapolate from animal test results to humans, because it will be known whether the animal embryo or fetus and the human embryo or fetus are exposed to a chemical at corresponding concentrations and times during development. Also, when the differences between animal and human DME activities are understood, mice can be genetically modified to make them more similar to humans in chemical metabolism. Studies on developmental components, such as signaling components and transcriptional regulators, that are similar to those discussed here for DMEs also should be conducted. Sequence information from the human and mouse genomes will facilitate these studies, as will studies on mice bearing targeted gene alterations.

- *Extrapolations from high to low doses.* Because exposure to a chemical at high doses might affect a variety of developmental processes, while exposure at low doses might affect only one critically sensitive pathway, studies using model test animals should be conducted to distinguish dose effects and, in particular, to distinguish effects that could potentially occur at exposure levels relevant to humans. Because a large number of chemicals cause apoptosis (cell death) in the embryo and fetus, the molecular-stress and checkpoint pathways should be given particular attention. Studies using sensitized model animals should be especially useful for defining low-dose responses.
- *Improved access to information.* To support the growth of knowledge in developmental toxicology and to organize information in a way that is useful for risk assessment, an inclusive national developmental toxicant database should be established, with entries from industry, academia, and government. The developmental toxicant database should include chemical toxicant information as well as information on known molecular targets and associations with developmental defects, both from animal tests and from humans. Steps should be taken to link this database with the databases of developmental biology (e.g., the database of phenotypes of mice with mutations in their signaling components, which are being generated by genetic modification techniques), and genomics. Databases describing metabolic pathways for drugs and environmental agents, and DME and transporter protein polymorphisms should be linked as well. Ideally, a separate relational database in which signaling pathways are grouped should be established and used when chemicals are identified as interacting with an element of the pathway. This relational database could help to suggest potential biological interactions of a chemical with other chemicals that affect components of the same pathway and record the involvement of signaling pathways in all aspects of development from a wide range of organisms.
- *Multidisciplinary outreach.* The challenges that investigators face when trying to work across fields, such as developmental biology, developmental toxicology, and risk assessment, are a key issue that the committee identified early in its deliberations. This issue previously impeded the successful application of the

new scientific information to improve developmental toxicity risk assessment. For the successful application of this report's findings, the committee believes that multidisciplinary educational and research programs must be conducted. Programs, such as workshops and professional meetings, should be organized so that researchers of developmental toxicology, developmental biology, genomics, medical genetics, epidemiology, and biostatistics can come together to exchange new insights, approaches, and techniques related to the analysis of developmental defects and to risk assessment. By accelerating the necessary research, cooperative research projects would move forward the recommendations of this report.

1

Introduction

Between 2% and 3% of all live-born infants are estimated to have a major developmental defect identified at birth (ICBD 1991; CDC 1995; Holmes 1997; March of Dimes 1999). The percentage increases substantially when all developmental defects—including nonstructural defects such as neurological and behavior problems that often are not detected until childhood or even adulthood—are considered. A developmental defect is defined as a structural or functional anomaly that results from an alteration in normal development.

The causes of most developmental defects are unknown. However, it is known that exposure to chemicals can result in developmental defects. In all, about 3% of developmental defects are attributable to an exposure of the mother to chemicals and physical agents, including environmental agents. A much larger fraction, perhaps 25%, are thought to be due to multifactorial causes resulting from the exposure of genetically predisposed individuals to environmental factors (e.g., infections, nutritional deficiencies and excesses, hyperthermia, ultraviolet radiation, X-rays, and manufactured and natural chemicals). There is concern that greater than 3% of developmental defects may be due to exposures to chemicals and physical agents. One reason for this concern is that only a fraction of the 60,000 to 90,000 chemicals in commercial use have been evaluated for their potential to cause developmental toxicity. Human-health concerns about environmental agents require that scientists and regulators attempt to understand and protect against the potential hazards of those agents on developing embryos, fetuses, and children.

In this committee's context of addressing the consequences of human prenatal exposure to environmental toxicants, a more inclusive and accurate term is "developmental defect" rather than "birth defect." The committee will use "developmental defect" throughout this report, because it includes the full range of kinds and severity of defects and the full range of times of detection—before, at, and after birth.

In recognition of the opportunity to use recent advances in developmental biology and genomics to elucidate further the role of environmental agents in human developmental defects, the NRC approved a project to evaluate the current understanding of the mechanism of action of toxicants that results in developmental defects, and make recommendations for the improvement of toxicant evaluation, ultimately in ways that would improve risk assessment. The specific tasks of the committee were as follows: evaluate the evidence supporting hypothesized mechanisms of developmental toxicity; evaluate the state of the science on testing for mechanisms of developmental effects; evaluate how that information can be used to improve qualitative and quantitative risk assessment for developmental effects; and develop recommendations for future research in developmental toxicology and developmental biology; focusing on those areas most likely to assist in risk assessment for developmental defects.

BACKGROUND

Awareness of developmental toxicants increased greatly in the early 1960s when the detrimental effect of thalidomide (used at that time as a sedative/hypnotic) primarily on human limb development was recognized (thalidomide causes other developmental defects as well). Before that time, various chemicals had been tested on adult animals but only intermittently on pregnant animals, and it was generally accepted that what was then thought of as the placental barrier protected the fetus from foreign agents. Since the recognition of prenatal vulnerability in the early 1960s, much has been done to detect potential developmental toxicants in the environment and to regulate human exposure to them. Adverse developmental effects of toxicants now are recognized to include not only malformations at birth but also growth retardation, death (including embryonic and fetal loss), and functional defects in the newborn. Over 1,200 specific compounds, pathogens, and conditions have been identified in experimental animals as causing adverse developmental effects, and the impact of human exposure to many of these agents is not understood (Shepard 1998).

Since the 1960s, the science of developmental toxicology—that is, the study of the impact of toxicants on critical processes of normal development—has advanced. The science of risk assessment of chemical effects on humans, which depends on the advances in toxicology, has also advanced. To predict risk, assessors rely primarily on two kinds of information: estimates of the level of human exposure to a particular chemical, and estimates of the chemical's toxicity for humans based on the developmental outcome of offspring from experimental pregnant animals exposed to that chemical. Occasionally, information is available from other sources, such as (1) structure-activity relationships relating the toxicity of a chemical to other members of its chemical family; (2) the results of in vitro tests of the chemical; and (3) human epidemiological observations about the effect of the chemical.

To make their evaluations of risk, assessors seek accurate mechanism-based empirical data—that is, data based on a solid understanding of the mechanism of a chemical's toxicity, as determined by developmental toxicologists. Such data are sparse. Several uncertainties limit the estimation of a chemical's potential for developmental toxicity. Animal bioassays, principally using mammals, currently are considered to provide the most reliable data for extrapolating toxicant effects to humans. Because these bioassays are expensive and time consuming, only a small fraction of the compounds in commerce and in the environment have been fully evaluated for their toxicity potential in animals. The many attempts to devise simpler, less costly test systems involving tissue explants, cell cultures, or purified biological molecules have so far proved to be of only limited value in predicting the actions of compounds on human embryonic and fetal development. Among the reasons for poor predictability are the inherent differences between the simple test systems and humans regarding the uptake, distribution, metabolism, and excretion of chemicals and the lack of understanding about the basic mechanisms of development.

In the absence of accurate mechanism-based empirical data, risk assessors often make four kinds of default assumptions when recommending the acceptable levels of exposure of humans to an environmental agent. First, they assume that animal test results are relevant for humans. Unless there is contradictory evidence, humans are assumed to be the most susceptible mammals, and a factor of 10 below the maximum no-effect exposure level in the animal's development serves as a basis for setting the acceptable human exposure level. Second, a further 10-fold reduction is introduced to take into account the possibility that the animal's developmental response, which frequently was obtained at subchronic exposure to the chemical, might not reflect human responses at prolonged (chronic) exposures. Third, a 10-fold reduction is introduced to cover the possibility that susceptibility varies among human individuals, some being inherently more sensitive to the chemical. Fourth, a 10-fold reduction is sometimes introduced if the toxicity database for a chemical is incomplete. Because of the susceptibility of developing systems, an additional child specific factor (usually a 10-fold reduction) is sometimes applied. Although many risk assessors would prefer to use mechanism-based empirical data instead of those defaults (up to a 10,000-fold compounded reduction in acceptable human exposure beyond that given by the animal test) to improve their risk assessments for environmental agents, the test data for the assessors' use often are sparse because of limited resources and are of unknown applicability to humans because of a lack of understanding of basic mechanisms of developmental toxicity and of differences in humans and animals.

Thus, risk assessors are challenged by the problems of extrapolation, interpretation, cost, and speed. Considering the large number of environmental chemicals (both manufactured and naturally occurring chemicals) that are not adequately tested for potential developmental toxicity, scientists have been asked to develop testing approaches that are based on our rapidly expanding knowledge of normal development to provide more timely information with improved predictions for human developmental outcomes. These issues are ongoing challenges in the effort to assess human risk from environmental toxicants.

RECENT ADVANCES IN DEVELOPMENTAL BIOLOGY AND THE PROMISE OF GENOMICS

Developmental biology is the study of normal developmental processes. It begins with descriptions of the sequential events of development, from the formation of the oocyte (the egg precursor) and sperm, to fertilization, then to cell division, morphogenesis (the transformation of egg organization into embryonic organization), organogenesis (the formation of organs), cell differentiation, and embryonic and fetal growth. In its full scope, developmental biology covers the development and growth of the infant, child, and adolescent to the time of reproductive maturity. Developmental biology also describes events in the organism's spatial dimension (the changing number and position of cells, tissues, and organs) in the vast multicellular population of the embryo and fetus (approximately 1 trillion cells in a newborn infant).

In the past 15 years, remarkable advances have been made in the knowledge of the components, mechanisms, and processes of normal development, primarily as the result of new insights into the molecular biology of development. Developmental biology has become a study of the mechanisms of development at the molecular level, particularly of the interaction of components of intracellular genetic regulatory circuits with components of intercelluar signaling pathways. To cite a few of those insights, it is now known that the trillions of cells of a large mammal such as a human have the same genetic composition (genetic blueprint). As recently reaffirmed by the cloning of Dolly the lamb (Wilmut et al. 1997), the Cumulina mouse family (Wakayama et al. 1998), and a nonhuman primate (Chan et al. 2000), the genetic content of almost all of the cells in an animal does not change from that of the single-celled fertilized egg from which it developed. Despite having the same genes, the cells in an animal differ widely in their appearance, functions, and responses to environmental impacts. At least 300 cell types are recognized in humans (e.g., red blood cells, Purkinje nerve cells, and smooth or striated muscle cells), and the number of cell subtypes at different stages of development and different parts of the body is perhaps tens of thousands. These cell types differ greatly in their ribonucleic acids (RNA) and proteins, reflecting the different combinations of genes they express from the same genomic repertoire). Development can be viewed as evolution's foremost accomplishment in gene regulation, entailing a complex orchestration of which cells will express which genes when and where in the embryo and fetus. Two major elements in that regulation are (1) a large variety of specific transcription factors that act in an even larger variety of combinations to control differential gene expression; and (2) the chemical communication between cells during development that allows cells to turn specific genes on and off in response to signals from their neighbors.

The following is now realized:

- Embryonic and fetal development involves repeated signaling among groups of cells, and the expression of particular genes in a cell depends on signaling inputs from other cells in the local environment.
- The number of signaling pathways used in development is limited. About 17 signaling pathways are now recognized, and probably only a few more remain to be discovered. Each pathway consists of an intercellular chemical signal, a specific receptor on or within the cell, and a set of molecular transducers that transmit each signal to targets, such as to components of the transcription machinery, within the cell. These 17 pathways are used repeatedly at different times and places in the developing embryo and fetus. The roles of these pathways in development are a major focus of current research in developmental biology.
- Surprisingly, given the morphological diversity of animal embryos and fetuses, the 17 signaling pathways are highly conserved across numerous animal phyla (e.g., nematode worms to arthropods to chordates). The molecular targets and responses within cells also are conserved across phyla, including specific gene expression, cell migration, and cell proliferation. Those signaling and responding aspects of development presumably were already present in the pre-Cambrian common ancestor of animals of modern phyla as diverse as the chordates (including humans), the arthropods (including fruit flies), and the nematodes. The differences in the development of various organisms mostly reflect differences in the particular times, places, and combinations of use of the conserved pathways and responses.

Those findings give new validity to the use of model organisms to learn more about basic development in mammals, to provide mechanistic clues about human variability, and to analyze and assess the risks of potential developmental toxicants.

With the transformation of developmental biology in the past decade, DNA sequence data from a variety of organisms have accumulated at an explosive rate. The large-scale projects initiated under the Human Genome Project include the complete sequencing of the genomes of several widely used model organisms, such as yeast, the nematode *Caenorhabditis elegans*, the fruit fly *Drosophila melanogaster*, the laboratory mouse, and humans. The sequencing of the yeast, *C. elegans*, and *Drosophilia* genomes has already been completed (Goffeau et al.

1996; *C. elegans* Sequencing Consortium 1998; Adams et al. 2000). The *C. elegans* genome was the first metazoan genome sequenced. The mouse and human genomes should be completed in 3 years, perhaps sooner. Out of these efforts, the field of genomics has emerged. It includes the identification of all genes of an organism, all RNA transcripts of those genes, all the rules for the time, place, and conditions of expression of those genes, and the sequence variants within the population of that organism. Proteomics is the study of all the proteins expressed from all RNA transcripts. The promise of genomics and proteomics is great, because all studies of physiological function, developmental change, and evolutionary diversification will draw upon it.

As DNA sequence data from various organisms become available, the need to manage and analyze vast amounts of sequence data will increase. That need has spawned a new field of science called bioinformatics. The ability of scientists to make use of genomic databases will become increasingly important for coordinating developmental biology, developmental toxicology, and genomics.

COMMITTEE'S APPROACH TO ITS CHARGE

The project was conducted in two phases. The first phase consisted of a symposium entitled "New Approaches for Assessing the Etiology and Risks of Developmental Abnormalities from Chemical Exposure," which was held December 11-12, 1995 in Washington, D.C. The proceedings from that symposium were published in *Reproductive Toxicology* (Kimmel et al. 1997) and were used as background information for the second phase of the project in which a multidisciplinary committee with expertise in developmental biology and developmental toxicology was asked to develop a consensus report (this report) that evaluates recent revolutionary advances in the understanding of normal development and gene-environment interactions and in the technology connected to the Human Genome Project and assesses whether these advances provide opportunities for innovation in developmental toxicology and risk assessment. In its report, the committee attempts to make broad-based interdisciplinary proposals by drawing on information from several fields of science—developmental toxicology, developmental biology, molecular biology, epidemiology, and genetics—all of which impinge on the understanding of the action of developmental toxicants.

ORGANIZATION OF THE REPORT

This report is organized into eight chapters in addition to this Introduction.

Chapter 2 describes the type and frequency of developmental defects in more detail, the problems of collecting accurate data on defects, and the general understanding of possible intrinsic and extrinsic causes.

Chapter 3 describes the current methods of risk assessment for the evaluation of developmental toxicity and the uncertainties in this assessment that make it necessary for assessors to introduce large default corrections in estimating allowable exposure levels.

Chapter 4 describes the history and current status of developmental toxicology, summarizing the attempts to identify mechanisms of action of toxicants, and the mechanisms of presentation of active toxicants to the embryo and fetus, detailing a few examples of well-understood developmental toxicants. The committee concludes that although much progress has been made in the analysis of toxicant action, much more remains to be done, probably facilitated by the recent advances in developmental biology and genomics.

Chapter 5 describes the fields of human genetics and genomics, including the role of molecular epidemiology in toxicant detection and the difficulties in the detection of complex genotype-environment interactions. The committee concludes that powerful new comprehensive methods from genomics will be of great value in developmental toxicology. Such a method is the newly found capacity to detect human genetic variation.

Chapter 6 describes the history of developmental biology and recent advances in that field, stressing the central role of cell-to-cell signaling in development and the repeated use of a small number of signaling pathways at different times and places in development. The committee concludes that the evolutionary conservation of these pathways, of a variety of genetic regulatory circuits and molecular-stress and checkpoint pathways, and of numerous other cellular activities makes it likely that informed use of model organisms to detect and analyze toxicant action will be valuable.

Chapter 7 discusses new approaches for using model organisms to test chemicals for developmental toxicity, stressing the value of using those organisms for which development is well understood and for which genetic manipulation can be performed to optimize their usefulness.

Chapter 8 outlines a novel multilevel, multidisciplinary approach to improve understanding of the mechanisms of action of toxicants and to improve developmental toxicity risk assessment by applying the recent advances in developmental biology and genomics. The capacity to understand organismal differences in development and toxicant metabolism is now possible, and the importance of that information for extrapolations of animal data to humans is emphasized.

The final chapter, Chapter 9, summarizes the committee's conclusions and recommendations. Here it is emphasized that the recent advances in developmental biology and genomics create an opportunity for improved detection and analysis of toxicants and for a better understanding of the meaning of assay results for risk assessment.

Four appendixes are included in the report. Appendix A contains a glossary of definitions of key terms used throughout the report. Appendix B contains descriptions of protein and genomic databases that can be useful to developmental toxicologists. Appendix C contains figures of the 17 known signal transduction pathways. Finally, Appendix D contains bibliographic information on the Committee on Developmental Toxicology.

2

Developmental Defects and Their Causes

Major developmental defects, also referred to as major congenital anomalies, occur in approximately 3% of live births, that is, in 120,000 of the approximately 4 million births per year in the United States (ICBD 1991; CDC 1995; Holmes 1997; March of Dimes 1999; NCHS 1998). These anomalies are defined as ones that are life threatening, require major surgery, or present a significant disability (Marden et al. 1964).

In 1995, major developmental defects accounted for approximately 70% of neonatal deaths (occurring before 1 month of age) and 22% of the 6,500 deaths of infants (before 15 months of age) in the United States (March of Dimes 1999). Approximately 30% of admissions to pediatric hospitals are for health problems associated with such defects.

For more than 20 years, major developmental defects have been the leading single cause of infant mortality in the United States (Petrini et al. 1997). Although infant mortality in the United States has declined by approximately 40% from 1968 to 1995, infant mortality attributable to major developmental defects has declined slightly less, by 34%, and, thus, the overall proportion of infant mortality due to developmental defects has increased from 14% to 22% from 1968 to 1995 (Ventura et al. 1997). In 1995, the leading defects associated with infant death were heart defects (31.4%), respiratory defects (14.5%), nervous system defects (13.1%), multiple anomalies associated with chromosomal aberrations (13.4%), and musculoskeletal anomalies (7.2%) (Petrini et al. 1997).

The personal costs of developmental defects, including emotional and mental stress, are impossible to measure. In national health considerations, it is standard practice to compare dollar costs. The 1992 estimated lifetime cost for 18 of the most significant developmental defects in the United States was $8 billion

(CDC 1995; Waitzman et al. 1994). The lifetime per-patient cost for spina bifida alone was estimated at $250,000, and the total annual cost for all surviving infants with spina bifida in the United States was $200 million (Sever et al. 1993). A recent study reported that the total lifetime costs for persons born in 1996 with mental retardation, autism, or cerebral palsy will be $47 billion, $4.9 billion, and $12 billion, respectively (Honeycutt et al. 1999). Major developmental defects are the fifth leading cause of years of potential life lost (YPLL) (CDC 1987). For comparison, loss attributed to heart disease before age 65 is 1,600,265 YPLL, loss attributed to cancer is 1,813,245 YPLL, and loss attributed to major congenital anomalies is 694,715 YPLL (CDC 1987).

Those major developmental defects represent only one class of the most socially and medically recognized developmental defects. Several other classes are identified below. Their prevalence has been harder to estimate. To begin with, at least one minor structural defect (e.g., preauricular sinus and syndactyly for toes 2-3) has been identified in 14.1% to 22.3% of live-born infants, a frequency that is 5 to 7 times higher than that for major defects (Leppig et al. 1987). The less-recognized defects are of lesser clinical and cosmetic importance, and the estimate of their birth prevalence varies considerably because of substantial differences in definition and detection and the lack of a national systematic database for this information.

Another class is made up of functional deficits—that is, deficits that are not accompanied by an overt structural defect but are expressed in a variety of ways ranging from delays in growth to deficits in behavioral and neurological development. Many of these deficits are only recognized in infancy or later in childhood (e.g., attention deficit hyperactivity disorder and dyslexia). Developmental defects with reproductive consequences might not be detected until much later. Finally, there is evidence of some mid-life health conditions (e.g., heart conditions) correlating with abnormal birth status (e.g., low birth weight) (Barker 1999). The costs and years of life lost have not been estimated for these more subtle developmental defects among live-born infants.

A further expanded view of developmental defects is gained by examining all pregnancy outcomes (Table 2-1), not only live-birth outcomes. The most common type of outcome in humans is early-pregnancy loss shortly after implantation (Zinaman et al. 1996; Wilcox et al. 1999). That occurs in 20-30% of pregnancies. Many of those losses are difficult to detect and enumerate because they occur prior to clinical recognition of the pregnancy. Spontaneous abortions of clinically recognized pregnancies (generally starting in the 8th week after the last menstrual period) occur in 10-20% of pregnancies (Hatasaka 1994), also a high frequency. Thus, these two categories dominate all other defects. In 40-50% of the spontaneous abortions examined, some type of chromosomal aberration was found, most frequently an extra or missing chromosome (Jacobs and Hassold 1995). Many chromosomally abnormal embryos have anatomical malformations. Fetal deaths (after 20 weeks of gestation) and stillbirths occur in 1-4% of preg-

TABLE 2-1 Frequency of a Variety of Developmental Outcomes

Outcome	Frequency	Reference
Early pregnancy loss (before 8 weeks)	20-30% of implantations	Zinaman et al. 1996; Wilcox et al. 1999
Spontaneous abortion (8-20 weeks)	10-20% of clinically recognized pregnancies	Hatasaka 1994
Chromosomal aberrations in spontaneous abortions (8-12 weeks)	40-50% of spontaneous abortions	Jacobs and Hassold 1995
Late fetal deaths after 20 weeks and stillbirths	1-4% of the sum of live births and late fetal deaths	Fretts et al. 1995
Major congenital anomalies at birth	2-3% of live births	Oakley 1986
Minor developmental defects at birth	14-22% of live births	Leppig et al. 1987
Major developmental defects leading to infant death (before age 15 months)	0.016% of live births	March of Dimes 1999
Chromosomal aberrations in live births	1% of live births	Oakley 1986
Severe mental retardation	0.4% of children to age 15	Mastroianni et al. 1994
Neural tube defects	0.001% of live births	Velie and Shaw 1996

nancies (Fretts et al. 1995). Chromosomal aberrations occur in approximately 1% of live births (Oakley 1986). Severe mental retardation is an example of a functional deficit that might not be recognized at birth but is recognized in approximately 0.4% of children before 15 years of age (Velie and Shaw 1996).

Developmental defects are often defined as those originating in the embryo and fetus, that is, in the prenatal period. A developmental toxicant is then a toxic agent or condition to which the pregnant mother is exposed. However, development goes on throughout the life cycle and includes for example, the continued growth and differentiation of the nervous, skeletal, and reproductive systems in the juvenile and adolescent, and the continuous renewal of cells of the skin, gut lining, and hematopoetic system of the adult. Thus, it is arbitrary to define developmental toxicants only as those affecting the embryo or fetus through maternal exposure in the period of pregnancy. In this report, the committee emphasizes developmental toxicants to which the mother may be exposed in the prenatal period. However, the division is not sharp, and it is to be expected that toxicant exposures at other periods could affect juvenile, adolescent, and adult development, and might affect gametes and reproductive organs in ways that are only expressed much later in the period of pregnancy.

TOXICANT EXPOSURE AND DEVELOPMENTAL DEFECTS

The current understanding of the various causes of developmental defects is incomplete. A crude distinction can be made between intrinsic and extrinsic causes. Intrinsic causes include genetic defects (mutations), endogenous chromosomal imbalances (e.g., meiotic nondisjunctions), endogenous metabolism (e.g., phenylketonurea), and perhaps failures in the complex developmental processes themselves. Extrinsic causes include the enormous variety of environmental inputs such as infection, nutritional deficiencies and excesses, life-style factors (e.g., alcohol), and closer to the concerns of this committee, the myriad agents—pharmaceuticals, synthetic chemicals, solvents, pesticides, fungicides, herbicides, cosmetics, food additives, natural plant and animal toxins and products, and other environmental chemicals—encountered by humans. Other environmental factors, such as hyperthermia, ultraviolet irradiation, and X-rays, should be included. As noted before, developmental defects comprise all structural and functional deficits detected in the implanted embryo, fetus, neonate, infant, or child.

The committee was asked to consider environmental agents that might cause developmental defects. Such agents include mercury, lead, and polychlorinated biphenyls. Natural plant and animal products and toxins have long been recognized as agents that can cause toxicity. They were some of the first environmental agents to be identified as teratogens. Agents can enter the environment by either deliberate (e.g., pesticide residues on food) or accidental (e.g., chemical spills) releases, and humans can be exposed through food, drinking water, or air. Pharmaceuticals and food additives generally would not be considered environmental agents; however, many of the issues under consideration for environmental agents can also apply to these agents. Additionally, it is possible that they incidentally enter the environment at significant concentrations and become environmental agents.

What fraction of developmental defects can be attributed to extrinisic or intrinsic causes? Wilson (1973) estimated that 25% of congenital anomalies in humans are attributable to genetic causes. Then, the author estimated that 65-75% of developmental defects are of unknown causation and attributed fewer than 10% of the anomalies to known environmental causes, including maternal diseases (e.g., diabetes and hypertension), infectious agents (e.g., rubella and syphilis), and mechanical problems (e.g., uterine deformations). Approximately 1% are known to be due to environmental toxicant exposures, including ionizing radiation and hyperthermia (Wilson 1973).

In a more recent evaluation, Nelson and Holmes (1989; also see Holmes 1997) collected data on 69,277 infants, of which 2.24% had at least one major congenital anomaly. The infants were in a surveillance program at a university hospital and were not from the population at large and, therefore, these percentages should be viewed cautiously. Nelson and Holmes estimated the causes of congenital anomalies to be genetic, 28%; multifactorial inheritance, 23%; uterine factors and twinning, 3%; toxicants, 3%; and unknown, 43%.

Multifactorial inheritance (23%), which is a category not distinguished by Wilson, has a genetic and an environmental component. The term is used when geneological studies indicate that a physical trait, disease, or developmental defect occurs at a higher rate within families than expected in the general population, but the patterns of inheritance do not follow strict Mendelian segregation rules. To explain the departure from Mendelian rules, the genetic variant of a gene is said to predispose the individual, but further circumstances, either environmental or other genetic factors, are needed for the production of the disease. An example of multifactorial inheritance is the relationship between maternal smoking, transforming growth factor (TGF) polymorphisms, and oral cleft (Hwang et al. 1995). This example is described in detail in Chapter 5.

Such a departure from Mendelian rules might be attributable to environmental factors, but the departure could as well be due to the requirement for a combination of particular alleles of two or more genes to produce the trait (a polygenic trait) or to genomic imprinting. Specific genes and environmental exposures have been associated in multifactorial inheritance in only a few instances, but increased information is becoming available. With the identification of the multifactorial inheritance category in the Nelson and Holmes study, Wilson's unknown-cause category of 65-75% is reduced to 42-52%, equaling the 43% unknown-cause category of Nelson and Holmes. As the Nelson and Holmes figures indicate, however, the knowledge about the causes and prevention of developmental defects continues to be limited (Mattison 1997).

Today, about 3% of the major developmental defects are estimated to be attributable to toxicant exposure (Oakley 1986; Kimmel 1997; March of Dimes 1999), but that figure is a rough approximation. It is generally recognized that 40-50 extrinsic agents probably have acted as human developmental toxicants and that more than 1,200 chemical and physical agents produce developmental defects in experimental animals (Shepard 1998; Schardein 2000). It should be noted that much of the developmental toxicity testing on experimental animals was conducted at up to maternally toxic doses and, therefore, observed effects at those doses might not be the same as effects observed after exposure to environmentally relevant doses. It is not known how many of the 1,200 agents actually produce developmental defects in humans, and the figure is not obtainable by direct testing in humans. In light of the experimental animal results, many of the agents have never entered the marketplace or environment, and others are handled with great caution according to preventive public-health and workplace-safety guidelines.

In all, only about 50 chemical and physical agents are known to cause developmental defects in humans (Friedman and Polifka 1994; Shepard 1998). They include so-called "life-style" chemicals, such as alcohol, accounting for 0.1-0.2% of defects in live-born infants and cocaine, a variety of pharmaceuticals, and several environmental agents (e.g., mercury, lead, and polychlorinated biphenyls). Table 2-2 lists several representative human developmental toxicants. There is information available on proposed mechanisms of action for these toxicants, however, it is infrequently synthesized into a cohesive and comprehensive mechanistic explanation. Over 80% of agents known to produce developmental defects in humans also cause developmental defects in at least one test animal (rat, mouse, or rabbit) (Shepard1998).

The actual percentage of environmental agents that are developmental toxicants in humans could be higher or lower than 3% for several reasons. For example, the epidemiological methods for identifying toxicants are inherently insensitive and depend on the systematic examination of large human populations. Such large-scale examination is difficult to do (Selevan 1985). Thalidomide was an exception. Because it caused such a distinctive outcome (bilateral limb shortening) of a rare human malformation, its effects were recognized in small patient groups. Even though the frequency of fetal alcohol syndrome is high compared with other developmental disorders, it took many years to identify alcohol as a human teratogen because the physical alterations are subtle, and the learning and social adjustment problems are sometimes not detectable until several years after birth. When a human exposure problem is suspected, epidemiological testing can be performed to assess toxicity, but few of the 1,200 agents have been so examined. Also, the number of agents that cause developmental toxicity might be higher if the multifactorial inheritance category of birth defects contains, as indeed is suspected, cases of human variants who are genetically more susceptible (predisposed) to particular environmental conditions than are others. Finally, the number might be higher if some toxicants (extrinsic causes) produce malformations as a consequence of their primary effect in causing genetic damage (intrinsic cause).

In conclusion, although it is recognized that environmental agents can, and some do, act as developmental toxicants, it is still unclear how large a role these agents play in producing human congenital anomalies relative to other sources of developmental toxicants such as pharmaceuticals and food additives, and relative to intrinsic causes such as genetic differences.

THE CHEMICAL UNIVERSE

The "chemical universe" refers to the collective variety of chemicals that humans encounter. This variety is theoretically infinite if no limit is set on the molecular size of chemicals, because new and more complex compounds can always be made by coupling together simpler chemical units. In practice, how-

TABLE 2-2 Representative Human Developmental Toxicants

Agent	Use	Developmentally Toxic Dosage	Adverse Effects
13-*cis* retinoic acid	Treatment of cystic acne	0.4-1.5 mg/kg/d	Craniofacial and cardiovascular malformations and intellectual deficits
Aminopterin	Folate antagonist	1-2 mg/kg/d	Abortion, central nervous system and craniofacial defects, and growth retardation
Angiotensin converting enzyme inhibitors	Antihypertensive	Therapeutic dose (differs for each)	Fetal death, stillbirth, oligohydramnios, growth retardation, hypotension, and renal failure
Cigarette smoke	Stimulant	>20/day	Growth retardation and facial defects
Coumarin derivatives	Anticoagulants	Therapeutic dose (differs for each)	Facial defects, limb anomalies, growth retardation, and neonatal respiratory distress
Cyclophosphamide	Antineoplastic	4 mg/kg/d	Limb and facial defects
Diethylstilbestrol	Synthetic estrogen	0.1-3 mg/kg/d	Reproductive tract malformations and vaginal cancer
Diphenylhydantoin	Anticonvulsant	8 mg/kg/d	Craniofacial defects, growth retardation, fetal loss, and intellectual deficit
Etretinate	Treatment of psoriasis	0.5-1 mg/kg/d	Limb, ear, cardiac, and thymic defects
Lead	Environmental contaminant	10-15 μg/dl/blood[a]	Abortion, growth retardation, and neurobehavioral deficits
Lithium	Bipolar disorder	3-5 mg/kg/d	Cardiac defects
Methylmercury	Environmental contaminant	10 μg/kg/d[b]	Central nervous system defects
Penicillamine	Chelator	20 mg/kg/d	Connective tissue defects
Polychlorinated biphenyls	Environmental contaminant		Growth retardation, hyperpigmentation, and neurobehavioral deficit[b]
Thalidomide	Sedative/ hypnotic	0.7-3 mg/kg/d	Structural malformations (particularly reduction defects of the limbs and ears)
Valproic acid	Anticonvulsant	5-10 mg/kg/d	Neural tube closure defects

[a] The dosage of lead needed to reach these blood levels is dependent on route of exposure. Typically, humans are exposed to lead through a combination of inhalation and oral ingestion.
[b] There is uncertainty about the dosage associated with adverse effects.

ever, chemists tend to produce synthetic chemicals or analyze natural chemicals that are between a molecular weight of a few hundred to a few thousand. Technological advances in chemistry will continue to increase the number of synthetic chemicals. Some of these chemicals will undoubtedly be of benefit to society; however, potentially harmful chemicals need to be identified so that human exposure is controlled or prevented. For risk assessment, a key goal is to determine which exposures to chemicals may be harmful to humans before exposure occurs.

Millions of synthetic chemicals are registered with the American Chemical Society, but fewer than 100,000 are currently in commercial or industrial use and, therefore, available for introduction into the environment (EPA 1997). Most of these chemicals have not been tested for developmental toxicity. For example, EPA (1998a) conducted a study assessing data availability on close to 3,000 chemicals that the United States produces or imports at more than 1 million pounds per year and concluded that only 23% of those chemicals had been tested for reproductive and developmental toxicity. Test data were considered available if any studies relevant to reproductive and developmental toxicity were located.

The number of synthetic chemicals is likely to increase greatly in the near future. Recent advances in combinatorial chemistry have made it possible for chemists to synthesize in parallel small amounts of a large number of chemicals (a "library," on the order of 10^4-10^6 kinds per application). Biologically active members of the library are selected by their performance in specific biological assays (usually in vitro assays rather than animal tests) based on recent insights into the workings of cellular, developmental, and pathological processes. These new synthetic and selective methods are expected to lead to the development of drugs that are more efficacious and have specific pharmacological activities. Although all but the most promising chemical candidates will be restricted to the laboratory, it will be a challenge to gain toxicity information about many of these so that not only the most efficacious but also the safest can advance to the next phase of drug development. As drug discovery and development approaches become more sophisticated, toxicity testing approaches must also become more sophisticated.

Toxins, such as chemicals from microorganisms, fungi, plants, and animals (e.g., sponges, coelenterates, and bryozoa), have not been analyzed systematically. Systematic analysis has shown that the variety of chemical constituents is known to be great in some naturally occurring substances. For example, over 400 chemicals have been identified in red wine, and over 1,000 chemicals have been found in tobacco or tobacco smoke. Naturally occurring substances sometimes have significant pharmacological and toxic properties (see, e.g., NRC 1996). Catalogs of known naturally occurring plant toxins, for example, include more than 2,000 entries, and the number with pharmacological activity is larger (Keeler and Tu 1991; Harborne and Baxter 1996).

Animals, including humans, have evolved enzymes and ligand-binding proteins to metabolize and eliminate many natural environmental chemicals. They have also evolved adaptive mechanisms, stress responses, and checkpoint pathways to prevent or correct damage from various environmental chemicals and physical conditions (e.g., heat shock), and hence to survive in their environment. Thanks to the broad specificities of these proteins and adaptive processes, animals also detoxify and adaptively respond to many synthetic chemicals as well, even though the animal has never seen these chemicals before in its evolution. Understanding detoxification and adaptation processes in animals and especially in humans has widespread implications for the field of developmental toxicology (for reviews, see Juchau 1980; Juchau et al. 1980; Shepard et al. 1983). Still, some small fraction of old and new chemicals, synthetic and natural, can elude the animal's defenses enough to impact components of its developmental processes, thereby leading to developmental defects.

SUMMARY

The frequency at which all classes of developmental defects occur is thought to be very high, perhaps exceeding half of initial pregnancies. However, the total frequency of developmental defects is only vaguely known, and the means of surveillance for defects are only approximate. It is thought that among newborns with major developmental defects, genetic transmission accounts for perhaps 25% of the cases. Lesser genetic defects, which are insufficient on their own to cause major defects but are sufficient in combination with environmental factors or other genetic factors, account for perhaps another 25% of major defects. Genotype combined with environmental causes is a class of developmental defect that is expected to receive incisive attention in the near future. Genotype is an important class because of its implications that some environmental agents might act as toxicants for some people (predisposed individuals) but not for others, making risk assessment a process requiring information about human diversity as well as toxicant action. A few percent (approximately 3%) of developmental defects are probably attributable to chemicals and physical agents alone and have no known genetic contribution. An unknown fraction of those are due to environmental toxicants. Finally, the causes of nearly half of the major defects immediately detected at birth are so poorly understood that they cannot even be classified as being caused by intrinsic or extrinsic factors or both. Presumably, some fraction of them have a complex environmental component. At the same time, there is a steadily expanding universe of chemicals and combinations of chemicals to which humans are exposed. Most of these chemicals have never been tested for developmental toxicity.

3

Current Practices for Assessing Risk for Developmental Defects and Their Limitations

Since the mid-1900s, various governmental agencies in the United States have taken responsibility for protecting the health of the public by regulating safe usage practices for drugs, food additives, pesticides, and environmental and occupational chemicals (Gallo 1996; Omenn and Faustman 1997). In the 1970s, risk assessment began as an organized activity of federal agencies to set acceptable exposure levels or tolerance levels. Earlier the American Conference of Governmental Industrial Hygienists had set threshold limit values for workers and the U.S. Food and Drug Administration (FDA) had established acceptable daily intakes for dietary pesticide residues and food additives. In 1983, the National Research Council published a report entitled *Risk Assessment in the Federal Government: Managing the Process* (often referred to as the "Red Book"), which provided a common framework for risk assessment (NRC 1983). In 1991, the U.S. Environmental Protection Agency (EPA) published risk assessment guidelines specific for developmental toxicity (EPA 1991).

In this chapter, the committee highlights risk assessment practices as they relate to the evaluation of chemicals as potential developmental toxicants and identifies limitations in the current risk assessment approaches.

THE DEVELOPMENTAL TOXICITY RISK ASSESSMENT PROCESS

"Human health risk assessment" refers to the process of systematically characterizing potential adverse health effects in humans that result from exposure to chemicals and physical agents (NRC 1983). For developmental toxicity, this assessment means evaluating the potential for chemical exposure to cause any of four types of adverse developmental end points: growth retardation; gross, skeletal, or visceral malformations; adverse functional outcomes; and lethality. Developmental toxicity risk assessment includes evaluating all available experimental animal and human toxicity data and the dose, route, duration, and timing of exposure to determine if an agent causes developmental toxicity (EPA 1991; Moore et al. 1995).

As discussed in the "Red Book," risk management, in contrast to risk assessment, is the application of risk assessment information in policy and decision-making processes to balance risks and benefits (e.g., for therapeutic applications); set target levels of acceptable risk (e.g., for food contaminants and water pollutants); set priorities for the program activities of regulatory agencies, manufacturers, and environmental and consumer organizations; and estimate residual risks after a risk-reduction effort has been taken (e.g., folic acid supplementation in food). Figure 3-1 shows the NRC paradigm for risk assessment and risk management. As shown in this figure, risk characterization refers to the synthesis of qualitative and quantitative information for both toxicity and exposure assessments (EPA 1995). It also usually includes a discussion of the uncertainties in the analysis.

The following sections describe some of the specific approaches used for toxicity assessment. Four types of informational methods that can be used for

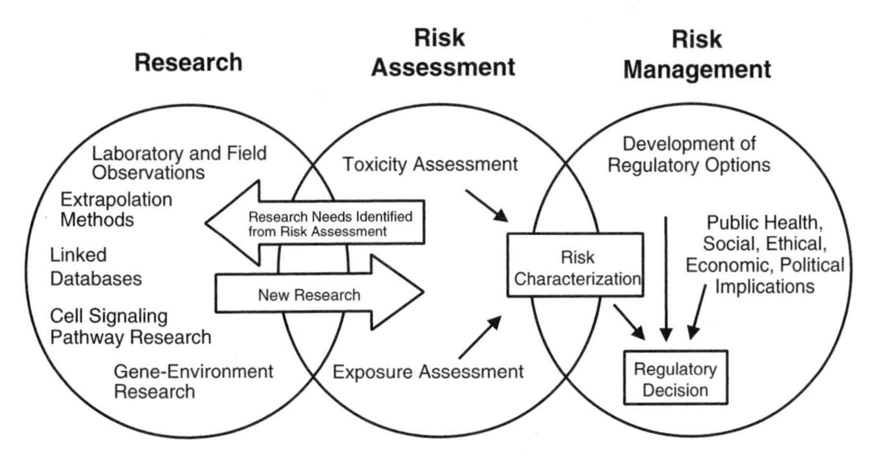

FIGURE 3-1 Risk assessment and risk management paradigm from the NRC modified for developmental toxicity risk assessments. In accordance with this committee's deliberations, the research section now includes a two-way arrow and specifically highlights emerging research on gene-environment interaction and developmental cell-signaling pathways. The iterative feedback loop between research and risk assessment is necessary to translate new findings in biology into scientifically based risk assessments. Source: Adapted from NRC 1983.

these assessments are chemical structure-activity information, in vitro assessments, in vivo animal bioassays, and epidemiological studies. Two additional steps in risk assessment, dose-response assessment and exposure assessment, are described. Finally, the use of toxicokinetic information and biomarkers in developmental toxicity risk assessment is discussed.

Chemical Structure-Activity Information

Information on a chemical's structure, stability, solubility, reactivity, and electrophilicity can provide useful clues to its potential to be absorbed and distributed throughout the body and to be reactive with biological tissues. In fact, despite early concepts of a true placental barrier, it is now appreciated that all lipid-soluble compounds have access to the developing cells of an embryo and fetus. Properties of lipid solubility and characteristics such as chemical size and pK_a can be used to predict the potential for chemicals to cross the placenta and have access to conceptus tissues (Slikker and Miller 1994).

Structure-activity relationships (SARs) are developed to show the relationship between the specific chemical structures or moieties of agents and their capacity to produce certain toxic effects. For glycol ethers, retinoic acid, valproic acid, and their derivatives and for several other commercial products and therapeutics, good SAR data exist for developmental effects. Recently, SARs were reported for valproic acid derivatives that activate the peroxisomal proliferation pathway and cause developmental toxicity (Lampen et al. 1999).

Early research on receptor binding identified SARs for environmental agents such as benzopyrene and dioxin. Toxicity equivalency factors (TEFs) have been developed that relate the relative toxicity of each compound to a reference compound, such as benzo[*a*]pyrene (B[*a*]P) or 2,3,7,8-tetrachlorodibenzo-*p*-dioxin (TCDD) for pyrenes and dioxins, respectively (Van den Berg et al. 1998). Complexities arise when different toxicity end points have different SARs. To be useful for developmental toxicity risk assessments, SARs (and TEFs) must be evaluated for each of the end points of developmental toxicity.

In Vitro Assessments

Alternatives to pregnant experimental mammals in studies of developmental toxicology have often been grouped together as in vitro approaches, but that is misleading, because they include not only ex vivo mammalian embryos, tissues, cells, and subcellular preparations but also embryos of nonmammalian species. Broadly, such alternatives have had two applications: to test chemicals for potential effects and to analyze mechanisms of effect.

Mechanistic uses of ex vivo methods have much in common with investigative studies in other areas of biology. They have made major contributions to understanding developmental toxicity, because of the manipulations possible in vitro, including the removal of the maternal environment, the ablation or transplantation of tissues and cells, labeling and tracking of cells or molecules, biochemical and gene manipulations by the use of inhibitors and anti-sense RNA, and real-time physiological monitoring of the embryo. The types of information generated include the identification of proximate developmental toxicants, exact tissue sites of accumulation, initial biochemical insults, gene expression changes, intrinsic SARs (of the parent compound), and identification of disrupted developmental pathways.

The search for alternatives for testing purposes is driven by the need to assess a larger number of chemicals than that allowed by using available resources for in vivo methods and also by the desire to reduce or replace the use of experimental mammals. Two levels of testing should be distinguished: secondary and primary. Secondary testing is the assessment of chemicals that have some known potential developmental toxicity. Most commonly, secondary testing involves analogs of prototype chemicals that have known in vivo developmental toxicity. The objective is to replicate the observed developmental toxicity in a simple system. The approach has been successful, especially for pharmaceuticals and particularly with the use of isolated mammalian embryos and embryonic cells in culture. For example, the approach has been used for testing retinoids (Kistler and Howard 1990) and triazoles (Flint and Boyle 1985). For that type of use, a universal validation of the method is not required. It is sufficient to show that the method replicates a specific in vivo effect for the particular chemicals under study.

Primary testing, in contrast, is the testing of chemicals that have no known potential toxicity, the aim being to predict in vivo actions. There must be confidence that the test outcome will accurately classify most chemicals by their potential to cause human developmental toxicity. Furthermore, the required sensitivity and selectivity will vary, depending on the purpose of the test. Sensitivity is the proportion of in vivo toxicants that are positive in the test, and selectivity is the proportion of inactive chemicals that are negative in the test. In some contexts, for example, in drug discovery by combinatorial chemistry, the aim is the early elimination of potential toxicants. False-positive results are not problematic, because there are many other chemicals from which to choose. Conversely, if the context is hazard identification and the aim is to set priorities for further in vivo testing, then a high rate of false-positive results would be inappropriate. Thus, there is a drive to validate tests for screening purposes by measuring their sensitivity and selectivity (Lave and Omenn 1986). Regardless of all the testing-related problems about to be discussed, it is worth bearing in mind that some countries have already banned the use of mammals for testing in certain situations, so there is an obligation to continue to refine in vitro approaches.

Alternative testing for developmental toxicity has a long history, encompassing regular international conferences (Ebert and Marois 1976; Kimmel et al. 1982; Schwetz 1993), comprehensive reviews (Brown and Freeman 1984; Faustman 1988; Welsch 1992; Brown et al. 1995), and much debate in print (Mirkes 1996;

Daston 1996). Alternative tests for development toxicity are not currently used by any regulatory agency.

Intrinsic Limitations

Alternatives to in vivo testing will never detect all the developmental toxicants that have actions in pregnant mammals. This is true for several reasons. First, some toxicants initiate their effects outside the embryo and in the maternal or placental compartments. Second, some effects are mediated by physiological changes only represented within the intact embryo (e.g., peripheral vascular perfusion). Third, known mechanisms of developmental toxicity are diverse, so it is unlikely that all targets will be present in a simple system. Fourth, some adverse outcomes are only observable as functional impairment postnatally. Finally, most alternative systems are static and have neither the dynamic changes in concentration associated with physiological disposition in vivo nor the metabolic transformation of the test agent.

Validation

Validation is complex and includes protocol standardization, interlaboratory consistency, and statistical prediction models, but the fundamental question remains: how well does the system mimic the susceptibility of human development? This has yet to be answered for any system, and there are a number of problems that are discussed below.

In Vitro Test for What Outcome? The type of adverse outcome induced by a chemical in vivo often varies between individuals, across species, and sometimes with routes or schedules of administration. Thus, although the initial aim of alternative tests was to predict the overall induction of congenital malformations, it is more appropriate to consider that in vitro tests can help to predict specified developmental toxicity and to identify potential mechanisms of disruption to particular cell-signaling and genetic regulatory pathways.

General Versus Specific Toxicity. Presumably, all chemicals would disturb development, if a high enough concentration were delivered to the embryo, even though such concentrations might be unattainable in mammals because of maternal toxicity. However, chemicals vary widely in their intrinsic hazard to development. For example, high-affinity ligands for some nuclear-hormone receptors cause irreversible developmental defects (see Chapter 4). It would be helpful to be able to discriminate such chemicals from those that affect development (D) only at exposures that are simultaneously toxic to the adult (A). The A/D ratio (the ratio of adult toxic dose to developmentally toxic dose) attempts to measure that specificity. However, use of that value has been tempered by the demonstration that A/D ratios are not necessarily consistent across species (Daston et al. 1991).

Which In Vivo Database? The database on humans is probably too heterogeneous to use for validation studies. For example, it is biased toward pharmaceuticals, and the exposure range is too small for most chemicals, so chemicals with reliable negative results are hard to identify. By default then, comparisons have been made with experimental mammal testing data. The information in this database is also heterogeneous in exposure times, routes, and doses; species; end points; and adverse outcomes. To avoid some of these problems but retain the use of existing data, perhaps the only option is to use data exclusively from orthodox segment II type tests in which animals are exposed during the period of major organogenesis. This approach eliminates many of the chemicals historically used in validation studies, because they have never been formally tested in vivo.

Chemicals for Validation. Much effort has been expended on the analysis of in vivo animal test data to produce a list of chemicals for use in validation studies. A prototype list produced by Smith et al. (1983) was subsequently found to be inadequate and an expert committee was set up to address that issue (Schwetz 1993). Because of the difficulty of the task, that committee was not able to complete its task. There has been considerable disagreement over what is and what is not developmentally toxic in vivo and over the severity of that action. There is currently no consensus on how to categorize, stratify, or quantify the developmental toxicity of chemicals. Most validation studies have used a binary classification: developmental toxicants or nontoxicants (Parsons et al. 1990; Uphill et al. 1990). This is a gross oversimplification of the richness of information available. More recently, chemicals have often been grouped informally into three or four categories: (1) toxic to development in all species, no maternal toxicity; (2) toxic to development in some species, no maternal toxicity; (3) toxic to development in some species, some maternal toxicity; (4) no evidence for developmental toxicity in any species tested. However, without formal definition of categories and consistent in vivo testing, there is disagreement in assigning chemicals to such groups (Wise et al. 1990; Daston et al. 1995; Newall and Beedles 1996; Spielmann et al. 1997). Many validation studies have been biased by the inherent toxicity inequality of chemicals selected (Brown 1987). It has been common to select chemicals of potent and general biological activity, such as antimetabolites, nucleotide or nucleoside analogs, and alkylating agents, as developmental toxicants. In contrast, the chosen nondevelopmental toxicants have frequently been endogenous intermediary biochemicals, such as acetate, glutamate, and lysine, or chemicals specifically designed to be nontoxic to mammalian cells, such as antibiotics, saccharin, and cyclamate. It comes as no surprise that developmental models respond differently to two such disparate groups. The proper strategy should be to select chemicals that are largely similar in their general toxicity and potency, but different in specific developmental hazard (Schwetz and Harris 1993). This has never been achieved.

An additional problem in categorizing chemicals, even those tested according to standard protocols, is that toxicokinetics and metabolism are rarely investigated sufficiently to indicate whether a negative outcome in vivo is a reflection of a true lack of inherent developmental toxicity potential or a low embryonic exposure. This outcome can lead to a situation in which a chemical is correctly identified as a potential developmental toxicant from an in vitro test, but the effective exposure can never be achieved in vivo.

Existing and Extinct Alternative Tests for Primary Screening

Because there are no known common mechanisms of developmental toxicity on which to base a design for a primary screening test, three other approaches have been taken. These are the use of (1) mammalian embryos or parts of embryos in culture, (2) free-living nonmammalian embryos, and (3) cell cultures in which processes thought to be required for normal development are assayed (e.g., proliferation, adhesion, communication, and differentiation). More than 30 test systems have been devised and preliminarily assessed (see Table 3-1). All those test systems that use embryos monitor gross morphological end points. Few tests are actively used for screening purposes (Brown et al. 1995). Rodent embryo culture, micromass, and stem-cell assays are currently being validated in a European Union-sponsored trial (Spielmann et al. 1998). The validation of the frog embryo teratogenesis assay in *Xenopus* (FETAX) is being reviewed by the U.S. National Toxicology Program Interagency Center for the Evaluation of Alternative Toxicological Methods (NIEHS 1997; Fort et al. 1998).

Rather than having been eliminated by objective criteria, most other systems were simply not adopted by scientists and were not pursued by their originators. For example, there have been no studies comparing several systems for relative performance or using more sophisticated molecular end points. A few systems have been eliminated by poor performance. The mouse ovarian tumor (MOT) cell-attachment method and the human embryonic palatal mesenchymal (HEPM) cell-proliferation method were simultaneously assessed by the U.S. National Toxicology Program (Steele at al. 1988) and shown to have a combined specificity of only 50%. The hydra assay is novel in having been designed specifically to estimate the A/D ratio. Although transiently popular, usage diminished with the demonstration that the A/D ratio is not consistent across species (Daston et al. 1991) and with other concerns about comparability with mammalian responses.

New Receptor-Based Tests

Endocrine disruptions by chemicals are beyond the scope of this report but are relevant in terms of the overlap in receptors involved and in the in vitro ap-

TABLE 3-1 Systems Proposed as Alternatives to Pregnant Mammals to Test for Developmental Toxicity[a]

System	End Point Monitored	References
Mammalian embryos ex utero		
Rodent whole-embryo culture	Morphogenesis	Webster et al. 1997
Rodent embryonic organ culture: limb; palate	Morphogenesis	Kochhar 1982; Abbott et al. 1992
Sub-mammalian embryos		
Avia: Chick embryotoxicity test (CHEST)	Morphogenesis	Peterka and Pexieder 1994
Amphibia: Frog embryo test (FETAX)	Morphogenesis	Fort et al. 1998
Fish: Mekada, zebrafish	Morphogenesis	Herrmann 1993
Arthropods: Cricket, artemia, *Drosophila*	Morphogenesis	Walton 1983; Sleet and Brendel 1985; Lynch et al. 1991
Flatworms: Planaria	Morphogenesis	Best and Morita 1991
Echinoderms: Sea urchin	Morphogenesis	Graillet et al. 1993
Coelenterates: Hydra	Regeneration	Johnson et al. 1988
Protista: Slime mold	Morphogenesis	Tillner et al. 1996
Cell cultures—primary		
Micromass (limb or mid-brain, rodent or chick embryo)	Differentiation	Flint 1993
Human amniotic and chorionic villus	Stress response	Honda et al. 1991
Human placental explants	Proliferation, differentiation	Genbacev et al. 1993
Drosophila embryo	Differentiation	Bournias Vardiabasis 1990
Chick embryo: neural retinal, neural crest, brain	Differentiation	Reinhardt 1993
Cell cultures—established lines		
Mouse ovarian tumor	Attachment	Braun et al. 1982
Human embryo palatal mesenchyme	Proliferation	Zhao et al. 1995
Neuroblastoma	Differentiation	Kisaalita 1997
V79/HEPM	Communication	Toraason et al. 1992
Pox virus in infected cells	Viral replication	Keller and Smith 1982
HELA	Proliferation	Freese 1982
EC cells	Differentiation	Hooghe and Ooms 1995
ES lines	Differentiation	Spielmann et al. 1997

[a] The reference given for each system is the most recent available, or one that will lead into the appropriate literature. Italicized systems are those still in active use in 1998.

proaches being pursued. Interference with estrogen, androgen, glucocorticoid, or thyroxine receptor function can result in developmental and endocrine toxicities. Major efforts are under way to devise screening methods to assess interference with those receptors (EPA 1998b). This task is complex, because chemicals could be agonists, partial agonists, antagonists, or negative antagonists (Limbird and Taylor 1998) or interact with other steps in the pathway. Caution is needed, therefore, in extrapolating from simple tests. Nevertheless, a variety of tests have been devised to assess receptor binding, activation of response elements, and cellular responses (e.g., proliferation). Similar approaches could be devised for other signaling pathway receptors involved in developmental toxicity.

Animal Bioassays

In vivo animal bioassays are a critical component in human health risk assessment. A basic underlying premise of risk assessment is that mammalian animal bioassays are predictive of potential adverse human health impacts. This assumption, and the assumption that humans are the most sensitive mammalian species, have served as the basis for human health risk assessment.

Several study protocols to test for developmental toxicity in animals are accepted and used by regulatory agencies such as EPA and FDA. Describing the various protocols goes beyond the scope of this report and the reader is referred to the original guidelines (EPA 1991, 1996a, 1998c,d,e; FDA 1994; OECD 1998) for detailed descriptions. T.F. Collins et al. (1998) contains a discussion and a comparison of EPA, FDA, and OECD guidelines.

Information obtained from in vivo bioassays includes the identification of potentially sensitive target organ systems; maternal toxicity; embryonic and fetal lethality; specific types of malformations including gross, visceral, and skeletal malformations; and altered birth weight and growth retardation. These assays can also provide information on reproductive effects, multigenerational effects, and prenatal and postnatal function. In vivo bioassays determine critical effects that are used for quantitative assessments by taking the no-observed-adverse-effect level (NOAEL) for the most sensitive effects.

The focus of animal bioassays primarily has been toxicity assessment, including hazard identification and dose-response assessment. The aim of such studies is to identify qualitatively what spectrum of effects a test chemical can produce and to put those effects in the context of dose-response relationships. Because there is uncertainty in extrapolation from animal studies to humans, several assumptions are made, including the following: (1) an agent that causes an adverse developmental effect in experimental animals might cause an effect in humans; (2) all end points (i.e., death, structural abnormalities, growth alterations, and functional deficits) of developmental toxicity are of potential concern; and (3) specific types of developmental effects observed in experimental animals might not be manifested in the exact same manner as those observed in humans.

Much of the literature before 1975 concerning studies of in-utero-induced adverse developmental outcome is troubled by small sample sizes, inappropriate routes and modes of exposure, inconsistent methodology, and excessively high dose or concentration exposures. Many of those deficiencies have been corrected by the regulatory mandate of adhering to Good Laboratory Practices (OECD 1987; FDA 1987; EPA 1990).

Several studies of concordance between the perturbed developmental outcomes in experimental animal studies and the human clinical experience have been made (Nisbet and Karch 1983; Kimmel et al. 1984; Francis et al. 1990; Hemminki and Vineis 1985; Newman et al. 1993). The most rigorous and earliest of those was done in the early 1980s and is contained in a technical report for the National Center for Toxicological Research (NCTR) (C.A. Kimmel, EPA, unpublished report, 1984).[1] In general, these studies concluded that there is concordance of developmental effects between animals and humans and that humans are as sensitive or more sensitive than the most sensitive animal species.

The NCTR study was notable because it employed criteria of acceptance for both human and experimental animal reports that included study design and statistical power considerations. Additionally, the authors held to the premise that adverse developmental effects represented a continuum of responses—or at least a number of interrelated effects—including in utero growth retardation, death of the products of conception, frank malformations, and functional deficits that manifest themselves in later stages in life. Hence, an effect on any one of these end points in experimental animals or human studies was considered a basis for concordance. Concordance did not require an exact mimicry of response among species. This was not required because exposure conditions (e.g., timing and duration of exposure and toxicokinetic differences) and tissue sensitivity (e.g., toxicodynamic differences) could differ enough between experimental animals and humans to result in a different type of effect.

Many different agents—mostly chemical agents but also physical agents—have been evaluated to determine their capacity to produce developmental toxicity in experimental animal models, such as the rat, mouse, and rabbit. Most of those studies have been conducted by private industry and federal government-funded research programs and involved test agents that had not yet entered the market. Schwetz and Harris (1993) provide a good review of 50 chemicals that the National Toxicology Program has evaluated for developmental toxicity using rodent bioassays.

As discussed in Chapter 2, humans were never exposed to many of the materials that have been evaluated in rodent bioassays and that have been shown to affect animal prenatal development adversely. Thus, it will never be known

[1] Dr. Kimmel presented this information to the committee during its meeting on October 6, 1997, in Washington, D.C.

whether comparable adverse effects would have been caused had similar human exposure occurred. Summary compilations from published data covering more than 4,000 different entities of exposure conditions indicate that more than 1,200 agents, predominantly chemical agents, have produced adverse developmental outcomes by the end point criteria stated above, often including congenital anomalies, in one or more species of experimental animals (Shepard 1998). Among this large number are about 50 agents (almost exclusively designated as drugs) that are known to cause adverse developmental effects in human beings. For most of the agents that were evaluated for developmental hazard potential in experimental animals, human exposures will never occur. Thus, public health was protected, but ascertainment of concordance of animal and human responses was undetermined for those agents. When exposures occur, rarely have human assessments been sufficient for definitive evaluation and establishment of cause-and-effect associations. Because of the background incidence of human developmental abnormalities (addressed in Chapter 2) and the difficulties in conducting epidemiology studies, such associations are extremely difficult to establish unless the outcome is unusual and striking, as was the case of thalidomide.

Among industrial chemicals and environmental contaminants that have been studied in pregnant animal models, often the estimated maximum tolerated dose (MTD) was repeatedly given in conformance with the testing guidelines. Internationally, regulatory authorities require in many instances that the MTD, even up to maternally toxic concentrations, be administered to ensure that no developmental toxicity occurs. Therefore, the underlying principle is that, if regulatory standards are set to protect against maternal toxicity, no adverse effects will occur in offspring. Unfortunately, all too frequently the focus of developmental toxicity testing has been to study the effects of an agent only at high doses that are most likely irrelevant to environmental and occupational exposures. For industrial and environmental chemicals, the dosing regimens at or even above MTDs, as applied in hazard identification studies, typically contrast sharply with anticipated human exposures that are commonly much lower in extent or magnitude, often uncertain, or even entirely unknown.

Because of the design of developmental hazard identification studies, the overwhelming majority of the more than 1,200 agents found to elicit adverse developmental outcomes in experimental animals were tested at doses many times higher than anticipated human exposures during pregnancy and have often elicited extreme maternal toxicity. Furthermore, exposure of the pregnant animals was sustained throughout all of organogenesis by daily repeated administrations, and minimal or no regard was taken for toxicokinetic considerations (see toxicokinetics section of this chapter for details).

Therefore, there are problems associated with the application of these assays for assessing human developmental toxicity potential. Repeated administration of an MTD might produce adverse results that are not indicative of risk from ambient exposure concentrations or intermittent exposures. It is a continuing challenge for test design and interpretation to minimize the problems noted above and, thereby, improve the predictiveness of laboratory animal toxicology protocols.

Epidemiology

Four approaches have traditionally been used to evaluate human developmental toxicity: (1) case series, (2) randomized controlled trials, (3) cohort studies, and (4) case-control studies.

Case series comprise an important first step in assessing relationships between exposures and adverse pregnancy outcomes. Many developmental toxicants are first recognized by astute clinicians who correlate specific patterns of developmental defects or developmental disabilities with specific exposures during pregnancy. Most notable among agents first identified this way as causing developmental defects are rubella and thalidomide (Gregg 1941; Lenz and Knapp 1962; for a review, see Rosa 1992). Case series can be useful when the outcome is distinctive, the exposed population is large enough that numerous cases are recognized, and the dose and timing are well described. Case series should be interpreted with caution, however, because the association can be due entirely to chance. They rarely permit identification of a causal link between exposure and outcome due to their anecdotal nature and the high background of adverse pregnancy outcomes in humans (see Table 2-1). Their greatest value is in the generation of hypotheses for further investigation.

Randomized controlled trials are the most widely accepted type of epidemiological study for assessing the relationship between an intervention and an outcome. Subjects are enrolled into a randomized trial based on pre-established criteria. They are randomly assigned to a reference group (placebo or alternate treatment) or a test group and administered a test agent under controlled conditions. Because the agent of interest is deliberately administered, this type of study is not appropriate for assessing risk of chemical exposures on pregnancy, or of adverse effects of chemicals in general. Randomized control trials have their widest use in tests of the efficacy of pharmaceuticals and other medical interventions.

Cohort studies are observational epidemiological studies in which individuals are assigned to groups (cohorts) on the basis of pre-existing exposure status and are followed to determine pregnancy outcome. The cohort study approach is limited to the investigation of few exposures, but allows for the assessment of numerous developmental endpoints. Considering the rarity of congenital anomalies, large studies are needed to detect differences between cohorts. For example, spina bifida occurs in 0.1% of most American populations and might not be detected even in a cohort of 1,000 pregnancies. Even though cohort studies allow for the next best determination of a causal association (after randomized controlled trials), they are often not practical. An additional problem is that some individuals from the cohort will no longer be available for follow-up of the pregnancy outcome. Postmarketing surveillance by the pharmaceutical industry can be viewed as a type of cohort study, although an unexposed cohort is often not studied concurrently.

The case-control study is the most common design used in assessing the association between exposure and pregnancy outcome. In this type of study, the pregnancy outcome is identified (usually congenital anomalies in live-born infants), and a retrospective evaluation is then conducted to determine the exposure pattern. In case-control studies, the number of developmental end points that can be assessed is small, but several exposures can be investigated. The case-control study is the most efficient study design for capturing rare events, such as congenital anomalies. Accurate ascertainment of exposure can be problematic for case-control studies. Recall bias can occur among women who deliver abnormal infants (i.e., exposures are recalled more extensively by women with abnormal infants than by those with normal infants). Selection of an appropriate control group, which ideally is identical to the case group except for the outcome of interest, can also be difficult.

There is no formula whereby a causal relationship can be established between an exposure and an adverse pregnancy outcome. Results from epidemiological studies should be interpreted with caution because associations found can be due to the following:

- Unmeasured confounding, particularly confounding by indication.
- Exposure misclassification (inability to pinpoint relevant dose and timing).
- Outcome misclassification (related to the heterogeneity of birth defects).
- Biological interactions (subgroups with differing genetic susceptibilities or presence of additional exposures).
- Differential prenatal survival (in studies evaluating live-born infants, spontaneous abortion or elective termination of abnormal fetuses should be taken into account).

Evidence from a number of sources, including human and experimental animal data, must be collectively considered to determine the strength of the association (Rothman 1986; Khoury et al. 1992).

There are many problems in identifying associations between exposures and adverse pregnancy outcomes using conventional epidemiological approaches. Weak or moderate associations (relative risks or odds ratios ranging from 1 to 3) are typically found between environmental exposures and pregnancy outcomes (Khoury et al. 1992). For example, maternal smoking is weakly associated with oral clefts (odds ratios between 1 and 2) (Khoury et al. 1989). Insulin-dependent diabetes is associated somewhat more strongly with major malformations (a relative risk of 7) (Becerra et al. 1990), and potent developmental toxicants, such as isotretinoin and thalidomide, are very strongly associated with major malformations (relative risks in the range of 25 and 300-400, respectively) (Lenz and Knapp 1962; Lammer et al. 1985). Although conventional epidemiological studies have been useful in quantifying the magnitude of risk produced by those potent agents, they were first identified as human developmental toxicants through case reports. Conventional epidemiological studies can be influenced sufficiently with biases, uncertainties, and methodological weaknesses that they may not be useful to detect accurately and assign significance to weak associations—those with relative risks in the range of 1 to 3, the range in which many environmental toxicants can be expected to act (Taubes 1995). In the context of risk assessment, such methodological limits mean that a 2- or 3-fold increase of risk, which amounts to a major health problem, would go undetected.

Another concern with conventional epidemiological studies on chemicals is that studies frequently rely on occupationally exposed cohorts under conditions in which exposure patterns are higher and potentially more consistent than environmental exposures of the general public.

Another potential complication in interpreting data from conventional epidemiological studies is that the complexities and variabilities of human activities, such as life-style factors and diet, cannot be controlled in human studies in the same manner as animal studies. Thus, interpretation of epidemiological study results requires sophisticated experimental designs and analyses to ascertain true relationships.

The ability of an epidemiological study to identify chemically related effects is dependent on the size of the study population, the variability of population effects, the study design, and the background incidence of the adverse health effect being studied. Such information is especially important for risk assessors when they are evaluating epidemiological studies with widely varying results. Understanding how much power a study with negative results has to detect an adverse outcome strengthens the utility of these studies for risk assessments.

Dose-Response Assessment

As part of the evaluation of dose-response relationships, a quantitative evaluation is conducted (EPA 1991; Moore et al. 1995). Doses or concentrations are identified that have no or minimal associated adverse developmental effects. A NOAEL or a lowest-observed-adverse-effect level (LOAEL) is chosen from one of the experimental doses or concentrations tested. These levels are identified for each human and experimental animal study and manifestation of developmental toxicity (i.e., death, structural abnormalities, growth alterations, and functional deficits). Using the NOAEL or other most sensitive effect levels (i.e., end points adversely affected at the lowest doses tested), the reference dose (RfD) or reference concentration (RfC) is determined. These values are an estimate of a daily exposure to the human population that is assumed to be without appreciable risk of adverse developmental effects (EPA 1991).

An acceptable daily intake (ADI) can also be determined from a NOAEL. ADI values are used for pesticides and food additives to define the daily intake of chemicals, which appear to be without appreciable risk of harm during an entire lifetime.

In an alternative approach, a specific effect dose or concentration, such as the ED_{05} (EC_{05}) or ED_{10} (EC_{10}) (the best estimate of the dose at a 5% or a 10% level of response) is determined for the dose-response curve based on rodent or human epidemiological studies (Crump 1984; Allen et al. 1994 a,b). That dose, which, unlike the NOAEL or LOAEL, does not have to be one of the experimental doses, represents the dose that results in a 5% or 10% response in the study population. The dose is determined from the experimental results by using a dose-response curve fitting program. Studies have confirmed that these levels of response (5-10%) represent the minimal level of effect that can statistically be resolved in a relatively robust bioassay (with the current design for detecting developmental toxicants). The benchmark dose (BMD) is frequently calculated using the lower confidence limit of the dose that results in a 5-10% response and thus represents with 95% confidence the lowest dose giving an increased 5% or 10% response in exposed populations over unexposed populations. Continuous responses from developmental toxicity bioassays include percentage of fetuses malformed, percentage of litters having one or more malformed fetuses, and birth weights (Kavlock et al. 1995).

RfDs and ADI values are derived from NOAELs or BMDs by dividing by uncertainty factors. Uncertainty factors, which are derived from animal and human data, generally involve dividing by a default value of 10 to account for uncertainties (EPA 1991). Uncertainty factors for developmental effects are applied to the NOAEL or BMD to include a 10-fold factor for interspecies extrapolation and a 10-fold factor for intraspecies variation. Additional 10-fold factors might be used to account for insufficiency in the database, extrapolation from subchronic to chronic exposures, and extrapolation from a LOAEL to a NOAEL. In practice, the aggregate product of all the default values is most often a factor of 100 to 1,000 (i.e., the acceptable human exposure concentration is 100- to 1,000-fold lower than the dose in the animal study that had little or no observable developmental effects). Because of increasing concern for susceptibility of children, the Food Quality Protection Act (Public Law 104-170; August 3, 1996) specifies an additional 10-fold default factor be applied under specific conditions. The default values are used unless there are research results that support the use of a different value. The need for uncertainty factors could be reduced with better data on comparative toxicokinetics, susceptible populations, and mechanisms of action. For example, if a NOAEL in a rat developmental toxicity study is used to set a RfD, then, in general, the NOAEL would be divided by 10 to account for extrapolation from animal studies to humans (interspecies extrapolation) and by another 10-fold factor to account for sensitive versus average human responses (intraspecies differences). Modifying factors can be used to change these default values of 10 and have been used to decrease the 10-fold factor used for interspecies extrapolation when there is sufficient knowledge about the similarities in toxicant kinetics in both rodents and humans (Moore et al. 1995).

Pharmaceutical agents almost always have a smaller difference between therapeutic and toxic dosages than is considered safe for environmental agents. A narrow therapeutic index is considered acceptable because pharmaceutical agents are given under the guidance of a health professional and because the therapeutic benefit has been determined to outweigh the risk. That is not the situation for agents such as environmental contaminants and food additives; hence, RfDs and ADIs are more conservative. It is worth stressing that the default uncertainty factors can be superceded when relevant exposure data are available to address key uncertainties. For example, the residual amounts of ethanol present in fruit juices, yeast breads, or vanilla ice cream are within a factor of 100-1,000 of the amount of alcohol (taken in alcoholic beverages) that results in fetal alcohol syndrome but do not constitute a risk to the fetus.

Renwick (1998) has discussed the concept of viewing uncertainty as being composed of kinetic and dynamic components and has proposed the concept of reducing the uncertainty associated with either or both of those components with additional mechanistic information. This structure has provided an initial framework by which mechanistic information can be used in the current risk assessment approaches. Chapters that follow will show how important new information on species differences and data on human variability can replace our reliance on such default approaches for developmental toxicity risk assessment.

Exposure Assessment

Exposure assessments are a critical component of the risk characterization process. Ideally, for developmental toxicity, one would like to have information about how much of the critical reactive species of the test chemical is present at the biological target throughout gestation. At the molecular level, that would mean knowing how much of and how long the toxic compound is bound to a specific target receptor. At the organism level, that would mean knowing how exposure occurred, when it occurred, how much of a compound was absorbed, what type of metabolism of the compound took place within the maternal compartment, and how that metabolism affected the exposure of the conceptus compartment. Information on how the conceptus metabolized and eliminated the compounds would also be important to know. Thus, an understanding of both time of exposure and dose is particularly important for characterizing the potential impacts of developmental toxicants. Numerous studies have shown dramatically different dose-response relationships when exposures occur even 8-12 hr apart due to the significant temporal differences in tissue susceptibility. Temporal differences make exposure assessments for developmental toxicants one of the most challenging of all exposure assessments. Although general toxicokinetic models exist for pregnancy, a refined exposure-modeling tool is not available for most chemicals. The lack of that tool is probably a key factor behind efforts to use exposure biomarkers and direct tissue measurements to improve estimates of conceptus exposures. Biomarkers of exposure have proved to be especially useful in addressing the limitations of exposure data in epidemiological assessments. Such biomarkers allow for a more accurate and representative exposure assessment and, when linked temporarily with biomarkers of effect, can frequently enhance the ability of a human epidemiological study to estimate human risk. New advances in biomarkers of susceptibility are allowing investigators to understand more fully variability in human response and have been proposed as improvements for human risk characterization.

One approach for using exposure information is to calculate a margin of exposure (MOE). The MOE is a ratio of the dose judged to be without effect to the anticipated levels of human exposure (Moore et al. 1995). However, because the calculation does not include the use of any default values to account for sources of uncertainty, it can only give a general indication of different levels of effects versus exposures levels for a quick exposure-scenario comparison.

The challenge for human health risk assessment is to convert exposure assessment information into relevant information for humans. The subsequent section on toxicokinetics will expand upon these concepts and will explain what information is needed to conduct relevant human exposure assessments.

Toxicokinetics

Toxicokinetics is the description of the absorption, distribution, metabolism, and excretion of a toxic chemical into and from the body (commonly referred to as ADME). The importance of chemical toxicokinetics for risk assessment is demonstrated by several example documents (California Environmental Protection Agency 1991; Moore et al. 1995; EPA 1996a; O'Flaherty 1997). These combined documents show a consensus that toxicokinetic data provide key elements to understanding species differences in response to developmental toxicants. Figure 3-2 shows an illustration of how exposure to a compound is then evaluated by using ADME. ADME controls how much of, when, and in what form a toxicant comes in contact with target organs. For developmental toxicants, these key questions are related to the amount and form of the toxicant that reaches tissues of the conceptus. Such knowledge can reduce the uncertainty in the extrapolation of results collected from experimental animals for the prediction of hazard associated with exposure of pregnant women. This section provides a brief discussion of a number of issues that require further investigation, as work continues to improve the scientific basis for risk assessment.

Decisions about toxicity hazard and risk to human development based on toxicokinetics from pregnant animals can rarely provide an unequivocal answer

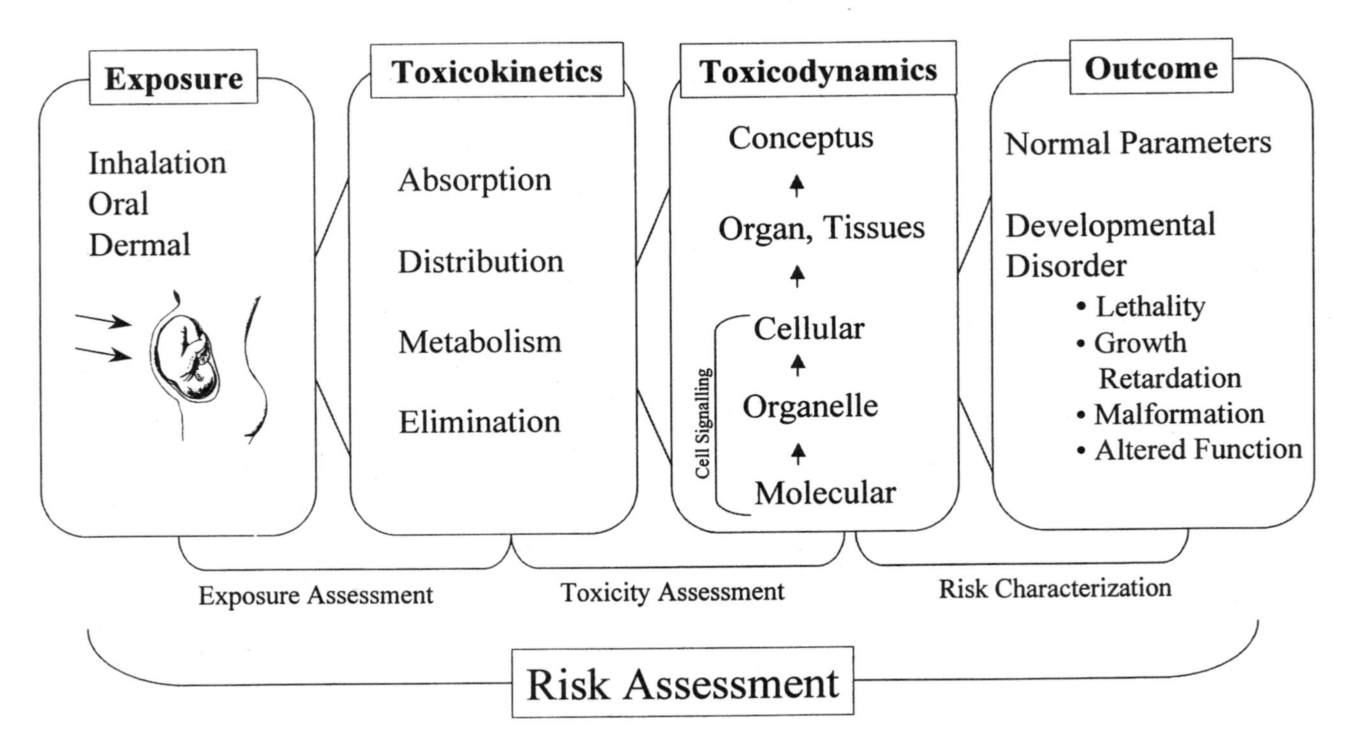

FIGURE 3-2 Overall framework to describe assessment of the effects of a toxicant on development.

of yes or no. The use of retinoid creams for dermal application can serve as an example. Retinoids were already known to cause developmental toxicity in pregnant animals of every test species examined. The regulatory decision concerning retinoid creams for dermal application was based on barely detectable changes in the blood concentrations of endogenous retinoids after dermal application of the drug. However, there is still minimal marketing of retinoid creams for dermal application. A similar case has been noted for vitamin A pills (retinyl palmitate) administered at doses greater than 30,000 IU per day (R.K. Miller et al. 1998).

If data show that a chemical is absorbed into the systemic circulation, then the next most important piece of toxicokinetic information is a determination of whether the biologically active toxicant is the parent compound, a metabolite, or both. Without such knowledge, the usefulness of other toxicokinetic data is diminished. If the active toxicant is not clearly delineated, then the qualitative and quantitative metabolic patterns that often vary between species for the agent of interest cannot be constructively applied in the characterization of hazard and the management of potential risks.

A determination of species concordance in susceptibility, in both basic research studies and developmental toxicity hazard assessment testing of chemical agents, is often the key element that provides the foundation for a generalization of the findings and extrapolations relevant for humans. Toxicokinetics are therefore studied to compare absorption, distribution, metabolism, and elimination of the test agent and its relevant metabolites. Toxicokinetic measurements can be used to determine the internal dose delivered to target tissues rather than relying on the administered dose, thereby taking into account species differences and individual variations in the extent and duration of systemic exposure in maternal and conceptus compartments. These interspecies variations and the interindividual differences in the same species indicate that each individual has a specific "fingerprint" of unique alleles of genes encoding drug-metabolizing enzymes (DMEs) as well as receptor and transcription factors that regulate the expression of genes encoding those DMEs. All DMEs appear to have endogenous substrates and are used in the biological functions of the normal animal or human, yet these enzymes have specificities that are sufficiently broad to metabolize endogenous substrates and environmental agents (Nebert 1994). Extrapolating toxicokinetic results from laboratory animals to nonhuman primates and humans is further complicated by the fact that the DMEs in the embryo and fetus differ considerably from one another with species-specific temporal patterns of expression observed throughout gestation and postnatal periods (Miller et al. 1996; Cresteil 1998). Such differences can allow drug-metabolizing reactions to occur in primates that are not yet functional in the common laboratory animal conceptus. The significance of such differences is magnified when considered with the differences in the susceptibility between the conceptus and adult tissues.

Few studies have been conducted in pregnant animals that have compared species-specific toxicokinetics. The data collected make it apparent that interspecies differences in susceptibility to developmentally toxic effects can frequently be due to differences in absorption, fate, or elimination of the agent of interest rather than to fundamental species-specific differences in biological response. Examples of embryotoxic drugs that have undergone detailed comparative evaluations are valproic acid (Nau 1986) and retinoids (Kraft et al. 1993; Kraft and Juchau 1993; Nau et al. 1994).

Regulatory authorities have required that the highest dose administered in animal toxicity studies in support of product registration be the estimated MTD. The administration of the MTD serves to maximize the likelihood of manifestation of a biological and possibly adverse response and ensure detection of all inherent toxicities. However, very large doses of environmental agents might be required in animals to reach the MTD, compared with anticipated human exposures at low concentrations. Such high doses of a test agent in animals can result in different kinetic and dynamic processes than those occurring at lower, more environmentally relevant exposures. For example, high doses can saturate elimination or repair processes or stimulate cell division or apoptosis, which might result in grossly exaggerated target organ concentrations and manifestations of toxicity. Indeed, toxicokinetic studies conducted in past decades have elucidated those phenomena for numerous therapeutic and environmental agents. The insights gained have led to the simplistic subdivision of linear and nonlinear classification of kinetics. Toxicokinetic data are essential to ascertain whether similar intervals or concentrations of the chemical or its metabolites result from different doses. High doses often result in nonlinear kinetics and subsequently elicit toxic effects that are not observed at low doses associated with linear toxicokinetics. Consideration of these kinetic and dose-response relationships is needed as new information is evaluated for use in risk assessment.

Toxicokinetic considerations are not only important in the interpretation of potential health effects and their relevance across species but they are also important in the determination of a developmental toxicity study design. For example, in one type of conventional developmental toxicology study design, studies initiate dosing at the beginning of organogenesis of the chosen test species. The study design might be flawed for compounds that have a long half-life, because steady-state concentrations are reached only after dosing for approximately four half-lives. Thus, a compound that has a half-life of 24 hr will not reach steady-state concentrations until 4 days after four consecutive daily administrations. Such a toxicokinetic property might miss the window of susceptibility to chemical perturbation of a specific developmental process in a long half-life test species. In animals with short gestation durations, such as the mouse or rat, the embryo might reach a developmental stage of decreased teratogenic sensitivity by the time the toxicologically critical steady-state concentration is achieved. For some agents, it may be possible to overcome this problem by starting exposure of the dam earlier in pregnancy. Toxicokinetic information can thus aid in the proper design of new studies or more accurately interpret results from investigations already completed. In this context, the committee needs to point out some serious shortcomings in the present approaches to interspecies extrapolations of toxicokinetic information, particularly those extrapolations that make the jump from laboratory animals to pregnant women. All too often, toxicokinetic data from pregnant animals are collected only for the maternal organism, and equally important aspects of the conceptus compartments are entirely lacking. As the examples of some chemicals studied in more detail have shown, the maternal-conceptus kinetics of chemicals and drugs change dynamically throughout gestation. Pharmacokinetic measurements in human pregnancy related to therapeutically used pharmaceutical agents are typically derived from blood and tissue samples collected from term deliveries and constitute only one or just a few time points after the drug's administration. Many significant changes occur throughout gestation that can make such term kinetic assessments less directly applicable for evaluation of first-trimester exposures.

The present dosing regimens in safety evaluations do not always consider the profound differences in elimination half-lives of chemicals between animal test species and humans (Nau 1986). The studies on valproic acid revealed dramatic species differences in the toxicokinetics of this drug that correlate with the species-specific teratogenic response (Nau 1986). Technical means of dosing that overcome the toxicokinetic differences between humans and pregnant animals exist and have been shown to be applicable and useful. For example, the studies on valproic acid were conducted by subcutaneously implanting osmotic minipumps that can deliver a chemical at a constant rate and produce maternal serum pharmacokinetic profiles with concentrations of the test chemical that resemble those occurring in humans much more closely than single or even repeated bolus administrations (Nau et al. 1981, 1985). However, conventional developmental toxicity testing designs still do not use that methodology routinely. It was critical for the committee to consider these factors in their deliberations on how to use new biological information for human risk assessment. Toward that end, it will be helpful to understand whether a developmental toxicant acts by exceeding a certain threshold peak concentration (Cmax) for a brief period of time, or whether an extended exposure to a certain concentration of the chemical over some period of time, is required to induce abnormal development. Such toxicokinetic relationships are graphically expressed as a plot of the concentration of the chemical of concern in maternal plasma (and preferably also in the embryo) against time. This visual display of the chemical-analytical presence of the substance and time is commonly known as area under the curve (AUC), as shown in Figure 3-3. In developmental toxicity studies, both the Cmax and the AUC concepts have been all too uncritically applied in assessing the value of toxicokinetics in pregnancy. The maternal AUC has often been used to draw conclusions about the chemical exposure of the conceptus without the availability of any toxicokinetic information from the conceptus compartments. The assessment of conceptus toxicant levels can be invaluable for understanding animal species differences in develop-

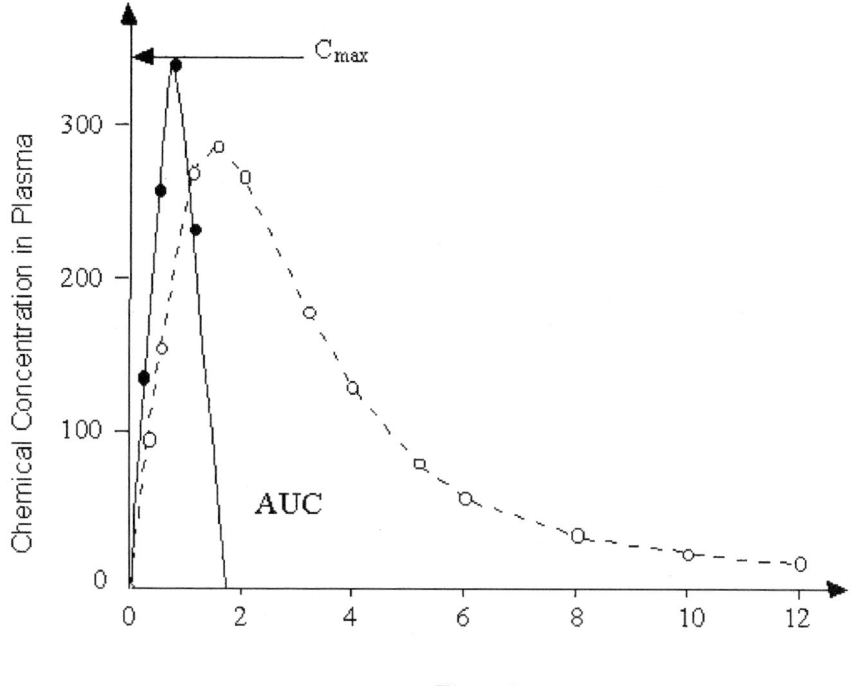

FIGURE 3-3 Two chemicals with different toxicokinetic properties are schematically illustrated. The concentration in maternal plasma of Chemical 1 (solid line) rises rapidly to reach its maximum (Cmax). Chemical 1 is then eliminated from the blood-plasma compartment in less than 2 hr after administration. In contrast, the plasma concentration of Chemical 2 (dashed line) rises more gradually and is slowly cleared from maternal plasma in an apparently biphasic fashion. The area under the curve (AUC) is defined by the plot of concentration of the chemical against time after administration.

mental toxicity and for assessing the validity of negative in vivo studies for compounds otherwise suspected to have high potential for developmental toxicity.

One chemical for which toxicokinetic data have been collected from maternal and conceptus compartments at two stages of pregnancy is 2-methoxyacetic acid, the proximate developmental toxicant derived from the maternal oxidation of 2-methoxyethanol, an ethylene glycol ether used as an industrial solvent. This chemical produces gross malformations in several test animal species examined, including nonhuman primates. Depending on the developmental age of an embryo at the time of exposure to sufficiently high concentrations of 2-methoxyacetic acid, the target tissues are either the developing anterior neuropore or the differentiating paw skeleton of the limbs, and exposure causes exencephaly or digit malformations, respectively (Terry et al. 1994). In the case of 2-methoxyethanol, the maternal plasma AUC of 2-methoxyacetic acid was highly indicative of that in the embryo and might serve as a surrogate of separate conceptus toxicokinetic measurements (Welsch et al. 1995, 1996).

Toxicokinetic information could be helpful in judging the extent of the hazard to humans from exposures if human kinetics are known (Yacobi et al. 1993). For example, the anticonvulsant drug valproic acid given to pregnant mice induces exencephaly in their embryos when a certain maternal plasma threshold concentration is surpassed for a very brief duration (Nau 1986). Larger total exposure over time (larger AUC) achieved by constant maternal drug infusion causes a dramatically lower incidence of exencephaly, indicating that the peak concentration (Cmax) rather than total exposure over time (AUC) induces the teratogenic response in mice. In contrast, clinical use of valproic acid for antiepileptic therapy requires the maintenance of valproic acid concentrations in an effective therapeutic range at which the required human doses produce serum Cmax values that are 6-10-fold lower than the teratogenic concentrations in mice (Nau 1986). A similar inference regarding Cmax as a cause of embryotoxic effects was made for caffeine in mice. A large single dose (100 mg/kg) induced a teratogenic response, whereas the same total amount divided into four separate administrations did not cause any malformations (Sullivan et al. 1987).

The embryotoxicity of other agents appears to depend on the total exposure over time (AUC). For example, the developmental toxicity of all-*trans* retinoic acid and cyclophosphamide (a chemotherapeutic alkylating agent) in the rat correlates best with duration of exposure (Tzimas et al. 1997; Reiners et al. 1987).

Caution in the interpretation of maternal AUC information without concomitant conceptus toxicokinetics is necessary because a single agent might act through both toxicokinetic exposure patterns, depending on the stage of development. 2-Methoxyacetic acid seems to induce mouse digit malformations best correlated with maternal and conceptus AUC (Clarke et al. 1992, 1993; Welsch et al. 1995, 1996). However, additional toxicokinetic data from both the maternal and the conceptus compartments at an earlier stage of mouse embryogenesis indicate that the agent induces neural tube defects that correlate best with Cmax in the conceptus tissues (Terry et al. 1994; Welsch et al. 1996). What is still lacking in these data is information on the toxicodynamic interaction of 2-methoxyacetic acid with a specific and still unknown recognition site (receptor) in the conceptus. The significance of considering both AUC and Cmax measurements for developmental toxicity risk assessment is especially important because of known temporal differences in tissue susceptibility. In cancer risk assessment, Haber's law (the product of concentration times time is equal to a constant) is used to normalize risk impacts. Such generalizing concepts cannot be applied in developmental toxicity risk assessment. A recent study by Weller et al. (1999) illustrated these differences for ethylene oxide developmental toxicity.

The toxicokinetic patterns that have been important in discriminating developmental toxicity are described here in terms of AUC and Cmax, and not in terms of metabolite profile (i.e., the qualitative similarities in a parent compound and its metabolites). Species are known to differ in the rates that they absorb, distribute, and excrete compounds (i.e., the metabolic rate manifested at AUC and Cmax). Pharmaceutical studies have demonstrated that metabolite profiles between species are often similar (Nau et al. 1994), and this similarity is one of the reasons that it is common practice to use various animal models to assess the potential toxicity of chemicals. The committee will later propose that human DME genes be introduced into model animals to further reduce differences in metabolism. These transgenic animals are likely to have similar metabolite profiles as humans but will be considerably different from humans in terms of metabolic rate.

In summary, the correct application of toxicokinetic information in the determination of hazard and in judgments concerning risk characterization requires a broad view of pharmacological, toxicological, and embryological principles. These principles have guided the committee in their considerations on how most effectively to incorporate recent advances in molecular and developmental biology in risk assessment.

BIOMARKERS

As the committee has outlined in the previous sections of this chapter, key challenges facing risk assessors include the need to understand critical initial events caused by toxicants (events that occur at low doses and early stages of toxicity) and to understand the implications of animal toxicity for human health. Ideally, appropriate biomarkers could serve as indicators to link exposure and early biological effects, and ultimately link those early effects with disease or pathogenesis. As numerous NRC reports have indicated, biomarkers of exposure, effects, and susceptibility are exactly the types of indicators that are needed to address these risk assessment challenges.

Specifically, biomarkers for developmental toxicity have been reviewed in the context of reproductive toxicology in a previous NRC (1989) report, *Biologic Markers in Reproductive Toxicology*. Three types of biomarkers have been defined (NRC 1989):

1. A *biologic marker of exposure* is an exogenous substance or its metabolite(s) or the product of an interaction between a foreign chemical and some target molecule or cell. The biomarker is measured in a compartment within an organism.
2. A *biologic marker of effect* represents a measurable biochemical, physiological, or other alteration within an organism that, depending on magnitude, can be recognized as causing an established or potential health impairment or disease.

3. A *biologic marker of susceptibility* is an indicator of an inherent or acquired limitation of an organism's ability to respond to the challenge of exposure to a specific foreign chemical substance.

It is easy to see from those definitions that biomarkers of exposure and effect should be useful for linking early, low-dose exposures with pathogenesis and providing a platform for cross-species and cross-compound comparisons. Likewise, it is easy to see how biomarkers of susceptibility could be especially useful for assessing differences in temporal sensitivity and developing tissues and for cross-species and intraspecies comparisons.

The validity of a biomarker for risk assessment depends on a demonstration that it is highly associated with the outcome, such as in this context, a developmental defect. At present, few biomarkers meet the test. When evaluating biomarkers, one must investigate the mechanistic basis of the association between the biomarker and the adverse events and then determine the reliability of the comparison in a large and varied population for specificity, sensitivity, and reproducibility. A biomarker does not have to be the definitive end point for defining the problem, although that is preferable, but even having a tool to identify candidate individuals for more definitive testing can be helpful. For example, the maternal serum α-fetoprotein levels are useful in clinical screens for neural tube defects, but serum α-fetoprotein is not a definitive test for such defects.

Temporal considerations are important for using biomarkers for developmental toxicity. When considering the validity of a screening test, the gestational age at the time of assessment and, more important, the gestational age at the time of exposure to the toxicant must be considered. Such issues have growing importance as fetal therapeutic interventions are increasingly available for use (Miller 1991). Accessibility to the biological material of interest is temporally determined. For example, invasive (e.g., percutaneous umbilical blood sampling, PUBS) and noninvasive biomarker procedures (e.g., ultrasound and Doppler) for assessing the developmental state of the fetus have made possible the use of interventions that have revolutionized the clinical capabilities to treat the affected fetus. At the same time, physicians can now predict, based upon patterns of uterine blood flow, which pregnancies have a greater risk for a poor reproductive outcome (Jaffe 1998).

Biomarkers of Exposure

The pre-eminent example in this category is methylmercury (MeHg). Maternal hair concentrations, as well as blood concentrations, of MeHg correlate with adverse developmental outcomes in the children exposed in utero (Clarkson 1987). Different threshold exposures have been observed in the adult and fetus for detrimental effects, based upon the hair analyses for MeHg. In fact, temporal records of MeHg exposure can be determined by measuring MeHg levels at various places in the human hair shaft.

Unfortunately, not all substances are comparable to MeHg in lending themselves to use as exposure biomarkers. For example, a debate continues concerning the dose of vitamin A, as retinyl esters versus retinol, that will produce malformations in humans. Of particular interest is the discussion about what dosages of vitamin A are needed to increase the blood concentrations of retinoic acid metabolites significantly above those seen in normal pregnant women. Doses of 30,000 international units of retinyl palmitate per day administered orally did not significantly increase the concentrations of retinoic acid in nonpregnant women above those concentrations already circulating in untreated pregnant women (R.K. Miller et al. 1998). Still, for many agents (e.g., ethanol, solvents, and retinoids) that cause developmental toxicity at or near adult toxic dosages, one might be able to monitor concentrations of the compound (or metabolites) in the exposed individual and thereby establish possible risk. Thus, biomarkers of exposure have the potential to be critical in establishing potential risk at a sensitive period during development.

For developmental toxicants that can produce developmental defects at dosages or concentrations not causing identifiable immediate adult toxicity (e.g., thalidomide and cigarette smoking), biomarkers of exposure that reveal actual concentrations of parent compounds or metabolites (e.g., cotinine as a nicotine-metabolite measure of cigarettes smoked) might be the only available indicators of risk.

It is believed that subtle changes in gene expression, as assayed by large-scale microassay analyses, are good examples of newly developing biomarkers of exposure. Those biomarkers still need to relate expression changes with early biological effects, occurring well before toxicity. In fact, there are extensive discussions to determine if these are truly "biomarkers of exposure" or "biomarkers of effect." Current efforts are under way to improve the detection of differences in patterns of gene expression for various chemical classes (e.g., peroxisomal proliferators and oxidants), with the aim of improving use of patterns rather than single changes as exposure biomarkers. In cases in which maternal toxic effects occur, the patterns of expression changes might be especially useful biomarkers to improve detection of developmental versus maternal toxicity.

Biomarkers of exposure often are used in occupational and molecular epidemiology. Aniline-hemoglobin adducts, benzo[*a*]pyrene-DNA adducts, aflatoxin B1-DNA adducts, elevated metallothionein, and elevated urinary 8-hydroxy-deoxyguanosine levels have been useful biomarkers for specific exposures. The cancer risk of exposure to dangerous concentrations of foreign or endogenous chemicals is assessed by the activation of a proto-oncogene or the inactivation of a tumor-suppressor gene (e.g., p53), reflecting the mutation of these genes in somatic tissues. In this usage, the biomarker of exposure also comes close to being a biomarker of effect, insofar as mutagenesis is thought to be an important step in carcinogenesis. Still other biomarkers, such as aryl hydrocarbon hydroxylase (AHH) and CYP1A1 at high concentrations, are taken to reflect induction of the enzymes by high internal concentrations of potentially toxic agents and are used to predict whether a population or individual might be at risk for perinatal morbidity or mortality.

Biomarkers of Effect

Biomarkers of effect at the molecular level are becoming as important as monitoring metabolites or a parent compound. Recently, Perera et al. (1998) confirmed an inverse relationship between concentrations of cotinine in plasma from newborns, a metabolite of nicotine, and birth weight and length. They also demonstrated a significant association between decreased body size at birth, body weight, and head circumference and increased concentrations of polycyclic aromatic hydrocarbon (PAH)-DNA adducts in umbilical cord blood above the median. This had previously been demonstrated for PAH-DNA adducts measured in the human placenta (Everson 1987; Everson et al. 1988). Such associations were related to cigarette smoking and environmental pollution. Those examples show that there can be a practical use of biomarkers of effect at the molecular level to assess exposure. Such measurements not only allow for epidemiological evaluations of environmental pollutants, such as cigarette smoke and air pollution, but they also allow those evaluations to help identify a subpopulation of individuals that might be at risk. Critical applications for such biomarkers in developmental toxicology are in the identification of those at risk, with hopes of reducing that risk by modifying exposure and by developing other intervention strategies to decrease the incidence of developmental defects. Other biomarkers include indicators of normal cell processes (e.g., cell proliferation that may occur at inappropriate times or at different levels of expression). Proliferation markers are often used for assessing immunological impacts where proliferation status is evaluated in the context of differentiation status. These immunological studies present similar issues to those in biomarker studies in developmental toxicology. Likewise, biomarkers of the apoptotic process (e.g., early biomarkers such as nexin, enzymatic changes in various caspase levels or types, and late biomarkers such as DNA fragmentation) can provide temporal, mechanistic biomarkers of effect that are also highly relevant for developmental toxicity assessments.

Other biomarkers of effect include increased concentrations of α-fetoprotein in amniotic fluid as indicative of neural tube defects, since delayed closure of the tube is thought to allow escape of this protein. Other biomarkers might be used in combination to enhance the collective ability to diagnosis or predict possible developmental anomalies (e.g., the triple assay of human chorionic gonadotropin, estriol, and α-fetoprotein for trisomy 21).

Biomarkers of Susceptibility

These biomarkers are used to identify either individuals or populations who might have a different risk based upon differences that are inherent (i.e., genetic) or acquired (i.e., from life history and conditions). The inherent category includes the polymorphisms for genes encoding DMEs and for genes for the receptor and transcription factors regulating the expression of the genes for DMEs, as discussed in a previous section of this chapter. The category also includes polymorphisms for genes encoding components of developmental processes, although the latter are still not well understood. The acquired category includes previous disease conditions, antibody immunity, nutrition, other chemical and pharmaceutical exposures, and various capacities for homeostasis.

As a monitor, the placenta has been a key test organ for identifying such sensitive populations and their responses to environmental exposures. For example, Welch et al. (1969) and Nebert et al. (1969) demonstrated that AHH is induced in the human placenta of cigarette smokers. With the ever-improving tools for investigation, biomarkers now have moved from proteins and enzyme activities induced by polycyclic aromatic hydrocarbons (e.g., benzo[*a*]pyrene) and dioxin (Manchester et al. 1984; Gurtoo et al. 1983) to biomarkers of combined effect and exposure, such as mRNAs (e.g., CYP1A1) plus DNA adducts (Everson et al. 1987,1988; Perera et al. 1998).

The molecular probes to identify such subpopulations are useful as biomarkers not only for identifying individuals at risk but also for exploring the underlying mechanisms by which those individuals or populations are at risk by demonstrating allelic polymorphisms in a particular gene. As discussed above, gene-environment interactions have been noted for the induction of cleft palate in humans (Hwang et al. 1995) through a combination of cigarette smoking and TGF genotype. Alone, neither variable demonstrates an association with cleft palate. Such an example demonstrates the possibility for understanding why only a small percentage of a population exposed to a developmental toxicant might be at risk, but more important, it identifies a biological association that might lead to a mechanistic understanding of how a particular developmental defect might occur.

There are serious concerns about using the term "biomarkers of susceptibility" to describe a person's particular set of alleles because these have a hereditary basis (e.g., slow acetylator activity, low G6PD activity, low 5,10-methylene tetrahydrofolate reductase activity, or high CYP1A1 activity). The committee emphasizes the need for a distinction between biomarkers of susceptibility reflecting inherent limitations versus those reflecting acquired limitations. The former require a full understanding of the complex genetic implications before they can be used. An allele encoding an altered DME, for example, might put the individual at increased risk for toxicity caused by one environmental chemical but at decreased risk for toxicity caused by another drug or environmental chemical. Combinations of alleles might also have exaggerating or compensating effects.

LIMITATIONS IN DEVELOPMENTAL TOXICITY RISK ASSESSMENTS

Although it can be argued that the current approach of risk assessment appears to have worked reasonably well for hazard identification, many assumptions must be made before it can be applied. One such default assumption is that outcomes for rodent tests are relevant for human risk prediction. Such assumptions are generically used because information on the mechanisms of action for specific developmental toxicants is inadequate and because the lack of mechanistic information results in the use of default uncertainty factors. The most important limitation is the paucity of human data, and the lack of methodology to adequately assess humans. Mechanism of action can be pursued in animal models, but it is also the lack of an understanding of human development that hampers risk assessment.

For risk characterization, the bioassays used for regulatory assessment have provided limited dose-response information. The information is limited because the focus is on the effects of high doses at or near maternal toxicity to emphasize identification of hazards. That focus has provided little quantitative information on the dose-response relationship in the low-dose region, the region of greatest importance for extrapolation in human risk assessment. The lack of useful dose-response data has had several impacts. As mentioned previously, conservative use of uncertainty factors predominates for converting NOAELs and BMDs to RfDs for determination of acceptable safe exposure levels. The dominance of animal testing at high doses has also had the unfortunate consequence of providing minimal useful mechanistic information, because assessments are frequently conducted at doses where homeostatic mechanisms are overwhelmed (Nebert 1994), and mechanistic clues about critical toxicant-induced changes are hidden.

The lack of mechanistic information has also resulted in assumptions about sensitivity among humans. Present practice in risk assessment almost always makes use of a default factor of 10 to take into account the variability in sensitivity (i.e., there is a 10-fold difference in susceptibility of the most sensitive individual and the average individual). This assumption has been experimentally addressed for relatively few chemicals (for a review, see Neumann and Kimmel 1998). However, the default assumption could change as researchers gain more information about the underlying basis for responses to toxicants. To date, the greatest progress in characterizing human variability is from research on DMEs. With time, there will also be data on other factors that influence susceptibility. For example, as discussed in detail in Chapter 5, a particular allele of transforming growth factor conveys more than a 10-fold increase in risk of oral clefts in infants whose mothers smoke cigarettes (Hwang et al. 1995; Shaw et al.1996).

As knowledge of human variation in responses increases from the results of the Human Genome Project, a risk-assessment framework is needed in which these default factors are replaced with mechanistic data on relevant toxicant-induced changes.

Molecular approaches should be useful in resolving the issue of extrapolation across species. There is general agreement that the molecular control of development is highly conserved, although the pattern of development of structures at higher levels of biological organization can be very dissimilar. The committee discusses such conservatism in Chapter 7 and suggests that models that assess a small number of those control points and pathways might be relevant for evaluating the potential for chemicals to impact the critical pathways in development, regardless of the species from which the model is derived. The same principle applies to extrapolation from rat or mouse to humans. The critical data to support predictions of ultimate effect in human embryos include a description of the pathogenetic steps that ensue from toxicant actions at the molecular level and lead to structural malformations. Pattern formation genes and signal transduction pathways are so highly conserved across groups of animals that actions of toxicants on those gene products and processes are likely to be comparable and have similar toxicodynamic impacts. The events that follow perturbations at the molecular level are likely to be more prone to interspecies variability, for example, the toxicokinetic differences observed with chemicals that require metabolic activation, as metabolic rates often vary markedly between species. At this point, the predictive value of hazard identification data in alternative models, particularly those that are phylogenetically removed from humans, becomes limited. Therefore, characterization of the pathogenetic events that result in dysmorphogenesis will lead to better prediction of (1) whether the critical events are present and when they are functional in humans, predisposing them to an adverse outcome; and (2) the kinds of adverse outcomes that are possible, based on the temporal and spatial locations of the critical events.

A better understanding of the molecular and cellular mechanisms involved in the pathogenesis of abnormal development might provide a method for answering such questions as to whether a residual level of risk exists at the RfD or ADI and what that level might be. A residual level of risk might also provide a method for determining what exposure concentration can be permitted before the probability of an adverse event begins to increase. The resolving power of animal studies to distinguish an increase in the rate of frank malformations is relatively weak. For example, in a study with 20 pregnant rats per dose group, an increase in the malformation rate must double from the background rate to be statistically significant. Mechanistic and pathogenetic events may prove to be much more sensitive and, therefore, provide a data-driven means to extend the dose-response curve below the NOAEL or BMD for malformations. Although these effects might not be adverse, they might be biomarkers or early indicators for the process of pathogenesis and might help to determine the shape and slope of the dose-response curve at doses that approach relevance for human exposure. It then becomes a matter of conducting further research to understand the magnitude of response on a molecular or cellular end point that is needed to produce a structural defect with adverse physiological, structural, or functional developmental effects. A method has been proposed for constructing dose-response curves that combine data on frank malformations with data on less-severe effects that are not considered adverse. For example, Allen et al. (1996) combined data on rib malformations with those on rib variations in rats prenatally exposed to boric acid. This method could easily be adapted to include molecular events.

Although it can be postulated that many molecular and cellular events that are the precedents of abnormal development are unlikely to have strictly linear dose-response curves, there is minimal information on developmental specific processes. There have been extensive discussions on the shape of receptor-versus-nonreceptor-based responses that were initiated directly from recent advances in the understanding of molecular events; yet, little is known about actual events at low-dose exposures, as opposed to generation of hypothesized dose-response relationships at low doses. Hypothetical biologically based dose-response models have been proposed on a toxicant-specific basis (Shuey et al. 1995; Leroux et al. 1996). What appears to be most relevant for this report is a call for increased understanding of toxicant-induced molecular changes and an investigation of how these early events are linked to manifestations of adverse developmental outcomes. Empirical work will be needed to establish the magnitude of response at each level of organization required to provoke a response at the next level. Such investigations should be conducted to obtain quantitative information on the kinetics of the toxicant and the dynamics of the toxicological interaction in the temporal context of development (Faustman et al. 1999; Faustman et al. 2000).

Current practices in developmental toxicity risk assessment recognize the concept of "critical windows of sensitivity" in development, but a fundamental understanding of applying the molecular and developmental biological events that define those windows is lacking. This lack of understanding again results in the application of additional child-specific uncertainty factors in efforts to address the sensitivity of the developing conceptus rather than emphasizing the search for the biological understanding of critical windows of susceptibility.

A corollary to the problem of low-dose extrapolation is the assumption that effects observed at high-dose concentrations in experimental animals are relevant to the prediction of risk of adverse effects at ambient exposure concentrations. As discussed previously, testing of chemicals has the inherent dilemma of requiring exaggerated doses and concentrations to maximize the chances of detecting the potential for adverse effects and requiring understanding that the interpretability of the results might be limited because of the possibility that physiological processes in the pregnant animal have been so overwhelmed that the observed responses are qualitatively different from the responses at lower doses. The uneasy resolution of the dilemma has been to assume that the high dose and concentration effects are predictive of all effects at low exposure doses and concentrations unless they are proved to be secondary to maternal toxicity. Some research has been conducted to demonstrate the existence of maternally mediated mechanisms of adverse development (Daston et al. 1991a, 1994). Examples are the induction of transitory zinc deficiencies in the dam by metallothionein inducers (Daston and Lehman-McKeeman 1996) and the overwhelming of acid-base buffering by acidic metabolites of ethylene glycol (Carney et al. 1996). Understanding the molecular processes that lead to specific developmental abnormalities will be useful in determining the low-dose relevance of high-dose effects. In those instances in which the high-dose effects are predictive of low-dose responses, the relevant molecular processes would be expected to increase with dose (i.e., to involve higher levels of gene expression or cellular response and involve more cells). For those instances in which the high-dose effects are the result of a secondary mechanism, the dose-response curve for the adverse effect and the underlying molecular perturbation would be expected to be steep, with an inflection at the dose where the maternal homeostasis was overwhelmed.

SUMMARY

This chapter has defined developmental toxicity risk assessment and outlined issues that regulators face as they strive to protect the human population from chemically induced birth defects. Each section also identified limitations in the current knowledge and methodologies. Biomarkers for developmental toxicity are also discussed. They hold great potential for epidemiological analysis of developmental defects, especially those defects due to complex gene-environment interactions. The information presented in this chapter, and in the next chapter on mechanisms of developmental toxicants (Chapter 4), will be used to define the current state of developmental toxicology and will provide a context for how advances in developmental biology and genomics can improve the approaches for protecting public health.

4

Mechanisms of Developmental Toxicity

This chapter presents a historical perspective of the field of developmental toxicology and then an analysis of various mechanisms by which agents cause developmental toxicity, as currently understood by developmental toxicologists. The chapter was prepared in response to the first charge to the committee to evaluate the evidence supporting hypothesized mechanisms of developmental toxicity.

HISTORY OF DEVELOPMENTAL TOXICOLOGY: GROWTH OF A NEW FIELD

Teratology, the study of abnormal development, has a long history, much of which is shared with developmental biology. Progress in experimentally determining the causes of abnormal development began in earnest in the nineteenth century and continued through the first half of the twentieth century. By then it was recognized that genetic, nutritional (e.g., cretinism), infectious (e.g., congenital rubella syndrome), and chemical factors caused congenital anomalies in humans and that such manifestations of perturbed development could also be experimentally elicited in various animal systems.

Much of the early research into chemical causes of abnormal development used the same animal models and approaches that were in common use in experimental developmental biology. For example, Dareste elucidated the concept of critical periods of susceptibility by treating chick embryos at various developmental stages with hypoxia, which was induced by coating various fractions of the egg's surface with wax (Dareste 1877, as cited in Wilson and Fraser 1977). The author concluded that the developmental stage during which treatment occurred is the critical factor in determining which organ was affected. The concept of developmental phase specificity and other core principles of experimental developmental toxicology grew out of that early work. Knowledge about normal developmental processes was applied to understand abnormal development, the pathogenesis resulting in malformations. In the course of such investigations, various chemical and physical insults were used to perturb development to elucidate the underlying normal processes.

In the 1950s, approximately at the time when basic scientific research in normal and abnormal embryo development was expanding, phocomelia and amelia of the arms and legs, two very rare manifestations of human limb defects, occurred in infants born in several European countries and Australia. Independently and almost simultaneously, two alert physicians, one in Australia and one in Germany, concluded that the sudden increase in those defects was attributable to treatment with the pharmaceutical thalidomide early in the pregnancies of the mothers of the affected babies (McBride 1961; Lenz 1961; Lenz and Knapp 1962). Until that time, thalidomide was thought to be a "nontoxic" sedative/hypnotic. The removal of thalidomide from the international market brought the phocomelia epidemic to an end. By various estimates, 7,000 to 10,000 babies were affected. Animal studies later confirmed that thalidomide has toxicological properties that affect development in some species (for reviews, see Neubert and Neubert 1997 and Stephens 1988).

Mass media coverage of the thalidomide tragedy brought the field of developmental toxicology to the attention of the public. For the first time, it was tragically demonstrated that a chemical agent had the potential to profoundly affect human development. Responsible scientists and members of the regulatory community recognized that developmental toxicity might not be unique to thalidomide. It was soon realized that the toxicity testing methods for pharmaceutical products in use at that time were focused primarily on the adult and were inadequate to predict the response or susceptibility of the embryo or fetus. An ad hoc committee of scientists, including many of the charter members of the Teratology Society, was assembled by the U.S. Food and Drug Administration (FDA) to develop guidelines for assessing the hazard potential of new therapeutic agents for developmental toxicity. Other regulatory agencies in the United States and around the world have since adopted similar guidelines, which are applied not only to pharmaceuticals but to all classes of chemicals with significant human exposure potential.

The recognition that environmental factors, whether chemical, physical, or biological, could elicit malformations resulted in a new emphasis on identifying and characterizing other agents that might cause adverse impacts. The disciplines of developmental biology, pharmacology, toxicology, and obstetrics and gynecology were brought together to address these research questions. As our ability to understand the cellular and macromolecular actions of chemicals grew, so did the field of developmental toxicology, which increasingly focuses on understanding the mechanisms underlying chemically induced developmental defects. Perhaps the most influential review and assessment of the state of knowledge published in that expansionary phase of teratology was J.G. Wilson's landmark book *Environment and Birth Defects* (1973). To this day, Wilson's book is useful as a summary of the principles of teratogenesis (described below) that were developed from the work of teratologists such as Josef Warkany, Lauri Saxen, Robert Brent, Jan Langman, and David Smith. This time period was also one in which numerous clinical discoveries were made of chemical agents that produce abnormal development in humans. Fetal alcohol syndrome, undoubtedly already a problem since antiquity, was first recognized and described in the scientific literature (Lemoine et al. 1968; Jones and Smith 1973; Jones et al. 1973). Unfortunately, in spite of the intervening years and educational efforts, this syndrome remains all too common and is currently estimated as affecting 1.95 of every 1,000 live births in the United States (Abel 1995). A retinoic-acid-induced human embryopathy was described soon after Accutane (13-*cis* retinoic acid) was introduced as an efficacious drug to treat severe cystic acne (Lammer et al. 1985). Anticonvulsant drugs were also recognized as associated with abnormal development (Finnell et al. 1997a). A large body of work on the heavy metal lead, a ubiquitous contaminant, has shown that it can produce subtle effects on neurobehavioral development (for a review, see Bellinger 1994), emphasizing the fact that abnormal development can manifest itself as subtle functional deficits and not just structural changes. Another heavy metal, mercury, also was identified as a human developmental toxicant after an epidemic of cerebral palsy with microcephaly in Minamata, Japan, was associated with the ingestion of fish contaminated with methyl mercury (Harada and Noda 1988).

PRINCIPLES OF TERATOLOGY

Prior to and shortly after the thalidomide crisis, a body of data had accumulated showing that many chemical, biological, and physical agents can induce malformations in mammalian species, such as mice, rats, rabbits, and guinea pigs. On the basis of those accumulated data, Wilson in the 1970s formulated six principles of teratology that have guided research in developmental toxicology to this day (Wilson 1973). These principles are relevant to cite because they provide the context for the specific mechanisms of toxicity considered later in the chapter.

1. "The access of adverse environmental influences to developing tissues depends on the nature of the influences (agent)." This is to say, developmental toxicants can be accessible to the conceptus (the embryo or fetus, plus the embryo-derived extra-embryonic tissues) in two ways, directly or indirectly. Examples of the former include ionizing radiation, microwaves, and ultrasound, which travel directly through maternal tissues without modification and then interact with the conceptus. Most known developmental toxicants gain access to

the conceptus indirectly. In the maternal body, they are subject to potential metabolic alterations (e.g., biotransformations in the liver), distribution, storage, and excretion that either enhance or diminish their potential to affect the conceptus adversely. The net result of all these interventions is that some level of active developmental toxicant is available to cross the placenta and eventually reaches target sites in the conceptus. Although it is frequently assumed that the developmental toxicant must reach targets in the conceptus to disrupt development, it should be noted that adverse effects on growth and development can be mediated indirectly through effects on accessory tissues, such as the yolk sac and placenta, or on maternal tissues.

2. "The final manifestations of abnormal development are death, malformation, growth retardation, and functional disorder." This principle highlights the now well-known fact that structural malformations are not the only possible outcome after the conceptus is exposed to a developmental toxicant. In fact, it is now known that in many cases the outcomes are interrelated. For example, at a relatively high dose of a developmental toxicant, the conceptus might suffer a high level of cell death that cannot be replenished by available repair and compensatory mechanisms. This, in turn could result in growth retardation if the induced cell death is widespread, and in death of the conceptus if the cell death compromises organ systems essential for viability of the conceptus. At lower doses particular malformations and functional disorders might occur. Which outcome, or combination of outcomes, will occur depends on the dose and chemical characteristics of the developmental toxicant (discussed in the third and fifth principle, respectively) and the developmental stage of the conceptus at the time of exposure (discussed in the fourth principle).

3. "Manifestations of deviant development increase in degree as dosage increases from the no-effect level to the totally lethal level." Sufficient evidence was available in the 1970s to support the relationship of dose with the incidence of structural malformations, death, and, to a lesser degree, growth retardation. Evidence accumulated since then extends it to functional deficits as well. It is also important to point out that the relationship between dose and response, although monotonic, does not have to be linear. It can be a steep S-shaped curve for developmental toxicants, sometimes going from a no-effect level to maximal effects within a doubling of the dose.

4. "Susceptibility to teratogenic agents varies with the developmental stage at the time of exposure." The change of susceptibility was originally published by Wilson as a hypothetical curve in which the degree of sensitivity to developmental toxicant-induced structural malformations was low during the pre-implantation phase, maximal during organogenesis, and low during fetal development. This shape of the developmental sensitivity curve reflects results from many stud-

ies of a variety of developmental toxicants, when developmental defects are scored at birth. The curve highlights the general conclusion that organ systems are most susceptible to pertubation by developmental toxicants, just prior to and during the overt phase of organ formation and differentiation, which occurs in mammals after the period of implantation, streak formation, and streak regression.

Several caveats need to be addressed, however. First, the pre-implantation period should not be viewed as a refractory period in terms of induced structural malformations (Rutledge 1997; Dwivedi and Iannaccone 1998). For example, ethylene oxide (EtO) can induce structural abnormalities in mice when administered during pre-implantation stages of embryogenesis (Generoso et al. 1987). Results from this study are particularly instructive because they show that agents can induce skeletal effects when administered to the pregnant dam at the zygote stage of development, long before skeletogenesis begins. Moreover, the spectrum of skeletal defects observed after exposure at the zygote stage differs from those observed after exposure during organogenesis. The mechanisms underlying that stage-specific effect of EtO on skeletal development are unknown.

The second caveat is that, although the susceptibility curve generally reflects the reality for structural defects, it is not a good generalization for developmental-toxicant-induced death, growth retardation, or functional deficits. For example, toxicant-induced death tends to occur most frequently at pre- and peri-implantation stages. As many as 30% of fertilized human oocytes are estimated to die during those early stages (see Chapter 2), and the role of developmental toxicants in that human loss is largely unknown.

The final point to be made is that development from fertilization to birth is a progressive process so that any adverse outcome (i.e., death, growth retardation, malformation, or functional deficits) after exposure to developmental toxicants will be dictated, in part, by the set of developmental processes active at the time of exposure.

5. "Teratogenic agents act through specific mechanisms on developing cells and tissues to initiate abnormal embryogenesis (pathogenesis)." Recent research focusing on how exogenous chemicals interact with endogenous molecular targets has increased our understanding of the mechanisms of action of toxicants. A detailed discussion on the mechanisms of action of toxicants follows this section.

6. "Susceptibility to teratogenesis depends on the genotype of the conceptus and the mother in which this interacts with environmental factors." This principle was originally based on the knowledge that different species and strains of animals respond differently to a developmental toxicant. For example, it was already known in the 1970s that mouse embryos are unusually susceptible to the induction of cleft palate by glucocorticoids, and other mammalian species are resistant. Also, some mouse strains are sensitive to hyperthermia-induced neural

tube defects and others are not. The particular genotype of individual offspring in the same mouse uterus was shown to be associated with benzo[a]pyrene-induced birth defects (Shum et al. 1979; Nebert 1989). A more recent example is the finding that oral clefting is more common in the offspring of mothers who smoke and who have a variant allele of the transforming growth factor (TGF) gene. The correlation implicates direct or indirect interactions between constituents in tobacco smoke and TGF, a secreted protein that binds to the epidermal-growth-factor receptor and is known to be expressed in palatal epithelium before and during palatal closure (Hwang et al. 1995; see Chapter 5). Examples such as these, together with a wealth of data indicating that many developmental defects of unknown etiology exhibit a multifactorial pattern of inheritance, have led to the conclusion that gene-gene and gene-environment interactions play a significant role in the etiology of many developmental defects.

MECHANISMS OF TOXICITY

An understanding of how exposure to a toxicant can result in an adverse developmental outcome is needed to develop intervention and preventive public health practices. Risk assessors seek to obtain mechanism-based toxicity results from animal tests in order to make justifiable extrapolations to humans. The process by which a toxicant can produce dysmorphogenesis, growth retardation, lethality, and functional alterations commonly is referred to as the "mechanism" by which developmental toxicity is produced. In general, it has been difficult to analyze mechanisms in sufficient detail and depth for risk assessment purposes. There are four reasons.

1. Normal development is extremely complex, and it is possible that there is a myriad of points at which a toxicant might interact with an important molecular component and cause developmental toxicity. Information about molecular components and processes of development has only been available in the past few years, largely through the study of developmental mutants of invertebrate model organisms, such as *Drosophila* and *Caenorhabditis elegans*. As highlighted by Wilson's principles, an understanding of mechanisms would be greatly enhanced by identifying critical key events altered by toxicants. Recent advances in research on signaling pathways and genetic regulatory circuits in development might have identified especially critical processes, ones that, if studied for their alteration by developmental toxicants, might provide exciting new clues for mechanistic investigations (see discussion in Chapters 6 and 7). For now, such insights are available in only a few cases, such as toxicant interactions with components of the nuclear hormone-receptor family of signal receptors and gene regulators.

2. Environmental toxicants include a wide range of chemical, physical, and biological agents that initiate a wide variety of mechanisms. Some agents are

specific for one or a few targets in the development or physiology of the conceptus, and others have a broad effect on many targets at different times and places in the conceptus and mother. Thus, the developmental toxicologist who focuses on these agents is probably faced with a wide variety of mechanisms.

3. Some toxicants might affect only a fraction of individuals in the population, probably because of genetic differences or differences in health history (diseases, nutrition, or other exposures). The differences add considerable complexity.

4. A mechanistic understanding of developmental toxicity involves understanding at several levels of biological organization. Once a toxicant interacts with a molecular component of the cell, it presumably affects its immediate function, so the function and alteration must be known. Then, the consequence for the altered function for the completion of a developmental process must be known. For example, in order to link specific branchiofacial defects with the action of a suspected toxicant, it is necessary to characterize the migratory events, proliferation control processes, and patterns of differentiation-promoting signal systems that affect neural crest cells from the time of their emigration from the neural tube. Other kinds of toxicants might alter specialized functions of organs of the fetus (e.g., the heart) and thus manifest impacts at the organ level. Yet other toxicants might cause cell death in the conceptus at a variety of times and locations and have multiple impacts.

A developmental toxicologist must understand the potential action of toxicants at many levels of biological organization to understand the overall processes of developmental toxicity (Faustman et al. 1997). Recognizing this dilemma, the U.S. Environmental Protection Agency (EPA) and the International Programme on Chemical Safety (IPCS) have defined chemical "modes of action" in addition to "mechanisms." Modes of action are described in the proposed EPA Guidelines for Carcinogen Risk Assessment (1996b), and in the IPCS guidelines for international applications (IPCS Workshop on Developing a Conceptual Framework for Cancer Risk Assessment, 16-18 February 1999, Lyon, France). In these definitions, ". . . mechanism is taken to infer detailed molecular knowledge of the initial events that result in an adverse response in the organism, whereas "mode of action" refers to the cascade of major changes that occurs during the development of the adverse event." Mode of action is contrasted to mechanism of action in that the latter usually implies a more detailed understanding of molecular and cellular events than the former. Furthermore, the altered process by which the initial molecular interactions lead to a structural or functional deficit is called pathogenesis. Pathogenesis can involve altered pathology at the cellular, tissue, and organ functional level.

To preserve the full range of causes and effects relevant to risk assessment of human developmental toxicity, the committee has sought to use "mechanism of action" in the most inclusive sense, to include all events from initial molecular interactions to the developmental defect itself. Such an explanation would include the following types of mechanistic information:

- The toxicant's kinetics and means of absorption, distribution, metabolism, and excretion within the mother and conceptus.
- Its interaction (or those of a metabolite derived from it) with specific molecular components of cellular or developmental processes in the conceptus or with maternal or extraembryonic components of processes supporting development.
- The consequences of the interactions on the function of the components in a cellular or developmental process.
- The consequences of the altered process on a developmental outcome, namely, the generation of a defect.

In Chapter 8, the committee discusses "levels of information" needed to understand inclusive mechanisms. The information is obtainable from various model systems (including in vitro and cell culture, nonmammalian animals, and mammals). Hypotheses about toxicant action in humans, based on the information from animal models, can then be strengthened or dismissed using information obtained from various levels of human data.

General Kinds of Initial Interactions of Toxicants with Cellular Molecules

Receptor-Ligand Interactions

Some chemicals interact directly with endogenous receptors for hormones, growth factors, cell-signaling molecules, and other endogenous compounds. They can activate the receptor inappropriately (agonists), inhibit the ability of the endogenous ligand to bind the receptor (antagonists), act in a way that activates the receptor but produces a less than maximal response (partial agonist), or act in a way that causes a decrease from the normal baseline in an activity under the control of the receptor (negative agonist). Receptors can be broadly classified as cytosolic/nuclear or membrane bound. Cytosolic/nuclear receptors reside within the cell and have ligands that are small and generally hydrophobic so that they can pass easily through the cell membrane. After the ligand binds to these receptors, the complex translocates to the nucleus where it interacts directly with specific sequences of DNA to activate or inactivate the expression of specific genes. Examples of cytosolic receptors are the estrogen receptors (ERa,ERb,ERR), retinoic acid receptors (RAR and RXR), and aryl hydrocarbon receptor (AHR). Agents that interact with one or more of these receptors and are known to produce abnormal development include retinoic acid and synthetic retinoids, glucocorticoids, androgens, and 2,3,7,8-tetrachlorodibenzo-*p*-dioxin (TCDD). A detailed description of what is currently known about the mechanisms by which retinoic acid and TCDD perturb development can be found below. Table 4-1 describes several cytosolic receptors (in this case, nuclear hormone receptors) that are involved in receptor-mediated developmental toxicity.

Membrane receptors (i.e., trans-membrane proteins) are diverse and interact with a wide variety of molecules, from small molecules, such as glutamate and acetylcholine, and small proteins, such as insulin, to large proteins, such as Wingless-Int (WNT), Sonic Hedgehog (SHH), TGFβ, and Delta signals (discussed in Chapter 6). Signaling molecules interact with a portion of the receptor on the cell's exterior. The binding of a ligand to a membrane receptor leads to a cascade of events within the cell membrane and cell known as signal transduction, which often involves five or more steps, including second messengers (intracellular signaling compounds) (described further in Chapter 6). It is conceivable that toxicants could affect any of these steps. For example, toxicants could interfere with receptor interactions or alter the activity of intermediates of the signal-transduction cascade. The number of agents known to exert developmental toxicity via interaction with membrane receptors is smaller than that for cytosolic receptors. Several membrane receptors—the Hedgehog receptor Patched, endothelin receptors A and B, and the cation channel delayed-rectifying Ikr—are known to play a role in mediating developmental toxicity and are highlighted in Table 4-1.

Despite the few examples of toxicant interactions with membrane receptors, the mechanism might be important in understanding how certain chemicals disrupt development. Most normal developmental processes involve cell-cell signaling and are mediated by trans-membrane receptors, including inductions, cell-matrix interactions, cell proliferation, cell movement, and autocrine and paracrine effects. The potential is great, therefore, that these mechanisms are significant in developmental toxicity.

Covalent Binding

Covalent binding occurs when the exogenous molecule chemically reacts with an endogenous molecule (e.g., forming a DNA or protein adduct). Among the kinds of reactive chemicals are aldehydes, epoxides, quinonimines, free radicals, acylating agents, and alkylating agents. Exposure to these chemicals might then result in abnormal transcription or replication of DNA, or abnormal function of the adducted protein. Phosphoramide mustard, a reactive metabolite of cyclophosphamide, is an example of a developmental toxicant that forms DNA adducts (alkylation) in embryos (Cushnir et al. 1990). Many chemicals that are not initially reactive are converted by DMEs (e.g., cytochromes P450) to "potentiated" reactive derivatives. An example of a developmental toxicant that forms both DNA and protein adducts in embryos is diphenylhydantoin, whose mechanism is described below.

TABLE 4-1 Receptor-Mediated Developmental Toxicity

Receptor (official name[a])	Endogenous Ligands	Developmentally Toxic Ligand and Modifier	Typical Effects	Recent References
Basic Helix-Loop-Helix Transcription Superfamily				
Aryl hydrocarbon AHR	Unknown	Agonists: TCDD and related polycyclics	Cleft palate, hydronephrosis	Couture et al. 1990; Mimura et al. 1997; Abbott et al. 1998
Nuclear Hormone Receptors				
Androgen AR (NR3C4)	Testosterone Dihydro-testosterone	Agonists: 17 α-ethinyl-testosterone and related progestins Antagonists[b]: Flutamide	Agonists: Masculinization of female external genitals Antagonists: Inhibition of Wolffian duct and prostate development and feminization of external genitals in males	Kassim et al. 1997
Estrogen ERa, ERb (NR3A1 and 2)	Estradiol	Agonist: DES Antagonist: tamoxifen, clomiphene—weak	Agonist: various genital-tract defects in males and females	Cunha et al. 1999
Glucocorticoid GR (NR3C1)	Cortisol	Agonists: cortisone, dexamethasone, triamcinolone	Cleft palate	Fawcett et al. 1996
Retinoic acid				
RARα, β, and γ (NR1B1, 2, and 3)	All-*trans* & 9-*cis* retinoic acids	Agonists: numerous natural and synthetic retinoids Antagonists: BMS493, AGN 193109 and others	Almost all organ systems can be affected	Collins and Mao 1999; Chazaud et al. 1999; Kochhar et al. 1998; Elmazar et al. 1997.
RXRα, β, and γ (NR2B1, 2, and 3)	9-*cis* retinoic acid			
Thyroid hormone TRa and b (NR1A1 and NR1A2)	Thyroxine (T4 and T3)	Antagonist: nitrophen	Lung, diaphragm, harderian-gland defects	Brandsma et al. 1994

continues

TABLE 4-1 Continued

Receptor (official name[a])	Endogenous Ligands	Developmentally Toxic Ligand and Modifier	Typical Effects	Recent References
Hedgehog Receptor				
Patched	Sonic, Desert, and Indian Hedgehogs	Veratrum alkaloids: cyclopamine (mechanism unclear)	Cyclopia, holoprosencephaly	Incardona et al. 1998; Cooper et al. 1998
Membrane				
Endothelin receptors A and B	Endothelins 1, 2, and 3	Antagonists: L-753, 037, SB-209670, SB-217242	Craniofacial, thyroid, cardiovascular, intestinal aganglionosis (Hirschsprung's disease)	Spence et al. 1999; Treinen et al. 1999; Gershon 1999a,b
Cation Channels				
Delayed-rectifying IKr	Potassium ion	Inhibitors: almokalant, dofetilide, d-sotalol	Digit, cardiovascular, orofacial clefts	Webster et al. 1996; Wellfelt et al. 1999

[a] Nuclear Receptors Committee 1999.
[b] Also, five α reductase inhibitors (e.g., finasteride) affect prostate and external genitals (Clarke et al. 1993).

Peroxidation of Lipids and Proteins

Some chemicals exist as free radicals or generate free radicals during their metabolism. Free radicals are highly reactive and will oxidize proteins or lipids, changing their structure. The developmental toxicity of hydroxyurea is at least partially mediated by free radicals (DeSesso 1979; DeSesso and Goeringer 1990; DeSesso et al. 1994) and that of niridazole appears to be entirely mediated by radical production (Barber and Fantel 1993). Physical agents such as ionizing radiation also produce this type of oxidative damage, as does the body itself during reperfusion after an ischemic episode. See recent reviews by Fantel (1996) and Wells and Winn (1996) for a more detailed discussion on this topic.

Interference with Sulfhydryl Groups

Sulfhydryl groups often play an important role in maintaining the tertiary structure and, therefore, the biological activity of proteins, especially in the disulfide linkages of secreted proteins. In some proteins, sulfhydryl groups are functional groups of the active (catalytic) site. Metals like mercury and cadmium are examples of developmental toxicants that cause oxidative stress and bind strongly to sulfhydryl groups and interfere with function (see reviews by Clarkson 1993; Stohs and Bagchi 1995; Quig 1998). The mechanism of one form of mercury, methylmercury, toxicity is described in detail below.

Inhibition of Protein Function

This is a broad category. Protein function occurs at catalytic sites (catalysis), regulatory sites (regulation of protein activity), macromolecule binding sites (such as specific DNA binding), or protein-protein association sites (as in aggregation of ribosomal proteins).

Some agents interfere with enzymes whose catalytic function is important in development, somewhat similar to an antagonist binding to a receptor. For example, methotrexate, a cancer chemotherapeutic agent mimics a substrate of dihydrofolate reductase, and its inhibitory binding results in a functional folate deficiency, which is developmentally adverse (DeSesso and Goeringer 1991, 1992). The mechanism is described in more detail below. Angiotensin-converting-enzyme (ACE) inhibitors are another example of agents that interfere with development by blocking enzyme action. These drugs block the conversion of angiotensin I to angiotensin II. Angiotensin II is a potent vasoconstrictive agent controlling blood pressure in adults. In the human fetus and neonate, it is needed to maintain renal perfusion and glomerular filtration. When angiotensin II levels are reduced in the fetus, glomerular filtration pressure and urine production are reduced, causing fetal hypotension and oligohydramnios (reduced volume of amniotic fluid). Those primary mechanisms lead to fetal death and stillbirth, middle- to late-trimester onset of oligohydramnios, and intrauterine growth restriction followed by delivery of infants with hypotension and renal failure (Barr 1997).

Other chemicals block protein polymerization, such as colchicine and colcimid blocking tubulin polymerization to microtubules or cytochalasins blocking actin aggregation to microfilaments. Those drugs bind to protein-protein association sites. There are examples of chemicals also binding at other kinds of sites. All fit the geometry and weak bonding properties of the site and competitively interfere with the binding of the normal cell component (substrate or ligand). Chelators of essential elements may interfere with protein function by limiting the availability of metal co-factors. Examples of proteins that require metals to function are metalloproteinases and several other enzymes, and zinc-finger transcription factors.

Maternally Mediated Effects

All the mechanisms discussed above occur within the embryo. However, there are examples in which developmental toxicity is the consequence of toxicity in the mother. Effects on the embryo occur secondarily, as a result of some effect on the pregnant mother. Effects include chemically induced maternal hypoxia or secondary nutritional deficiencies. An example of the former is the case of diflunisal (5-(2,4-difluorophenyl) salicylic acid, a nonsteroidal anti-inflammatory drug), which causes hemolytic anemia in pregnant rabbits. The anemia leads to adverse developmental effects (Clark et al. 1984). An example of secondary nutritional deficiencies is the functional zinc deficiency brought about by substantial induction of metallothionein in maternal liver as part of a systemic acute-phase response to a wide array of chemicals that have little in common other than their capacity to induce an acute-phase response, including de novo expression of metallothionein in the liver (Daston and Lehman-McKeeman 1996). The events that take place within the embryo after toxicant-induced zinc deficiency are equivalent to those occurring during dietary deficiency, but the salient point for developmental toxicology and risk assessment is the recognition that maternal factors might contribute substantially to embryonic response.

Other Mechanistic Considerations

There are other mechanisms that might be found to affect development. These might include such events as DNA intercalation, interaction with as yet unidentified targets, or complicated interactions that involve multiple changes, each of which is necessary—but not by itself sufficient—to initiate a pathogenetic cascade.

General Kinds of Pathogenesis

Once the toxicant has interacted with an endogenous molecule, the function of the endogenous molecule will be altered to an extent depending on the dose and duration. In the developing embryo, the function of the endogenous molecule can be seen as having a cellular and a developmental role, simply because the embryo is composed of cells whose activities are directed to a developmental outcome. From the current knowledge of cell biology, various general classes of function can be cited as susceptible to alteration. Any of these might be affected as part of a toxicant-induced pathogenic process. The classes of function include altered

- gene expression,
- patterns of apoptosis (programmed cell death),
- replication, cell cycle, cell proliferation,
- secretion, endocytosis, uptake, migration, adhesion, and
- signal transduction.

Certain chemicals might affect more than one of these processes. There is mechanistic value in knowing which of these cell biological processes is affected.

All cells of all stages of development engage in the cell activities listed above. For an understanding of developmental consequences, however, more specific information is needed about which particular molecular components of which particular processes of development are affected. Some cell biological effects, such as failed cell proliferation or failed cell migration, might be several steps removed from the initial effect of the toxicant and several steps from the final effect of the altered cell behavior on development (e.g., a craniofacial defect). The challenge in recent years has been to identify particular molecular components of cellular and developmental processes, discern their activities, and understand the toxicant-caused alteration of activity. The recent information from cell and developmental biology has been essential for the progress in the understanding of mechanisms of toxicity.

Known Mechanistic Information on Selected Chemicals

The remainder of this chapter reviews the current hypotheses of mechanisms by which chemicals are thought to cause developmental toxicity. Eleven chemicals, listed alphabetically, are used to illustrate different mechanisms. For some chemicals, a great deal of evidence has been gathered supporting certain aspects of the mechanisms. For others, data are sparse and the understanding of the mechanism is incomplete. Experimental approaches used to study mechanisms of developmental toxicity are highlighted. In the near future, these approaches can be used in conjunction with new approaches from developmental biology and genetics to elucidate further the mechanisms by which chemicals cause developmental toxicity.

Class III Antiarrhythmic Drugs

An example of a chemical interaction with a membrane receptor that leads to adverse developmental outcome is the case of class III antiarrhythmic drugs, such as dofetilide and almokalant. In rat embryos, these drugs block specific potassium channels in myocardial cells, presumably by high-affinity interactions with adrenergic or muscarinic receptors. This action on potassium flux leads to a profound bradycardia, which in turn progresses to malformations (probably through hypoxia in the affected structures) or death in utero (Webster et al. 1996). This example also illustrates the process of pathogenesis, as the molecular interaction produces an effect on embryonal organ function, which in turn affects the further development and viability of the embryo.

Cyclopamine

Binns et al. (1963) reported that an epidemic of cyclopia with associated holoprosencephaly in sheep was caused by teratogenic compounds present in the subalpine lily, *Veratrum californicum*. Subsequent work by Keeler and Binns (1968) showed that the active teratogenic agents in this plant were cyclopamine, its glycoside alkaloid X, jervine, and veratosine. Of these teratogenic agents, cyclopamine and jervine are the most active. In addition, it was noted that these two compounds closely resemble cholesterol in structure. In an early study, Roux and Aubry (1966) showed that alterations in cholesterol metabolism induced by AY-9944, an inhibitor of the final step in cholesterol synthesis, induced holoprosencephaly in rats. Almost 30 years then passed before additional insights were gained into the teratogenic mechanism of cyclopamine-induced holoprosencephaly.

In the 1990s, new data from several unrelated fields converged to suggest that cyclopamine-induced holoprosencephaly was caused by interference with cholesterol metabolism and SHH signaling. First, SHH is synthesized as a precursor that must be cleaved and covalently linked with cholesterol to be active (Roelink et al. 1995; Porter et al. 1996). Second, SHH is necessary and sufficient for patterning the ventral neural tube (Tanabe and Jessell 1996). Third, mutations in SHH were shown to cause holoprosencephaly in mice (Chiang et al. 1996). Fourth, holoprosencephaly associated with Smith-Lemli-Opitz syndrome results from a genetic decrease in Δ7-DHC (Δ7-dehydrocholesterol) reductase activity (Kelley et al. 1996; Tint et al. 1994). These discoveries, coupled with earlier studies linking cylopamine to holoprosencephaly, led to the hypothesis that cyclopamine causes holoprosencephaly by interfering with SHH signaling. Direct confirmation of this hypothesis came in 1998 from two independent laboratories (Cooper et al. 1998; Incardona et al. 1998). Although many of the mechanistic details are lacking, both groups showed that cyclopamine inhibits SHH signaling.

Diethylstilbesterol (DES)

DES was prescribed from the 1940s to the 1970s to prevent pregnancy loss. DES was found to be a transplacental carcinogen and teratogen (affecting the hypothalamo-hypophysial axis and reproductive organs) in humans approximately 25 years after its introduction into women's health care (Herbst et al. 1971; Herbst 1981; Kaufman et al. 1980; Goldberg and Falcone 1999). Even more time passed before it was demonstrated to be a transplacental carcinogen in the mouse (McLachlan et al. 1980; C. Miller et al. 1998; Walker and Haven 1997), rat (Baggs et al. 1991; Henry and Miller 1986), and hamster (Khan et al. 1998).

Mechanistically, DES has produced a rich field for investigating mechanisms of teratogenic and carcinogenic action. Such animal and human observations place DES among the agents that can modify not only the estrogen receptor activity but also expression of uterine lactoferrin through signal transduction mechanisms (Newbold et al. 1997). More recent evidence has implicated chromosomes 3 and 6 as sites for gene control resulting in not only carcinogenesis but also teratogenesis (Hanselaar et al. 1997).

Recent investigations have coupled the effects of DES in the developing mouse female reproductive tract with downregulation of WNT7A, resulting in abnormal smooth-muscle proliferation (C. Miller et al. 1998). WNT7A is normally expressed in the luminal epithelium of the uterus. Following DES exposure in utero, low levels of WNT7A transcripts were detected at birth. Such alterations in the reproductive tract following DES exposure are consistent with knockout mice lacking *Wnt7a* having malformed female reproductive tracts (Miller and Sassoon 1998).

All of these investigations implicate the role of gene control and modification by estrogenic agents that might be more effective not only because of their estrogenic properties but also because of their pharmacokinetics and metabolism (Miller et al. 1982; Henry et al. 1984; Henry and Miller 1986). Thus, in the human, further questions are being raised about the gene-environment interactions based on the collection of experience with the use of DES during pregnancy (Hanselaar et al. 1997).

Diphenylhydantoin

Diphenylhydantoin (DPH; common name, phenytoin) is an anticonvulsant used to treat epilepsy. It produces abnormal development in fetuses whose mothers take the drug during pregnancy. Abnormalities include facial dysmorphism (epicanthal folds, hypertelorism, broad, flat bridge of the nose, upturned tip of nose, and prominent lips), distal digital hypoplasia, intrauterine growth retardation, and mental retardation. This cluster of defects has been termed the fetal hydantoin syndrome and occurs in about 11-17% of pregnancies in which the mother has taken the drug (Hanson et al. 1976; van Dyke et al. 1988).

It appears necessary for DPH to be metabolized by cytochrome P450 (CYP) enzymes to reactive intermediates that form adducts with DNA or protein within the embryo (for a review, see Wells et al. 1997). The most likely intermediate is an arene oxide. An alternative hypothesis suggests that DPH is metabolized by prostaglandin synthetase to a teratogenic intermediate. This hypothesis is supported by the observation that DPH teratogenicity in mice can be mitigated by cotreatment with aspirin, an inhibitor of prostaglandin synthetase (Wells et al. 1989). It has been observed that DPH treatment in rodents decreases the expression of the mRNAs for a number of important growth factors, including TGFβ, NT3 and WNT1 (Musselman et al. 1994). Whether the decrease is due to an effect on gene expression or a degradation of RNA by reactive intermediates of DPH is not known.

Methotrexate

Methotrexate is a cancer chemotherapeutic drug. It is a competitive inhibitor of dihydrofolate reductase, which converts folate to tetrahydrofolate. Tetrahydrofolate is then metabolized to various coenzymes that play a role in the synthesis of purines and amino acids and conversion of deoxyuridylate to thymidylate. Exposure of rabbits to methotrexate during gestation causes craniofacial defects, limb deformities, and decreased fetal weight in the offspring (DeSesso and Goeringer 1991, 1992). Similar defects have been observed clinically in babies of mothers who had been given methotrexate between 35 and 50 days of gestation (Milunsky et al. 1968; Warkany 1978). Using a metabolic derivative of folinic acid, the authors (DeSesso and Goeringer 1991, 1992) demonstrated that methotrexate causes developmental toxicity by inhibition of dihydrofolate reductase. The metabolic derivative replaced the normal product of the inhibited enzyme and eradicated the developmental toxicity.

Methylmercury

Methylmercury (MeHg) is an environmental toxicant that primarily affects the central nervous system (CNS) and, to a lesser extent, the liver and kidneys. To reach the brain, it must cross the blood-brain barrier by traversing the brain capillary endothelial cells. MeHg possesses a high affinity for thiol groups and will bind to endogenous sulfhydryl-containing ligands, such as proteins, and low-molecular-weight compounds, such as glutathione, found in blood and tissue (for a review, see Clarkson 1993). Hirayama (1980, 1985) reported that intravenous injection of MeHg chloride and cysteine in rats increased the rate of

MeHg uptake into brain tissue, an effect that was reversed by administration of a neutral amino acid. The author concluded that MeHg was transported across the blood-brain barrier by an amino acid carrier. Subsequent studies by Kerper et al. (1992, 1996) and Mokrzan et al. (1995) showed that the amino acid carrier is an L (leucine-preferring) amino acid transporter and that MeHg is released from the brain capillary endothelial cells into the brain interstitial space as a glutathione complex.

The brains of both humans and experimental animals (rodents and primates) exposed in utero to MeHg show changes in neuronal migration and distribution patterns, cell loss, low neuronal abundance, and microcephaly, changes consistent with effects on the microtubular cytoskeleton and inhibition of cell-cycle progression (Burbacher et al. 1990). The effects of MeHg on mitotic activity and mitotic spindle function both in vivo and in cell culture have been characterized (Imura et al. 1980; Rodier et al. 1984; Brown et al. 1988; Wasteneys et al. 1988). It has been shown that MeHg directly binds to tubulin and inhibits microtubule formation (Vogel et al. 1986). Ponce et al. (1994) conducted in vitro studies using primary embryo neuronal cells to characterize MeHg's effect on cell cycling and its role in developmental toxicity. Exposure at concentrations of 2 µM MeHg causes G2/M phase cell-cycle inhibition, and at 4 µM, all cell-cycle phases are inhibited, suggesting that the cytoskeleton and mitotic spindles might be particularly sensitive to MeHg.

Two recent studies have further characterized steps in the mechanism by which MeHg affects the cell cycle in embryos. Ou et al. (1997) used primary rodent embryonic neuronal cells to determine mRNA expression levels of two genes involved in a checkpoint pathway of cell-cycle arrest, *Gadd45* and *Gadd153*, in response to MeHg. Exposure at 2 µM caused both GADD45 and GADD153 mRNA levels to increase. The authors concluded that activation of these Gadd genes could be a mechanism by which MeHg causes cell-cycle arrest in embryos. The same laboratory investigated the involvement of *p21* (a cell-cycle regulatory gene of a checkpoint pathway of arrest of G1 and G2 phases of the cell cycle) in primary embryonic cells exposed to MeHg (Ou et al. 1999). The embryonic cells responded to MeHg exposure with a concentration-dependent increase in p21 mRNA, indicating that activation of cell-cycle regulatory genes could be one mechanism by which MeHg disrupts the cell cycle in embryos.

Retinoic Acid

Vitamin A (retinol) and the structurally related retinoids have a special place in developmental toxicology, both currently and historically. Nutritional deficiency of vitamin A was the first chemical manipulation to produce congenital malformations in a mammal (Hale 1933) and thus began the whole field of experimental mammalian teratology. Now, the biologically active metabolites of vitamin A, the retinoic acids (RAs), and their synthetic derivatives have become the most thoroughly studied of all teratogens. Because natural retinoids are signaling molecules, critical to many developmental processes, exogenous retinoids or vitamin A deficiencies are teratogenic in all animals studied, including humans (for review, see Collins and Mao 1999).

In higher animals, vitamin A is an essential vitamin, requiring absorption from the diet or synthesis from dietary retinyl esters, β-carotene, or other carotenoids. The all-*trans* form of retinol is most abundant, but there are a number of isomers, which generate the corresponding active retinoid isomers, including 9-*cis* and 11-*cis*, following metabolism. The absorption and distribution of retinol involves serum (RBP) and cellular (CRBP-I and CRBP-II) binding proteins. A number of enzymes are capable of converting retinol to retinoic acid, including CYP monooxygenases, alcohol dehydrogenases, and aldehyde dehydrogenases. Mutation of the mouse NAD-dependent retinaldehyde dehydrogenase-2 (ALDH2) (Niederreither et al. 1999) causes severe developmental defects, a result that shows this enzyme to be essential for embryonic RA synthesis. Further metabolism of RA is complex, involving multiple oxidation and conjugation pathways, some also CYP dependent (e.g., CYP26) (Kraft and Juchua 1993; Kraft et al. 1993; Nau et al. 1994; Trofimova-Griffin et al. 2000). Cellular binding proteins (CRABP-I and -II) are thought to influence intracellular levels of RA, but their exact role is unclear. The mouse knockout of CRABP-I is without phenotype, and CRABP-II null mice have polydactyly (Lampron et al. 1995). This phenotype is also the phenotype of the double-knockout mice, which do not, however, differ from wild-type animals in sensitivity to RA teratogenicity (Lampron et al. 1995).

Although there are some strain and species variations in developmental sensitivity and responses to exogenous retinoids, they are usually not profound. The effective oral dose of all-*trans*-RA in all mammals tested is broadly similar. In contrast, the potency of 13-*cis*-RA varies by two orders of magnitude. The explanation for this difference lies in species differences in metabolism, coupled with metabolite-specific placental transfer (see Collins and Mao 1999), and is a good illustration of the importance of toxicokinetics.

Some of the dysmorphogenic effects of retinoids are very well conserved across species. For example, RA-induced truncation of the forebrain, with posteriorization in the hindbrain, has been observed in mammals, birds, amphibia, and fish. In addition to CNS and craniofacial malformations, RA also affects the limbs, cardiovascular system, gut, and thymus; the predominant defects depend upon the phase of organogenesis exposed (Collins and Mao 1999). In mice, preorganogenesis RA treatment around the time of implantation induces body axis duplication and supernumerary limbs (Rutledge et al. 1994), whereas fetal exposures can cause functional and behavioral abnormalities (Nolen 1986). Human exposure to the pharmaceutical retinoids 13-*cis*-RA (isotretinoin) or etretinate predominantly affect CNS and cranial neural-crest development (Coberly et al. 1996).

A large number of whole-animal and in vitro studies have characterized the relationships between structure and developmental toxicity of retinoids. As for any chemical, the in vivo potency results from a combination of pharmacokinetic and pharmacodynamic properties. These studies suggest that the teratogenicity of retinoids is receptor-mediated, but there are other possibilities (see below). The structural requirements for teratogenicity have been reviewed (Willhite et al. 1989; Collins and Mao 1999) and show a wide diversity in the polyene side-chain and β-cyclogeranylidene ring modifications that still retain activity, although an acidic polar terminus appears indispensable. Some aromatic retinoids (arotinoids), such as TTNPB, are 1,000-fold more potent in vivo teratogens than RA. This potency appears to be due predominantly to slower elimination and reduced affinity for CRABPs (Pignatello et al. 1999). Another aromatic retinoid, etretinate, has a very long half-life in humans after multiple exposures, with measurable concentrations in serum 2 years after cessation of intake, probably because of storage and slow release from adipose tissue (Eisenhardt and Bickel 1994). Experimental studies show that the critical pharmacokinetic characteristic for retinoid teratogenicity is the area under the curve (AUC) (Tzimas et al. 1997), rather than a transient high dose.

The receptors for retinoids are of the nuclear hormone ligand-dependent transcription-factor superfamily (Nuclear Receptors Committee 1999). They are of two types: RARs (subclass NR1B) and RXRs (subclass NR2B) (see Chambon 1996). For each type, there are three receptors, α, β, and γ (NR1B1, -2, and -3 and NR2B1, -2, and -3), each encoded by a separate gene. For all these genes, with the exception of RXR, multiple isoforms have been detected (e.g., NR1B2a, -b, -c, -d), generated by differential promoter usage and alternative splicing. Most of the embryonic effects of retinoids seem to be mediated by RAR-RXR heterodimers, but RXRs can form homodimers and can also form heterodimers with a number of other nuclear receptors, the most important being those for thyroid hormones and for peroxisome proliferators (Mangelsdorf and Evans 1995). Several isomers of RA are agonists for RARs, including all-*trans*-RA, 9-*cis*-RA, 4-*oxo*-RA, and 3,4-didehydro-RA, and 9-*cis*-RA seems to be the predominant RXR agonist (Collins and Mao 1999).

Each receptor, and in some cases each isoform, has been knocked out in mice to test for its function in development. Many combinations of knockouts have also been generated. Loss of RARβ (all isoforms), RAR 1, or RAR 2 has no phenotypic effect (Li et al. 1993; Lohnes et al. 1993; Luo et al. 1995). In contrast, disruption of all isoforms of RAR or RAR causes many of the effects of vitamin A deficiency, including growth deficiencies and male sterility (Lohnes et al. 1993; Lufkin et al. 1993). Compound RAR null mice display all the malformations induced by vitamin A deficiency, including defects of the eyes, limbs, and heart and the craniofacial, urogenital, and reproductive systems (Lohnes et al. 1994; Mendelsohn et al. 1994). An interesting recent example is the compound RAR-RARβ null mouse, which causes syndactyly and demonstrates a role of RA in interdigit cell death (Dupe et al. 1999). Mice lacking RXR have a hypoplastic ventricular myocardium and placental defects (Sucov et al. 1994; Kastner et al. 1994). Combining RAR and RXR mutants causes considerably more severe defects, suggesting that normal embryonic retinoid signaling is mediated by RAR-RXR heterodimers (Kastner et al. 1997a,b).

Null mutant mice have been used to correlate individual receptors with specific teratogenic effects of retinoids. RAR appears to be essential for the RA-induced defects of truncation of the posterior axial skeleton and is partially required for neural-tube and craniofacial defects (Lohnes et al. 1993; Iulianella and Lohnes 1997). In contrast, RXR is required for RA-induced limb defects (Sucov et al. 1995). It is intriguing that in both cases the receptor is not required for normal development of the affected tissues but does mediate the teratogenic action, a result indicating that the receptor, when activated by exogenously added RA, is affecting gene expression at abnormal times and places, as compared with that done by endogenous retinoids. RARβ does not appear to directly mediate any teratogenic action of retinoids (Luo et al. 1995). Expression of a constitutively active RAR mimics the action of excess RA. For example, expression of active RAR 1 in limb causes the same limb defects as exogenous RA (Cash et al. 1997).

In general, the receptor specificity of retinoids correlates with their teratogenic actions; RAR agonists are potent teratogens and RXR agonists are ineffective and mixed agonists having intermediate activity (Kochhar et al. 1996). Although the lack of action of RXR agonists shows that RXR homodimers are not involved in RA teratogenicity, such ligands can potentiate some of the teratogenic effects of RAR agonists. For example, an RXR agonist increased the effects of an RAR agonist on some organs but not others (Elmazar et al. 1997). Comparative studies of selective RAR, β, and agonists showed a correlation of potency with receptor affinity and transactiviation as well as receptor-organ specificity (Elmazar et al. 1996). As expected, retinoid-receptor antagonists can reduce the developmental effects of agonists (Elmazar et al. 1997). In addition, pan-receptor antagonists, capable of blocking all types of retinoid receptor, reproduce the actions of vitamin A deficiency (Kochhar et al. 1998; Chazaud et al. 1999).

Of direct relevance to this report is that the relationship between structure and teratogenic activity of retinoids is accurately reflected in several in vitro systems. Indeed, rodent limb-bud micromass cultures have been used extensively to screen for the activities of retinoids (e.g., Kistler and Howard 1990). In addition, the dysmorphogenic action of the retinoids in vivo is faithfully reproduced in mammalian whole-embryo culture, and this system has contributed much to our current understanding of mechanism (e.g., Bechter et al. 1992; Chazaud et al. 1999).

The DNA target sequences for RAR-RXR heterodimers are termed RAREs (retinoic acid response elements) and are usually two direct sequence repeats, spaced by 1 to 5 bases (DR1-5), although there are other arrangements (Harmon et al. 1995; Chambon 1996). For example, a functional inverted repeat with zero nucleotides in the spacer (IR0) was recently described (Lee and Wei 1999). A large number of genes containing RAREs have now been identified, and the transcription of many has been shown to be modulated by RA treatment (Chambon 1996; Collins and Mao 1999). Similarly, the expression of numerous genes has been observed to change following embryonic RA exposure. How many of these changes are directly controlled by RAREs and which are critical for teratogenicity is largely unknown. The *HOX* genes, however, represent one class of functionally important downstream RA targets in teratogenesis.

The homeobox-containing HOX transcription factors are involved in patterning of the CNS, limbs, axial skeleton, and other organ systems, where their expression encodes positional identity. As discussed in Chapter 6, *HOX* genes are arranged on chromosomes in clusters in which the genes are colinear with their expression domains (Duboule 1998). For example, in the early CNS and somites, 3′ *HOX* genes are expressed rostrally and 5′ caudally. This colinearity is also manifest in responsiveness to RA, the induction of 3′ *HOX* genes being more rapid and abundant (Marshall et al. 1996). Endogenous retinoid signaling might be responsible for the progressive expression of *HOX* genes in vivo. The primitive (Hensen's) node, for example, is a site of RA synthesis and might pattern the paraxial mesoderm as it egresses through the primitive streak (Hogan et al. 1992). Exogenous retinoids can induce ectopic, expanded, or reduced HOX expression domains, which then establish abnormally arranged compartments of positional identity. These abnormal compartments then result in abnormal cell fate and morphogeneisis (Marshall et al. 1996).

There is good evidence for HOX-mediated retinoid teratogenicity in the axial skeleton, craniofacies, and limb, but perhaps the best understood example is the developmental effect in the hindbrain. The different *HOX* genes encode transcription factors that control the different identities of the rhombomeres (r) of the hindbrain. *Hoxb2* has a rostral expression boundary at r2-3, *Hoxb3* at r4-5, *Hoxb4* at r6-7, and *Hoxb1* in a band at r4 (Marshall et al. 1996; Studer et al. 1996). At early neural-plate stages in the mouse, exogenous retinoid treatment results in rostral expansion of these domains. In some cases, the treatment results in a transformation of r1-3 to an r4 identity with expansion of r5 (Conlon and Rossant 1992). In other cases, r2-3 is transformed into r4-5, and both the trigeminal motor nucleus and adjacent trigeminal ganglion are transformed into structures having a facial nucleus or ganglion appearance (Marshall et al. 1992). Analyses of the 3′ *Hox* genes reveal multiple RAREs that cooperate with other positive and negative regulatory elements to regulate spatial and temporal expression of the *HOX* genes (Marshall et al. 1996).

Retinoid-induced changes in *HOX* gene expression probably result in abnormally arranged compartments of HOX expression. The misexpressed *HOX* gene then activates and represses many other genes in abnormal places and thereby initiates abnormal development. Pathogenetic changes observed in retinoid-treated embryos include abnormal differentiation, migration, proliferation and apoptosis (Collins and Mao 1999). Some of these effects might not be receptor-mediated. For example, disruption of membranes, changes in phosphorylation, and increases in reactive oxygen species might play roles in retinoid teratogenicity.

The significance of the retinoids in developmental toxicology might extend further, because a range of chemicals likely mediate their developmental toxicities by interfering with endogenous retinoid signaling. Ethanol can act as a competitive inhibitor of retinol dehydrogenase activity, thus lowering retinoid synthesis, and this effect might be a component of fetal alcohol effects (Duester 1991). Similar mechanisms have been proposed for the anticonvulsants phenytoin, phenobarbital, carbamazepine, ethosuximide, and valproic acid (Nau et al. 1995; Fex et al. 1995). Inhibition of retinoid catabolism has also been implicated. Metabolism of retinoic acid via NADPH-dependent cytochrome P450 is inhibited by azole antifungal drugs (Vanden Bossche et al. 1988; Schwartz et al. 1995), which can induce retinoid-like craniofacial defects (Wang and Brown 1994). None of these examples is wholly convincing, but the overall concept is plausible, particularly because the retinoids are unusual as secreted signals in being small lipophilic molecules. Because many synthetic chemicals are also small and lipophilic, the potential for interaction might be higher than that for other peptide-based signaling pathways.

*TCDD (2,3,7,8-tetrachlorodibenzo-*p*-dioxin)*

The mechanism by which TCDD induces developmental toxicity has been studied extensively (for a review, see Wilson and Safe 1998) and is one of the best understood. It is summarized here to provide an example of how a chemical interacts with an endogenous cytoplasmic receptor (in this case, a basic helix-loop-helix receptor (bHLH)) and alters the expression of several dozen genes, one or more of which might result in an adverse developmental outcome.

TCDD, an environmental pollutant, is a byproduct of the production of chlorinated products such as herbicides and wood preservatives, and is developmentally toxic in many species. Exposure in utero to TCDD causes increased mortality and growth retardation, and structural and behavioral abnormalities, including the induction of cleft palate and hydronephrosis in mice. Evidence supports the hypothesis that TCDD binds the aryl hydrocarbon receptor (AHR), allowing the receptor to bind with AH-responsive elements on DNA and leading to changes in gene expression. For example, mice with wild-type high-affinity AHR exposed to TCDD have a higher incidence of developmental abnormalities than do mice with low-affinity receptors. (The mouse *Ahr* gene contains a mutation that lowers the affinity of the encoded protein for DNA (Chang et al. 1993; Poland et al. 1994)). Large amounts of AHR have been localized to the palatal shelves in normal mice susceptible to the effects of TCDD (i.e., mice having high-affinity receptors) and much lower amounts in resistant mice (i.e., mice having low-affinity receptors). Additional evidence of the relationship between TCDD and the AHR comes from a study in which mRNA and protein levels for the AHR were found to be decreased in TCDD-exposed mouse embryos compared with that in control embryos (Abbott et al. 1994). Changes in gene expression of epidermal growth factor (EGF), transforming growth factor α (TGFα), EGF receptor, transforming growth factor β1 (TGFβ1), and TGFβ2 have been reported in conjunction with TCDD-induced cleft palate (Abbott et al. 1989; Abbott and Birnbaum 1990). Genetically modified AHR-deficient mice are reported to be relatively unaffected by doses of TCDD that are 10-fold higher than the dose found to induce toxic and pathological effects in mice expressing functional AHR (Fernandez-Salguero et al. 1996). The results suggest that the TCDD-induced toxic effects are mediated by AHR.

Recently, comparative experiments by Abbott et al. (1998; 1999a,b) showed that formation of the palate, which divides the oral cavity from the nasal cavity in mouse and human embryos, involves homologous processes at the morphological, cellular, and molecular level. In organ culture, developing mouse and human embryo palates respond similarly to TCDD. Exposure to the chemical causes excessive epithelial cell proliferation, via several steps, which interferes with fusion of the opposing palatal shelves. Several factors that regulate proliferation in this fusion process have been identified, among them EGF, EGF receptor, TGFα, TGFβ1, TGFβ2, and TGFβ3. Expression patterns of protein and mRNA of each of the growth factor and receptor genes were examined during palatogenesis for mouse and human palates maintained in organ culture. The effects of TCDD exposure on expression of those genes as well as the AHR- and the AH-receptor nuclear translocator (ARNT) and the glucocorticoid receptor (GR), were also defined in both species. To compare the human and mouse palatal responses to TCDD, the induction of a dioxin-responsive gene, CYP1A1, was quantified and compared across species. This comparison required the generation of tissue-level concentration-response profiles across dose and time for both mouse and human profiles. Quantification of AHR and ARNT levels revealed differences between species in the expression levels of the effector molecules. The mouse and human responses at the same target-tissue concentration could thus be compared, and a sensitive molecular marker gene (CYP1A1 induction) could be correlated with gross morphological outcomes and changes in growth-factor expression. Overall, the data indicate that human palates expressed all of these regulatory genes, that responses to TCDD were detected, and that comparison between mouse and human palates revealed interspecies variation that might be a factor in each species' response to TCDD.

Comparison of in vitro exposure levels between human and mouse palate tissues have revealed profound species differences. At comparable stages of development, human embryo palates are much less sensitive to TCDD than mouse embryos. These studies allow the conclusion that human embryonic palatal tissue expresses much lower levels of AHR and ARNT, or has the poor affinity AHR, and the expression and induction of CYP1A1 were lower. A situation emerges in which approximately 350 times fewer receptors are expressed in the human embryo than in the mouse, approximately 200 times more TCDD is required in human embryo tissue than in mouse tissue to produce the critical effects, and the response of a transcriptionally regulated gene (CYP1A1) is approximately 1,500 times lower in human embryo palates than in mouse palates under identical exposure conditions.

Valproic Acid

Valproic acid (VPA) is an anticonvulsant drug used to treat epilepsy with the major side-effect of hepatotoxicity. VPA is unusual in that its human teratogenicity was predicted from laboratory animal studies, without any knowledge of mechanism (Brown et al. 1980; Kao et al. 1981). In all species, including humans, neural-tube-closure defects are a consistent component of the teratogenic effects, but many other organ systems are affected and their sensitivity varies among species (Kao et al. 1981; Robert 1992). The pharmacological effect of VPA appears to involve several mechanisms, including actions on γ-aminobutyric acid (GABA) synthesis and release; the release of γ-hydroxybutyric acid; attenuation of N-methyl-D-aspartate- (NMDA) type glutamate receptors, and direct effects on excitable membranes (Löscher 1999). The mechanisms leading to hepatotoxicity and teratogenicity are distinct and also differ from the pharmacological mechanisms. The exact mechanism of teratogenicity is unclear, but it too might be multifaceted. Suggested actions include effects on the cytoskeleton and cell motility (Walmod et al. 1998, 1999); several aspects of zinc, folate, methionine, homocysteine, and glutathione metabolism (Alonso-Aperte et al. 1999; Hishida and Nau 1998; Bui et al. 1998; Finnell et al. 1997b); peroxisome proliferation-activated receptor δ interaction (Lampen et al. 1999); and gene expression (Wlodarczyk et al. 1996). Despite the initial site of action ("receptor") being unknown, there is a wealth of information on structure-teratogenicity relationships.

VPA is a simple short-chain carboxylic acid, 2-propylpentanoic acid. The following features affect teratogenicity (Nau 1994): (1) A free carboxylic acid is required. Amides, such as valpromide, are inactive (Spiegelstein et al. 1999; Radatz et al. 1998), as are stable esters. (2) The C2 carbon must be bonded to one hydrogen and two alkyl chains, as well as the carboxyl group. Substituting the hydrogen with any group abolishes activity, and a single chain or unsaturated derivatives (e.g., 2-en-VPA) are also inactive. (3) Activity is greatest when the two alkyl chains are unbranched (Bojic et al. 1996) and contain three carbons (Bojic et al. 1998). (4) Introducing a side-chain double or triple bond terminally (between C3 and C4) enhances teratogenicity but, in any other position, abolishes activity. (5) When one side-chain has a terminal unsaturation, C2 is asymmetric and the enantiomers have markedly different potencies. In the cases of both 4-en-VPA and 4-yn-VPA, the *S*-enantiomer is more potent than the racemate and the *R*-enantiomer is virtually inactive (Andrews et al. 1997, 1995; Hauck and Nau 1992).

These structure-activity relationships are not due to pharmacokinetic differences, as shown by direct measurements of tissue levels and by the activities of VPA and analogs in embryo culture (Brown et al. 1987; Nau 1994; Andrews et al. 1997, 1995). They are also consistent across species (Andrews et al. 1997, 1995). The overall impression is that the teratogenic effect of valproids requires an interaction with a specific site, at which one alkyl chain becomes located in a hydrophobic pocket, thus enabling ionic bonding of the carboxyl group and the interaction of the second chain with a region that favors the high electron density of terminal unsaturation (Bojic et al. 1998).

Chemicals That Might Induce Apoptosis

More than 1,000 agents have been identified as teratogens in animal studies (Shepard 1998); moreover, a variety of studies has now shown that cell death is an early, common event in the teratogenic process initiated by many, if not all, teratogens (Scott 1977; Knudsen 1997). Often, teratogen-induced cell death occurs preferentially in areas of normal programmed cell death, suggesting that there might be a mechanistic link between programmed and teratogen-induced cell death (Alles and Sulik 1989). The importance of an appropriate amount of programmed cell death to normal development is highlighted by mouse mutants, such as Hammertoe, in which insufficient programmed cell death underlies abnormal limb development (Zakeri et al. 1994). Likewise, excessive teratogen-induced cell death is directly linked to abnormal development by the finding that 2-chloro-2′-deoxyadenosine-induced eye defects are associated with excessive teratogen-induced cell death (Wubah et al. 1996).

Recent research has shown that cell death induced by a variety of stimuli occurs by a process termed apoptosis. Although many of the details are still lacking, it is known that apoptosis is a tightly controlled process, triggered either internally or externally, by which a cell self-destructs in a manner that does not lead to destruction of neighboring cells. Key components in the execution phase of the apoptotic pathway are the intracellular cysteinyl-aspartate proteases known as caspases, particularly caspase-3 (Colussi and Kumar 1999). These enzymes are normally present in all cells as inactive precursors that become activated by cleavage at specific internal motifs. Once activated, these caspases function to degrade specific target substrates, such as poly(ADP-ribose)polymerase (PARP), DNA-PKs, and lamins. A recent report shows that developmental toxicants, such as hyperthermia, cyclophosphamide (an alkylating agent), and sodium arsenite (a thiol oxidant), induce increased cell death characterized by activation of caspase-3, cleavage of PARP, and fragmentation of DNA (Mirkes and Little 1998). Although some of the downstream events in the apoptotic pathway activated by developmental toxicants are now known, upstream events that initiate and regulate this pathway in mammalian postimplantation embryos are unknown but oxidative stress clearly plays a role (Nebert et al. 2000). In particular, it is not known how cells in the embryo perceive exposure to a developmental toxicant and then respond to the insult. The ability of cells to perceive a stimulus or perturbation and then transduce these events into appropriate intracellular responses is commonly referred to as signal transduction. Although little is known about the interaction between developmental toxicants and signaling pathways in mammalian embryos, a recent report showed that heat shock can rapidly activate the stress-activated protein kinase pathways mediated by c-Jun terminal kinase (JNK) and p38 (Wilson et al. 1999), as well as the unfolded protein stress pathway involving a variety of chaperon proteins (Welch 1991; Sidrauski et al. 1998).

Chemicals That Might Induce Autism and Their Role in Understanding the Disorder

Recent discoveries regarding mutations responsible for the genetic susceptibility to autism spectrum disorders (ASDs) are an example of how studies of developmental toxicity can lead to breakthroughs in understanding the etiology of human birth defects, even when the defects are ones with strong genetic components (for a review, see Rodier 2000).

ASDs are among the most common congenital anomalies, occurring at a rate of 2-5 per 1,000 births (Bryson et al. 1988; Bryson and Smith 1998). Until recently, little was known about the causes and even less about the nature of the CNS injury underlying the symptoms. Family studies had indicated that unknown genetic factors account for about 90% of the variance (Bailey et al. 1995), but linkage studies provided few regions, and no genes, unambiguously linked to autism (Myers et al. 1998; Philippe et al. 1999). The family studies suggested that environmental factors are also involved (Le Couteur et al. 1996) but could not identify any of those factors. In 1994, it was discovered that exposure of the closing neural tube of the human embryo to thalidomide, a well-known teratogen, could produce autism at a high rate (Strömland et al. 1994). Valproic acid (Christianson et al. 1994) and ethanol (Nanson 1992) have also been implicated as teratogens that increase the risk of autism. The critical period for induction of autism by thalidomide was determined from the somatic defects of the patients with autism, each of whom also had malformed ears and hearing deficits. This stage of development—days 20-24 of gestation, which is the period of neural tube closure—is much earlier than the periods usually considered in studies of neuroteratology, because only a few neurons of the brain stem form so early. Most of these are motor neurons for cranial-nerve nuclei (Bayer et al. 1993), and cranial-nerve dysfunctions are indeed present in the thalidomide cases. Figure 4-1 shows a comparison of brain-stem neuroanatomy of a control and a patient with autism.

Using the information about the critical period, Rodier and colleagues were

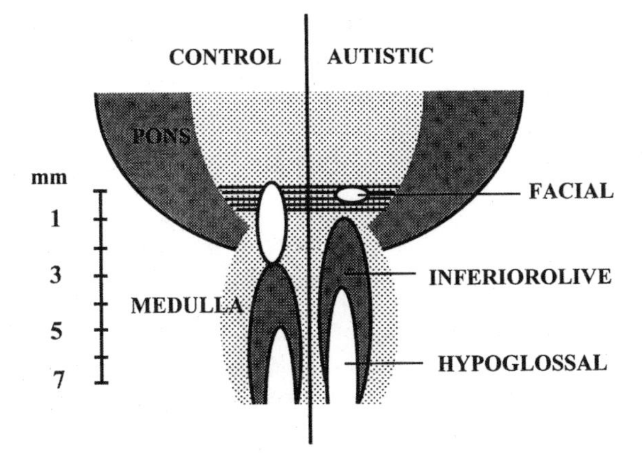

FIGURE 4-1 Comparison of brain-stem neuroanatomy of a control and a patient with autism. In the brain on the right, the number of neurons in the facial nucleus is greatly reduced, and a region caudal to it appears to be missing. The more caudal nuclei are shifted rostrally. The missing area is similar to the embryonic fifth rhombomere, from which most of the facial nucleus arises. Source: Adapted from Rodier et al. (1996).

able to describe alterations of the human brain stem related to autism and to create an animal model of the initial insult using exposure to valproic acid (Rodier et al. 1996). In addition, in cases of autism of unknown cause, they documented the existence of minor physical anomalies that were similar to those reported in the thalidomide-exposed cases (Rodier et al. 1997). The craniofacial symptoms had been reported before in the autism literature, but ignored because they seemed trivial in comparison to the disabling behavioral symptoms. However, the craniofacial defects speak directly to the embryological origin of the disorder. The anatomical studies described effects almost identical to those seen in mice with null mutations of the gene *HoxA1* (Chisaka et al. 1992; Carpenter et al. 1993), which is essential to brain-stem and ear development and expressed only during the period of neural tube closure. No variants of *HoxA1* had ever been detected in any mammalian species, but the teratological findings suggested the hypothesis that defective versions of the gene must exist and must contribute to the genetic etiology of autism.

Remarkably, an alternate allele of *HoxA1* was discovered in a substantial number of people diagnosed with ASD (Rodier 1998). The variant allele not only appeared significantly more frequently in familial cases than in historical controls or parent controls, but the number of homozygotes was significantly reduced from the expected value in all groups, suggesting that homozygosity for the variant reduces viability. (The mouse knockout of the same gene is lethal soon after birth.) A second variant of *HoxA1* has since been detected in cases of autism and is under investigation (Stodgell et al. 1999). Further, it has now been determined that valproic acid selectively alters the expression of *Hoxa1* in the embryo, providing an explanation for the similarity between the phenotypes of teratological cases and genetic cases of autism (Ingram and Rodier 1998). The increased understanding of ASDs is an example of how several research paths have converged to provide evidence supporting an interesting genotype-environment hypothesis.

SUMMARY

Early researchers of the causes of abnormal development used many of the same methods and animal models as developmental biologists studying normal development. Knowledge about normal developmental processes was essential to understand the developmental pathogenesis induced by chemical and physical agents, and, reciprocally, such agents at that time were used to disrupt normal development in order to understand the processes.

Our ability to understand the mechanisms by which chemicals act at the cellular and molecular levels to affect development has improved greatly during the past 2 decades. The improvement has occurred in concert with the advances in cell and developmental biology. As discussed in Chapter 6, progress in developmental biology came from the systematic analysis of developmental mutants of model animals, and progress in cell biology came from biochemical and molecular biological techniques, particularly gene cloning and sequence analysis.

Mechanism is an inclusive term for developmental toxicologists. To be complete, it should include information about (1) the toxicant's kinetics of uptake, distribution, storage, metabolism, and excretion as it gains access to the conceptus; (2) its interaction with a molecular component of a cellular or developmental process; (3) the consequence of that interaction for the component's function; and (4) the consequence of that altered function for the operation of cellular and developmental processes (pathogenesis), leading to a structural or functional developmental defect. Thus, a full description of a mechanism of toxicity would draw on molecular, cellular, and developmental knowledge.

The current hypotheses of the mechanisms of toxicity and the evidence supporting the hypotheses for 11 toxicants, or groups of toxicants, were reviewed. Of these, TCDD and retinoids are the exemplars at present. The hypotheses for the toxicity are quite complete and substantiated. The toxicants interact with known proteins, specific members of the bHLH and nuclear hormone receptor families, respectively, which are signal transduction components as well as genetic regulatory components. The liganded receptors activate (or perhaps repress) the expression of certain genes at abnormal times and places. The full range of genes is not yet known, but some genes are known. The products of the misexpressed genes affect developmental processes, such as cell migration, cell responses, and the expression of yet other genes. The connection of altered gene expression to defects of organogenesis is still somewhat weak, but the outlines have been established. Some of the altered expressions include genes encoding signaling components involved in cell-cell interactions during organogenesis.

For the case of cyclopamine, a transmembrane signaling pathway is known to be disrupted by the toxicant, leading to failure of an induction needed to pattern the eye field of the diencephalon. This is an example of a transmembrane signaling pathway that is a direct target for a toxicant.

For several other toxicants (e.g., methylmercury and methotrexate), the molecular target is not known but is probably some component of a process of cellular proliferation (e.g., DNA damage, block to DNA synthesis, disrupted spindle formation, and energy depletion). In many cases, high-dose disruption of the process results in cell death by apoptosis, preceded by the cell's attempts to restore viability by way of molecular stress and checkpoint pathways, some of which have been partially defined. Excessive cell death results in disrupted development. The developmental effects of these toxicants has been broadly defined, but most details are missing.

For a few toxicants, fetal organ function is probably compromised by processes similar to the agent's pharmaceutical-physiological effect on the mother (e.g., ACE inhibitors). Functional defects are in the early stages of elucidation (e.g., the connection of autism to altered HOX gene expression and altered rhombomere development in the hindbrain) and the possible role of some toxicants (including thalidomide) in altering the expression of those genes). Finally, several others act by mechanisms still largely unknown, despite knowledge of the final developmental defect (e.g., diphenylhydantoin and valproic acid).

In conclusion, there are only a few examples where the molecular, cellular, and developmental information is complete enough to say the hypothesis of the mechanism of toxicity is well substantiated. In no case is the mechanism of cellular and developmental toxicity fully known both toxicokinetically and toxicodynamically. However, it should be appreciated how broad and deep the scientific understanding has to be in order to have all the facets of a hypothesized mechanism distinguished and substantiated. The variety of mechanisms by which environmental toxicants probably work should be noted: mechanisms for toxicity are cellular, developmental, or physiological. Some involve two or more of these three. Some mechanisms occur at embryonic stages, fetal stages, or both, and some affect the conceptus, the mother, or both. Recent advances in the understanding of normal development (e.g., signaling pathways and transcriptional regulatory circuits) and cell biology (e.g., the cell cycle and checkpoint pathways) have identified critical processes, which, if investigated for their alteration by developmental toxicants, can provide exciting new advances in mechanistic investigations.

5

Human Genetics and the Human Genome Project

Genes are the fundamental units of heredity, and the genome is the organism's ensemble of genes. The genotype is the individual organism's unique set of all the genes. In a complex manner, the genotype governs the phenotype, which is the ensemble of all traits of the organism's appearance, function, and behavior. Genes are now known to be deoxyribonucleic acid (DNA) sequences from which ribonucleic acid (RNA) is transcribed. The transcripts of most, but not all, genes are then translated into proteins, which are composed of amino acid sequences and which perform most of the cell's functions by virtue of their catalytic activity and the interactions occurring at their specific binding sites. Hence, the gene is required for a phenotypic trait, because it encodes a protein involved in the generation of the trait.

It is not known precisely how many genes the human genome contains, but estimates range from 61,000 to 140,000 (Dickson 1999; Dunham et al. 1999). By comparison, the nematode *Caenorhabditis elegans* has 19,000 genes. In humans, only 5% of the DNA of the genome actually encodes proteins. The rest serves either as regulatory sequences that specify the conditions under which a gene will be transcribed, as introns (sequences that are transcribed but not translated), or as spacer DNA of yet unknown function. Each gene is located at a particular site on a chromosome, and in the diploid phase of the life cycle of humans and other metazoa, there are two chromosomes with that gene site in each nucleus. These two gene copies are called alleles. Particularly relevant to this report, many variant alleles of each gene have arisen during human evolution, and different alleles often confer slight or great differences in some particular trait of the organism, when members of a population are compared.

In this chapter, the committee describes the fields of human genetics and genomics. The role of molecular epidemiology in detecting developmental toxicants is discussed, as well as the difficulties in the detection of complex genotype-environment interactions.

GENOTYPE, PHENOTYPE, AND MULTIFACTORIAL INHERITANCE

In his classic experiments of the mid-nineteenth century, Gregor Mendel (1865) chose the pea plant (*Pisum sativum*) in which to study the segregation and assortment of particulate determinants of phenotypic traits. He was fortunate to choose several traits, each of which was controlled by a single genetic locus. The alleles at each locus, when inherited, acted in either a dominant or recessive manner, and their action was not significantly influenced by other genes or by environmental factors under his conditions of testing. Consequently, he observed precise and interpretable mathematical ratios for the phenotypes of the progeny in each breeding experiment. Traits of phenotype that show such easily interpretable patterns of inheritance are called simple, or Mendelian, traits, and these generally are governed by a single genetic locus.

However, the relationship between genotype and phenotype is almost always very complex. Even when scientists consider one particular gene and know its particular allelic form, its effect on phenotype is often subject to either or both of two variables: (1) the different alleles of various other major and modifier genes in the organism's genome, and (2) various environmental conditions. Such traits display a multifactoral pattern of inheritance (also called complex or non-Mendelian inheritance) and are termed complex traits or multiplex phenotypes (for a recent review, see Lander and Schork 1994). Multifactorial inheritance is much more common than simple inheritance. Such traits entail the interaction of two or more genes (a polygenic trait). The genes can contribute to the phenotypic trait in a quantitative and additive manner (e.g., genes *A*, *B*, and *C* might contribute 20%, 30%, and 50%, respectively, to a trait such as birth weight). These genes are called "quantitative trait loci," and genetic methods for analyzing their contributions are powerful. Alleles of the *BRCA1* gene, for example, appear to contribute about 5% to the overall risk of breast cancer (for a review, see Brody and Biesecker 1998), but several other contributors, which are analytically believed to exist, have not yet been identified. Still more complex patterns of inheritance can be traced to multiple genes acting in nonadditive manners. Segregated alleles might be neither dominant nor recessive. Finally, a gene might show incomplete penetrance (only some members of the population show the trait) or variable expressivity (members of the population vary in the extent of the trait) or both.

Other traits are modifiable by the environment. Such traits are not at all unusual and might overlap with polygenic traits. Studies in model organisms, such as the fruit fly *Drosophila melanogaster*, have long shown that a gene's effect on a trait can be modified by such extrinsic factors as temperature, chemicals, nutrition, and crowding. Geneticists who are particularly interested in evolution have argued that gene-environment interactions are so pervasive and important that one should not speak of a "phenotypic trait" of the organism but of its "norm of reaction," which is a set of phenotypes produced by an individual genotype when it is exposed to different environmental conditions (Stearnes 1989). The relationship between genotype, environment, and phenotype, which is sometimes called the gene-environment interaction, can be expressed as

$$\text{Genotype} + \text{Environment} \rightarrow \text{Phenotype}.$$

Although multifactorial inheritance is a nuisance to geneticists, it describes most human heritable diseases and virtually all susceptibilities. As mentioned in Chapter 2, approximately 25% of human developmental defects possibly follow multifactorial inheritance. Humans and experimental animals are notoriously heterogeneous in their responses to drugs or environmental pollutants. The favored explanation at present is that the heterogeneity reflects a combination of the heterogeneous exposure circumstances (extrinisic conditions) and heterogeneous genotypes for susceptibility (intrinsic conditions). Examples of exposure plus susceptibility would be the age of onset of lung cancer in cigarette smokers or the likelihood of asthma induced by urban pollution. The gene-environment relationship is further confounded in developmental toxicology by the need to consider the genotype of both the mother and the embryo or fetus, how and where a toxicant is metabolized, and the developmental stage at which a toxicant crosses the placenta. Gene-environment interactions are obviously relevant to the fields of molecular epidemiology and developmental toxicology.

POLYMORPHISMS

A polymorphism denotes the presence of two or more alleles of a particular gene within a population of organisms; the minority allele is present at a gene frequency of at least 1% (Hartl and Taubes 1998). That frequency is a somewhat arbitrary cutoff set by population geneticists and minority alleles of still lower frequency are called "rare alleles." In keeping with the Hardy-Weinberg distribution ($p^2 + 2pq + q^2$) for two alleles at a single locus, if the minority allele is present at a 1% gene frequency, it is then present in heterozygous form in about 2% of the members of that population and in homozygous form in 0.01% of the population.

Whatever the frequency, alleles are now defined in the most general way, namely, as different nucleotide sequences of the same gene—that is, as changes of one or more bases (adenine, thymine, cytosine, and guanine) relative to the reference DNA base sequence. However, finding such a difference does not in itself reveal much about an effect on phenotype. If the sequence difference occurs in a coding region of the gene, protein activity or stability might be affected. If the change is synonymous (i.e., the amino acid is not altered), conservative, or located in a region of the protein where any of several amino acids is acceptable, protein activity or stability might not be affected. If a DNA sequence change occurs in the transcribed region of the gene but not in the coding region, it might affect the reading frame, splicing, mRNA stability, translation efficiency, or transcriptional regulation. If outside the transcription region of the gene, the change still might affect the time, place, and level of expression of the gene, although not the protein's sequence. Additional work has to be done to identify the effect of the particular DNA change on protein function or level.

Polymorphisms reaching the 1% allelic frequency level are generally expected to have a selectively advantageous phenotypic consequence of altered protein activity or amount. However, a variant might be represented below the 1% level or close to 1%, because it arose in a small founder group of organisms that, because of local fortuitous circumstances, proliferated to a large population relative to other members of that species. Such a polymorphism might have no effect on protein activity or amount. It would just be a marker of that lineage of organisms. It might even have a negative selective effect.

Modern sequencing methods have greatly increased the capacity of researchers to detect alleles. For a particular gene sequence, any two unrelated people within a population are likely to have sequence differences. A gene sequence is taken to include all regulatory and transcribed regions of the DNA. When a base change first arises, due to oxidative hits, replication errors, ultraviolet-induced thymine dimers or other forms of DNA damage, one round of new DNA synthesis is usually required to become "fixed" in a double stranded form that is immune to repair and to count as a mutation. Before this synthesis, the base change often results in a mismatch in the DNA double helix, and a number of mismatch repair enzymes remove such errors (Snow 1997). However, out of every million or more DNA sites that become damaged, an error occasionally escapes uncorrected. Unrepaired mutations are thought to occur naturally at frequencies of once per 10^6-10^8 bases per generation. Because humans have such a large genome, roughly 75 new mutations accumulate per human individual per lifetime. Most of these are probably not deleterious. Many do not occur in protein coding regions (5% of the human DNA sequence) or, if they do, do not change the particular amino acid (synonymous substitutions). Some are deleterious, however, and the deleterious mutation rate in humans (nonsynonymous amino acid changes affecting activity) has recently been estimated to be at least 1.6 new deleterious mutations per diploid genome per generation. The authors conclude that this rate "is close to the upper rate tolerable for a species such as humans that has a low reproductive rate" (Eyre-Walker and Keightley 1999). It is likely that the human population is full of genetic variation, and this variation must be considered and appraised in any evaluation of an individual's susceptibility to developmental toxicants.

THE HUMAN GENOME PROJECT

The genome of an organism is the total genetic content of the organism, or more broadly, it is the organism's entire DNA content—including nontranscribed, non-*cis*-regulatory regions of DNA such as centromeres and telomeres. The study of the genomes of organisms, which is called genomics, includes areas of research determining the genetic and physical maps of genomes, the DNA sequences of genomes, the functions of genes and proteins, the *cis*-regulatory elements of genes, and the time, place, and conditions of expression of genes. A prominent part of genomics has become the managing of the massive amount of gathered information (a field referred to as bioinformatics) and the analysis of data with regard to, for example, aspects of the organization of the genome, the comparison of genomes of different organisms, and the global patterns of expression of genes.

The Human Genome Project (HGP) was launched in October 1990 by the National Institutes of Health (NIH) as a federally funded initiative. The immediate goal was then, as it is now, to complete the accurate sequencing of the approximately 3.5 billion human DNA base pairs (the haploid amount) by the end of 2003 (F.S. Collins et al. 1998). A "rough draft sequence", comprising approximately 90% of the human genome, was completed in mid-2000 (www.ornl.gov/hgmis/project/progress.html). In the longer term, a goal is to identify all human genes. Identification is difficult. In an organism such as yeast, which is favorable for the identification of genes by mutational genetic analysis, more than half the genes had gone undetected until the genome sequence became available (Brown and Botstein 1999). The lack of detection was in part due to large redundant regions of the yeast genome. In vertebrates, mutational genetic analysis is much more difficult, and redundancy might be more widespread. Therefore, initial gene identification by sequencing is the approach of choice. A gene is initially identified as an open reading frame (ORF), which can be discerned directly by looking at the sequence, or it is initially identified as an expressed sequence tag (EST) site, a sequence complementary to a known piece of transcribed RNA (see below). Thereafter, the goal is to identify each gene as a sequence encoding a full-length RNA and a protein of known function. The functions of nontranscribed regions, such as the numerous large *cis*-regulatory regions setting conditions for gene expression, will have to be elucidated as well. This task is still more difficult, currently involving a number of approaches, including the construction of transgenic animal lines carrying portions of the regulatory region in conjunction with a reporter gene (e.g., green fluorescent protein, GFP).

The functional analysis of the genome, in terms of the time and place of expression of genes and the functions of the gene products, is sometimes called "functional genomics" or even "post-genomics." The analysis of a protein's function might be simple if the protein sequence resembles that of another well-understood protein and might be difficult if sequence motifs are absent. The analysis of the organism's protein composition (called the "proteome") and function is sometimes called "proteomics." Targeted areas of the HGP currently include genetic and physical mapping of the human genome, DNA sequencing, analysis of the genomes of numerous important nonhuman organisms, informatics to handle the tremendous increase in the rate of information generated, resource and technology development, and the ethical, legal, and social implications (ELSI) of genetic research for individuals and for society (F.S. Collins et al. 1998).

The HGP is supported by NIH and U.S. Department of Energy (DOE) at 22 specialized genome research centers in the United States and in many university, national, and private-sector laboratories. At NIH, the name was changed to the National Center for Human Genome Research in 1993 and, since late 1996, has become the National Human Genome Research Institute (NHGRI). At least 14 other countries also have programs for analyzing the genomes of various organisms—ranging from microbes and economically important plants and animals to humans.

The explosion of genomics information has occurred sooner than the most daring scientists would have predicted. Following the first complete genome sequence, that of *Haemophilus influenzae* in 1995, seven more genomes were completed in the next 18 months, namely, four more eubacterial genomes, two archaebacterial genomes, and one unicellular eukaryote genome—that of the yeast *Saccharomyces cerevisiae*. In December 1998, the genome of the first multicellular eukaryote, *Caenorhabditis elegans*, was completed (100 kilobases of DNA sequence and 19,000 genes identified at least as ORFs). As of the end of 1999, more than 30,000 human genes had been partially identified, located, and sequenced. Human chromosome 22 has been sequenced and is projected to contain at least 679 genes (Dunham et al. 1999). In the mouse, at least 14,000 genes have been described. The fruit fly (*Drosophila melanogaster*) sequence was completed in 1999 (Adams et al. 2000). In mid-2000, an approximate ("working draft") human sequence was completed. By 2003, numerous nonhuman genomes will be sequenced as well, including the mouse *Mus musculus*, the zebrafish *Danio rerio*, the silkworm *Bombyx mori*, the rat, dog, cat, chicken, rice, corn, wheat, barley, cotton, the plant *Arabidopsis thaliana*, and probably also the cow, sheep, pig, and horse. The sequencing of the mouse genome is running well ahead of schedule.

New technologies, resources, and applications have become increasingly available to researchers of many diverse scientific fields, including cancer research, drug discovery, medical genetics, and environmental genetics, and their availability should also accelerate numerous major advances in developmental toxicology in the next decade, as discussed later in this chapter.

Functional Genomics and Microarray Technology

From the outset, it was expected that the completion of sequencing of the human genome would mark but a first step in the HGP. The information about the sequence and location of genes in the genome will greatly facilitate further studies—not only of human genetic variability but also of functional genomics. As noted above, the latter is the comprehensive analysis of gene expression and gene-product function. In the cases of yeast and *C. elegans* for which the entire genome sequence is already known, projects are under way to assess systematically the function of every gene product, for example, by knocking out yeast genes (causing a loss of function) one at a time and by associating an identified messenger (m) RNA with every ORF.

Some of this functional analysis can go forward even before a genome is sequenced. In the case of humans, the study of ESTs has been an important step of such analysis. mRNAs can be isolated from the organism and converted to complementary (c) DNAs, by reverse transcription (RT) and the polymerase chain reaction (PCR). The cDNAs are then cloned and sequenced to prepare a large and well-defined library of ESTs. The information is entered in a database. These sequences represent genes expressed in the human. For example, more than 1 million human ESTs are now available, representing greater than 50,000 genes. Each EST reflects an mRNA piece, not a full-length sequence. The most comprehensive libraries are prepared from a wide range of tissues and times of development in an effort to include all expressed genes. (Unfortunately for developmental toxicologists, although the initial sources of RNA included placenta, they were underrepresented in the variety of early embryonic tissue.) These sequences are useful in the course of genome sequencing to identify DNA regions that actually encode proteins (only 5% of the human genome sequence is thought to show up in processed mRNA sequences). New methods have become available to obtain full-length cDNAs from transcripts, and these will be more useful than fragments.

A further step of analysis of genome function is the determination of the time, place, and conditions of expression of each gene. Until recently, this analysis has been done one gene at a time. DNA microarray techniques recently have made possible the description of simultaneous changes of thousands of genes as cells and tissues undergo development or various changes of environmental conditions. In the study of toxicant effects on the organism, the analysis sometimes is called "toxicogenomics" or, in the study of the effects of pharmaceuticals, "pharmacogenomics." DNA microarray approaches are gaining widespread use (see Nuwaysir et al. 1999 for a discussion of its use in toxicology).

The technology is now suitable for simultaneously comparing the amounts of thousands of kinds of mRNA in two tissues or cell samples (e.g., a normal control tissue versus a tissue treated with a teratogenic agent). To do the comparison, thousands of different DNA sequences (e.g., each an oligomer of at least 25 nucleotides) are robotically spotted onto a microslide, and each sequence is placed on a known spot to make a DNA microarray. The DNA adheres to the glass, and each DNA spot is typically 20 micrometers (m) in diameter. For example, microarrays of 6,200 cDNA sequences, representing all the genes of yeast, have been fitted on a single slide or a few slides, and 8,900 cDNA sequences representing about 10% of human genes (mostly EST sequences) have been put on a few slides (Iyer et al. 1999). The expense and time of producing such slides are modest enough that a hundred or so can be prepared, each serving for the analysis of one comparison condition (e.g., analysis of several time points and concentration conditions). Each slide is an array of DNA "probes" by which the amounts of thousands of kinds of mRNA in cells, tissues, or embryos can be visualized simultaneously at each time and condition.

In the procedure of Brown and Botstein (1999), the tissue samples for comparison are separately extracted and the mRNAs are labeled with different fluorescent dye molecules, say green for the control and red for the treated tissue (see Figure 5-1). The RNAs are then mixed and added to the microarray slide (in their case, carrying 8,900 human DNA sequences) under conditions suitable for RNA-DNA hybridization, and then washed to remove unbound RNA. The slide is then read in a fluorescence microscope to see if each particular DNA spot has bound more of the green or rRNA. The ratio of red to green tells whether a particular gene is expressed more or less than normal under the treatment condition. Yellow is seen when equal red and green mRNA has hybridized. This technique has been applied recently to human cells cultured in the presence or absence of serum. Indeed, hundreds of genes changed expression, including many genes encoding stress-related proteins seen in wound healing (Iyer et al. 1998), a fact not previously realized in the years of study of the serum response of cultured cells. The technique also has been applied to yeast cells progressing along the sporulation pathway (Chu et al. 1998), yeast cells progressing through the cell cycle (Spellman et al. 1998), and yeast cells in a haploid, diploid, or tetraploid state (Galitski et al. 1999). Recently, it has been used to discover the response of a single kind of cell to two signaling ligands, each acting through a different receptor tyrosine kinase (Fambrough et al. 1999). In all cases, the expression of hundreds of genes changed. Viewing global patterns reveals that each gene does not seem to behave individually; instead, concerted expression of large batteries of genes seems to occur under various conditions. These results give credence to the value of analysis of the global patterns of gene expression. Previous studies of individual genes might have missed large-scale patterns. However, much remains to be done in the interpretation of the manifold changes of gene expression. As mentioned above, hundreds of gene expression changes are observed even with seemingly modest changes in a cell's circumstances, such as its ploidy.

In the analysis of toxicant effects, it is expected that cells or organisms could be treated with toxicants of unknown mechanism of action, and the changes of gene expression could be profiled by the DNA microarray method. If enough were known about the function and interaction of proteins encoded by the genes undergoing changes of expression, sound deductions might be made about the mechanism of action of the toxicant. In the future, it is expected that DNA microarray methods will allow rapid and detailed characterization of a cell or organism's response to a toxicant. As more information is collected, different

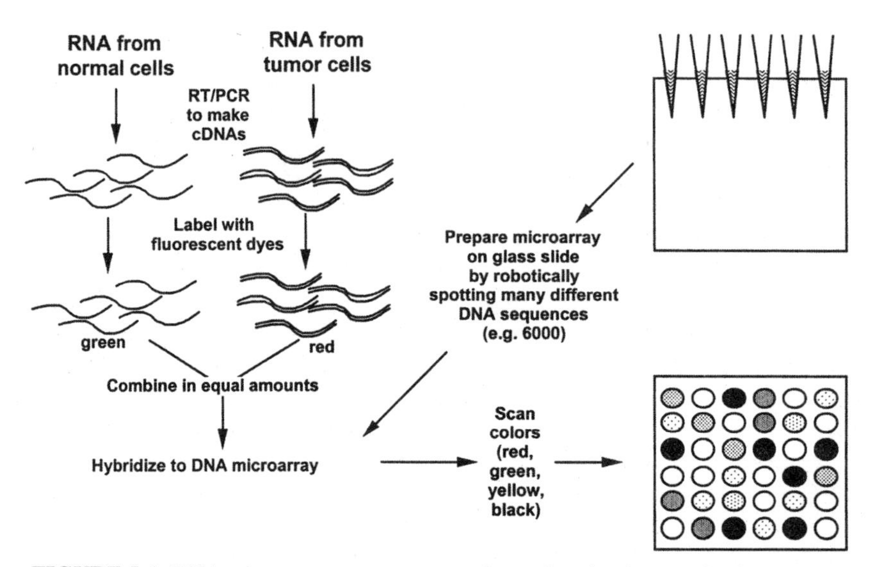

FIGURE 5-1 DNA microarrays as a means to determine simultaneously the amounts of thousands of kinds of mRNA in a cell or tissue. As shown in the upper left, mRNA is prepared separately from two kinds of cells, such as normal and tumorous, and each mRNA is converted to cDNA by reverse transcription and the polymerase chain reaction (RT-PCR) (see text). The separate cDNA samples are reacted with fluorescent dyes (e.g., green for the normal cDNA sample and red for the tumorous cDNA sample). The two are mixed in equal amounts. To prepare the DNA microarray, thousands of different DNA sequences are spotted robotically on a glass slide, each sequence in a known spot. The DNA adheres to the glass. The fluorescent cDNA mix is spread over the slide under hybridization conditions so that cDNAs stick to specific spots based on their complementary base sequence. Unbound cDNA is washed away and the slide is scanned in a fluorescence microscope to determine the green-to-red ratio of each spot. Pure red means that the RNA is present only in normal cells. Pure green means that the mRNA is present only in tumorous cells. Yellow means that the mRNA is present in equal amounts in both cells, and more red or green reflects more or less of the mRNA in the normal or tumorous cell. Black means that there is no mRNA in either kind of cell. Source: Adapted from the National Human Genome Research Institute's Glossary of Genetic Terms. Available via the Internet at <http://www.nhgri.nih.gov/DIR/VIP/Glossary/Illustration/microarray_technology.htm>.

---

toxicants can be grouped by their similarities of effect, and the analysis of toxicant action can be pursued on a more systematic basis. The DNA microarray method is already in use to compare normal cells and cancer cells. Many of the questions of interpretation of gene expression differences are being explored in that case as well.

Although it is preferrable to have a large set of cloned and sequenced DNAs (representing different identified genes) for the microarray, as is the case for yeast, it is not necessary. ESTs have been useful already for human and mouse studies, as in the serum study mentioned above (Iyer et al. 1999). If expression of a particular sequence, known only by its EST, is found to change greatly in the test condition, it might then qualify as interesting enough to deserve full-length cloning, sequencing, and further analysis of function.

Vast amounts of data accumulate in such comparisons (e.g., when 6,200 yeast genes (the entire yeast genome) or 8,900 human sequences are expressed differently under several conditions at several time points). The multidimensional data sets have challenged applied mathematicians to find means to express them in ways useful to biologists (e.g., see the methods of two-dimensional clustering analysis in Eisen et al. 1998; Alon et al. 1999; Tamayo et al. 1999). Yet, larger data sets loom on the horizon (e.g., the expression of perhaps 100,000 mouse or human genes at all times and places in development, not to mention with different toxicant exposures). As described below, the demand is great for managers and analysts of these data sets.

Although the various microarray techniques promise to reveal exciting new information about where, when, and under what conditions the genes of the genome are transcribed, this approach will not provide information concerning the translation and post-translational modification of proteins encoded by these mRNAs—that is, information about when and where the proteins are present and active. Protein function is almost always the immediate cause of cell function. To provide such functional information is the goal of proteomics. Proteomics has been defined (Anderson and Anderson 1998) as "the use of quantitative protein-level measurements of gene expression to characterize biological processes (e.g., disease processes and drug effects) and decipher the mechanisms of gene expression control." At the core of proteomics is the Human Protein Index—that is, the systematic identification of all human proteins (Anderson and Anderson 1982) using high-throughput, high-resolution, two-dimensional (2D)-gel electrophoresis to generate a gel with as many as 1,000 separate protein spots on it. A large amount of 2D-gel information is stored in the Proteomics database (see Appendix B for the Internet address). Plans have been made to identify every protein spot on a 2D gel, because the nanogram amount of protein in a spot is sufficient to determine a partial amino acid sequence by tandem mass spectrometry (Yates 1998). The partial sequence can then be looked up in the genome database and the protein identified. When proteins are modified by phosphorylation, acylation, glycosylation, farnesylation, limited proteolysis, or any of the other 30 or so covalent post-translational alterations, their migration on a 2D gel changes, thus allowing a correlation to be made with their activity or inactivity in the tissue. Such activity information cannot be gained from DNA microarray measurements of mRNA amounts.

Finally, many proteins are required to associate with other proteins in order to achieve activity, and there are various nondenaturing gel-electrophoresis methods to detect such associations. Current and future efforts can be expected to produce new technological advances for the analysis of proteins and their functions. In developmental toxicology, the combination of genomics and proteomics offers the possibility of assessing developmental toxicants not only for their capacity to alter gene expression but also for their capacity to alter protein function.

Applications of Genomic Technologies

Researchers have recently made use of the genomic technologies to identify and sequence genes with a role in disease etiology. It is probable that these genomic technologies will be applied to the study of the effects of chemicals on development in the near future. Two models already exist that demonstrate the application of new genomic technologies: the Cancer Genome Anatomy Project (CGAP), which began in 1997 and is administered by the National Cancer Institute, and the Environmental Genome Project (EGP), which began in 1998 and is administered by the National Institute of Environmental Health Sciences.

The goal of CGAP is to provide a complete catalogue of all genes whose expression changes in cancer cells relative to normal cells for all types of cancer (Pennisi 1997). In the past, a major barrier to the analysis of cancer cells has been the mixture of cell types (normal and cancerous) present in a typical tumor. It is difficult to get an accurate picture of gene expression in cancer cells if the RNA extracted from the whole tumor includes sequences from many different cell types. Using recently developed methods, CGAP researchers minimize that problem by microscopically selecting a small homogeneous group of cells, either normal, precancerous, or malignant (Emmert-Buck et al. 1996; Simone et al. 1998), which are then isolated by adhering them to a special laser-sensitive film. RNA sequences then are extracted from the isolated cells and amplified by RT-PCR to make cDNA libraries (Peterson et al. 1998). About 5,000 cells of homogeneous morphology have been required to obtain a wide mRNA representation. As few as 500 cells are needed to obtain cDNAs of abundant mRNA species. Such libraries will be important tools for describing the progression of cancer and providing diagnostic markers of the disease. They also provide sequences for functional analysis by way of DNA microarrays. As data are obtained, they are entered on the CGAP Web site for ready access and use by other investigators and for integration with other data in the large repository.

New technologies are expected to reduce the minimum number of cells needed to 1,000 or fewer and still get wide RNA representation. The lower the minimum number of cells required, the better for future developmental toxicological studies, because many developing cell types are present in small numbers during embryogenesis and fetal development.

The goal of EGP is to study the interaction between genetic susceptibility, environmental exposures, and disease. Specifically, researchers working on this 5-year project are attempting to identify allelic variants (i.e., polymorphisms) of 200 genes important in environmental diseases, to develop a centralized database of polymorphisms for those genes, and to foster epidemiological studies of gene-environment interactions in disease etiology. As discussed later in this chapter, evidence is particularly strong for increased chemical susceptibilities of individuals with polymorphisms of genes encoding drug-metabolizing enzymes (DMEs). EGP is using genomic technologies, such as high-throughput sequence analysis developed for the HGP, which will facilitate epidemiological studies of gene-environment interactions in disease.

The type of genomic technologies used in CGAP and EGP, and many of the data themselves, are anticipated to have an enormous impact on future research in developmental toxicology. Ideally, a developmental toxicology counterpart to such programs as CGAP and EGP would focus on obtaining data on gene expression at all times and in all tissues during normal development. Comparisons then could be made between embryos and fetuses from pregnant control (normal) animals and pregnant treated animals to look for differences in gene expression. Changes could be identified in the expression of genes encoding proteins known to function in cell signaling, transcriptional regulation, cell division, cell motility, cell adhesion, apoptosis, differentiation, metabolism, repair, electrolyte balance, homeostasis, or transport. Such studies will help to elucidate mechanisms by which extrinsic chemicals (potential developmental toxicants) act as agonists or antagonists to receptor- and enzyme-mediated subcellular processes during embryogenesis and fetal development. Such research will be a major force in merging the fields of developmental biology, genomics, and developmental toxicology.

Management of Genome Sequence and Functional Genomics Data

The explosion of molecular biology in the past two decades has led to enormous advances in DNA sequencing, which in turn has led to the increasingly rapid identification of genes as ORFs and as EST sites and the identification of the function of gene products by sequence motifs (e.g., homeodomains, zinc-finger domains, kinase domains, and SH2 and SH3 domains). There are four major nucleotide sequence databases: GenBank and the Genome Sequence Database (GSDB) in the United States, the European Molecular Biology Library (EMBL), and the DNA Data Bank of Japan (DDBJ). All groups exchange new and updated sequences electronically and usually on the same day of submission.

There is a two-decade history of these databases. The goal of the Los Alamos Sequence Library in 1979 at the Department of Energy (DOE)-sponsored Los Alamos National Laboratory (LANL) was to store DNA sequence data in electronic form. Within the same year, a similar database was also established at the EMBL in Heidelberg. In 1982, it was agreed that any data submitted or entered by one group would be forwarded immediately to the other, thereby avoiding duplication of effort. In 1982, the LANL database became GenBank when Bolt, Beranek and Newman (BBN) became the primary contractor for distribution of data-and-user support, and LANL was changed to a subcontractor of BBN. Sequence data activities at BBN and LANL were sponsored by the National Institute of General Medical Sciences (NIGMS), as well as DOE and other agencies. At the end of the first 5-year contract, IntelliGenetics became the primary contractor, and LANL again became the subcontractor in charge of designing and building the database. In October 1992, at the end of the second 5-year contract, NIGMS transferred GenBank to the National Center for Biotechnology Information (NCBI) at the National Library of Medicine. In August 1993, LANL and NCBI database resources became independent of one another. GenBank remained at NCBI, and LANL took the new name of Genome Sequence Database (GSDB) and moved to the National Center for Genome Resources (NCGR) in Santa Fe, New Mexico. The EMBL Nucleotide Sequence Database, which originated in 1982, is now maintained by the European Bioinformatics Institute (EBI), located near Cambridge, England—which also oversees the SWISS-PROT Protein Sequence Database and more than 30 other specialty databases. DDBJ, created in 1984 and sponsored by the Japanese Ministry of Education, Science, and Culture since 1986, accumulates nucleotide sequence data, mostly from Japanese scientists, and through electronic transfer makes more than a dozen other databases available.

Within a decade the genomes of at least 200 organisms, from numerous bacterial species to humans, will have been sequenced. By then, expression assays using high-throughput microarrays of DNA or cDNA or protein microarrays will be commonplace and provide us with overwhelming amounts of new information on the time and location (i.e., tissue and cell type) of expression of various genes and on the changes of gene expression in the organism's development and response to different exposure conditions (Reichhardt 1999). There will be a tremendous need for departments, or divisions, of bioinformatics in universities and industries to keep track of the data and to analyze it with respect to interesting questions about genome organization and function (see commentary by Reichhardt 1999). Additional information is likely to arise from the comparison of genomes of different organisms. The need to train large numbers of people in the new field of bioinformatics will be great. Information readily available on the Internet should facilitate the integration of the fields of developmental toxicology, human genetics, genomics, and developmental biology. In the future, developmental toxicologists will certainly benefit in many ways from ready and immediate access to this new information, but it should be appreciated that training will be required before the vast amounts of information can be used effectively. It is also unspecified at present how best to organize the data so that those involved in risk assessment can obtain what is most relevant.

RECENT DEVELOPMENTS IN MOLECULAR EPIDEMIOLOGY

Molecular tools have been used recently to identify interactions between genetic and environmental factors in the causation of complex diseases such as developmental defects. The majority of adverse pregnancy outcomes in humans are of unknown etiology and are viewed as complex traits in which exogenous agents might interact with particular combinations of allelic variants of genes controlling development and differentiation to produce adverse pregnancy outcomes. Studies of gene-environment interactions during embryogenesis and fetal development have become commonplace and increasingly appreciated. Categories of disease etiology can be viewed as spanning the range from totally genetic in causation to totally environmental in causation. These categories include single gene causation, chromosomal causation, multifactorial causation with high heritability, multifactorial causation with low heritability, infectious causes, and environmental causes (Khoury et al. 1993a).

Many single-gene disorders are characterized by a low frequency of the disease allele in the general population (allelic frequency, 1% or less) and high penetrance (a high proportion of individuals with the disease allele develop the disorder). Susceptibility genes for these single-gene disorders typically demonstrate Mendelian patterns of inheritance and are associated with high disease risk. Although individually rare, single-gene Mendelian disorders contribute significantly to infant morbidity and mortality. Approximately 4-7% of pediatric hospital admissions are made for recognized Mendelian diseases (Khoury et al. 1993b). From more than 2,000 likely single-gene developmental-defect syndromes in humans, the gene has been isolated and mapped for 100 of these syndromes and mapped but not yet isolated for another 100 (Winter 1996). In the mouse, for comparison, there are approximately 500 spontaneously occurring single-gene defects associated with developmental defects. Approximately 75 of these genes have been isolated, and greater than 400 have been mapped (Winter 1996). Furthermore, more than 1,000 mouse mutants have been prepared with known gene defects (many in components of signaling pathways), and many of these have phenotypes that fully qualify them as mouse single-gene developmental defects. Despite the availability of mouse mutants and the identification of genes important in development of *C. elegans*, *Drosophila*, and zebrafish (see Chapters 6 and 7), the study of single-gene defects contributing to developmental defects in humans has not yet received much experimental attention. Genes identified as important for development in *C. elegans*, *Drosophila*, and zebrafish, however, provide a rich source of information for identification of potential susceptibility genes in humans. That strikes this committee as an underutilized resource.

Multifactorial disorders, or complex diseases, are characterized by genetic complexity and probable gene-environment interactions (Ellsworth et al. 1997). These diseases tend to aggregate within families but are not inherited in simple Mendelian fashion. They are typically found in a higher proportion within affected families than expected in the general population. In contrast to single-gene disorders, susceptibility genes for complex disorders tend to be common in the population (allelic frequency more than 1%) and can be considered polymorphisms. Susceptibility genes are usually associated with low risk to the individual but high attributable risk in the population. In cases of high heritability, the disease trait might have alleles of several major genes and of several modifier genes contributing to its penetrance and expressivity. In those with low heritability, there might be alleles of several genes as well as specific environmental factors that together increase the risk of disease in a population.

Recent examples in which gene-environment interactions have been sucessfully elucidated for developmental defects are the following:

• Transforming growth factor (*TGF*α) polymorphisms and oral clefts: Evidence for an association between maternal smoking and oral clefts has been equivocal (Hwang et al. 1995). In this study, there was not an overall significant association between maternal smoking and oral clefts in the newborn. However, if the newborn had a variant allele (*TAQL* C2) of the *TGF*α gene, the odds ratio for oral clefts in infants of smoking mothers (more than 10 cigarettes per day) was 8.7, a 10-fold increase compared with infants of smoking mothers who did not have this variant allele. The variant allele alone was not associated with increased risk for oral clefts. *TGF*α is a ligand of a tyrosine kinase receptor.

• Homeobox gene *MSX1*, limb deficiencies, and smoking: Frequencies of rare alleles at the *MSX1* locus are slightly higher in infants with limb deficiencies compared with infants having other types of developmental defects (odds ratio 2.4). Infants carrying the rare alleles had a 2-fold increased risk of a limb deficiency when the mother smoked during pregnancy (odds ratio 4.8) compared with infants harboring the rare allele whose mothers did not smoke. Smoking alone was not associated with increased risk for limb deficiencies in this study (Hwang et al. 1998). MSX1 is a transcription factor whose activity often depends on BMP2,4 signals.

• Variable human susceptibility to developmental defects due to diphenylhydantoin (DPH or phenytoin): 10-20% of the offspring of epileptic women taking phenytoin during pregnancy have the fetal hydantoin syndrome (Hanson et al. 1976; van Dyke et al. 1988). Phenytoin is thought to be converted to a reactive intermediate to have teratogenic effects (Martz et al. 1977; for a review, see Finnell et al. 1997a). The population variability in response to phenytoin possibly reflects a heterogeneity of DME genotypes. In a pair of twin births in which only one twin had dysmorphologies of the hydantoin syndrome, the mother and the affected twin had decreased activity of the enzyme epoxide hydroxylase compared with the unaffected twin (Buehler 1984). Buehler et al. (1990) subsequently showed that children with the hydantoin syndrome indeed have lower activity of epoxide hydrolase. Epoxide hydrolase would serve to detoxify an arene oxide intermediate of phenytoin, and it has been suggested that reduced activity of this enzyme is responsible for increased susceptibility to phenytoin. Hassett et al. (1994) reported a polymorphism in the epoxide hydrolase gene that markedly decreases enzyme activity. Conversely, the DPH parent compound might be teratogenic, and genetic defects in CYP2C9 and CYP2C19 (the two enzymes that metabolize DPH) could result in an accumulation of the DPH teratogen. The frequencies of CYP2C9 and CYP2C19 poor metabolizing polymorphisms range between 10% and 25% of individuals in different populations—very similar to the percent of women taking DHP who have children with the fetal hydantoin syndrome.

- Aldehyde dehydrogenase 2, alcohol dehydrogenase 2, and the susceptibility to fetal alcohol syndrome: In the metabolism of ethyl alcohol, alcohol dehydrogenase (ADH) catalyzes the conversion of alcohol to acetaldyhyde, and acetylaldehyde dehydrogenase (ALDH) oxidizes the conversion of this product to acetic acid. Alcohol and acetaldehyde, but not acetic acid, are thought to have the potential for deleterious effects. Humans possess at least seven ADH genes and 13 ALDH genes. Crabb (1990) pointed out that the single base mutation in ALDH2 (the mitochondrial as opposed to the cytosolic ALDH), which is responsible for acute alcohol-flushing reaction and alcohol intolerance mostly in Asians, is the best-characterized genetic factor influencing alcohol drinking behavior (lower activity correlating with intolerance). He raised the possibility that polymorphisms in the several alcohol dehydrogenase genes might be related to risk of fetal alcohol syndrome (FAS). A genetic influence in fetal alcohol syndrome is suggested by twin studies: Streissguth and Dehaene (1993) established that the rate of concordance for the diagnosis of fetal alcohol syndrome was 5 out of 5 for monozygotic and 7 out of 11 for dizygotic twins. In two dizygotic pairs, one twin had FAS, and the other had fetal alcohol effects (FAE). In two other dizygotic pairs, one twin had no evident abnormality, and the other had FAE. Intelligence Quotient scores were most similar within pairs of monozygotic twins and least similar within pairs of dizygotic twins discordant for diagnosis. Johnson et al. (1996) documented the central nervous system (CNS) anomalies of FAS by magnetic resonance imaging. CNS and craniofacial abnormalities were predominantly symmetric and central or midline. The authors stated that the association emphasized the concept of the midline as a special developmental field. The CNS is vulnerable to adverse factors during embryogenesis and fetal growth and development.

As those four examples indicate, further investigation of gene-environment interactions using the tools of molecular epidemiology is likely to yield important new information on multifactorial causes of developmental defects. Two of the above-cited examples concern polymorphisms of genes encoding enzymes involved in the metabolism of an agent, namely, phenytoin or alcohol, and two of the examples concern polymorphisms of genes encoding protein intermediates of signal transduction pathways and genetic regulatory circuits (*TGF*α or *MSX1*), which are components of developmental processes.

The examination of gene-environment interactions is particularly advanced for disease conditions related to the DMEs, and this area of study is called ecogenetics or pharmacogenetics. There are phase I and phase II metabolizing enzymes. Phase I enzymes catalyze a conversion of the exogenous agent to a modified form, often an oxidized form. In some cases, the exogenous agent is toxic and the intermediate is not, but in other cases, the agent is nontoxic and the intermediate is toxic, an example of metabolic potentiation or activation. There probably are several hundred kinds of phase I enzymes (and genes encoding them) in mammals (including humans). The majority of them are members of the large cytochrome P450 monooxygenase family. Three or four kinds of P450 enzymes are thought to metabolize 70-80% of the prescription drugs taken by patients, and defects in phase I enzymes correlate with drug sensitivities and hazardous side effects. Phase II enzymes subsequently catalyze the conjugation of the modified intermediate to an endogenous harmless metabolite, such as a sugar or amino acid, and the conjugated form, which is usually nontoxic, is then excreted. In several well-analyzed cases, patients with high levels of phase I enzyme (hence, producing high amounts of a toxic intermediate) and low levels of phase II enzyme (hence, unable to get rid of that intermediate) were found to be particularly at risk from chemical exposures. Thus, human variants with altered levels of enzymes of one group or the other, or both, can have abnormal drug responses, as much as a 20- or 30-fold increase in drug sensitivity.

At least 60 ecogenetic or pharmacogenetic differences are now known; many are listed in Table 5-1. In this research, epidemiological methods and genomic methods are complementary, and progress in the near future seems assured. It seems likely that the fetus is at increased risk of developmental defects, because either the mother or the fetus cannot metabolize chemicals as well as others can or because they metabolize them better.

The other large area to investigate for the correlation of polymorphisms with developmental defects is that of the components of the developmental processes themselves, namely, key components of developmental processes, such as those of signal transduction pathways and genetic regulatory circuits. These components are the targets of exogenous agents that elude detoxification or are potentiated by phase I enzymes. The examples of *TGF*α with smoking and *MSX1* with limb defects are two that have been clarified. At this time, however, there are few good examples, perhaps simply because information about developmental processes has not been available until recently. Because the developmental components are conserved across phyla and have been well described in *Drosophila*, *C. elegans*, and now the mouse, the means are available to obtain related human sequences and search for polymorphisms. This research will be further discussed in Chapters 8 and 9. The phenotypes of mouse null mutants generated by the embryonic stem-cell-(ES) knockout technology have already contributed substantially to our knowledge about how alterations in those genes and pathways impact development. This kind of research is progressing rapidly. Complete deletion of some components of a variety of pathways fundamental to development results in embryo lethality, but in other cases for which there is a gene redundancy for the component, the deletion of the component leads to mice born with developmental

TABLE 5-1 Classification of a Partial List of Human Pharmacogenetic or Ecogenetic Differences[a]

Less enzyme or a defective protein
- N-acetylation polymorphisms (NAT2, NAT1)
- Increased susceptibility to chemical-induced hemolysis (G6PD deficiency) (G6PD)
- Hereditary methemoglobinemias; hemoglobinopathies
- P450 monooxygenase polymorphisms (oxidation deficiencies). Debrisoquine (CYP2D6), S-mephenytoin (CYP2C19 & 2C9), phenytoin (CYP2C9 & 2C19), nifedipine (CYP3A4), coumarin and nicotine (CYP2A6), theophylline (CYP1A2), acetaminophen (CYP2E1)
- Null mutants of glutathione transferase, mu class (GSTM1); theta class (GSTT1)
- Thiopurine methyltransferase (TPMT)
- Paraoxonase deficiency, sarinase (PON1)
- UDP glucuronosyltransferase (Gilbert's disease, UGT1A1; (S)-oxazepam, UGT2B7)
- NAD(P)H:quinone oxidoreductase (NQO1)
- Epoxide hydrolase (HYL1)
- Atypical alcohol dehydrogenase (ADH)
- Atypical or absent aldehyde dehydrogenase (ALDH2)
- Defect in converting aldophosphamide to carboxyphosphamide
- α1-antitrypsin (PI)
- α1-antichymotrypsin (ACT)
- Angiotensin-converting enzyme (DCP1, ACE)
- Acatalesemia (CAT)
- Dihydropyrimidine dehydrogenase (DPD)
- Succinyl sensitivity, atypical or absent serum cholinesterase (CHE1)
- Cholesteryl ester transfer protein (CETP)
- Butyrylcholinesterase (BCE1)
- Fish odor syndrome (FMO3)
- Glucocorticoid-remediable aldosteronism (CYP11B1, CYP11B2)
- Dubin-Johnson syndrome; multispecific organic anion transporter (MOAT, MRP)
- Altered serotonin transporter (5HHT)
- Altered dopamine transporter (DAT)
- Dopamine receptors (D2DR, D4DR)
- Defective drug transporters (e.g., MDR1), resistance to chemotherapeutic agents
- Licorice-induced pseudoaldosteronism (HSD11B1)
- Mineralocorticoid excess with hypertension (HSD11B2)
- Pyridoxine (vitamin B6)-responsive anemia (ALAS2)

Increased resistance to chemicals
- Inability to taste phenylthiourea
- Coumarin anticoagulant resistance
- Androgen resistance
- Estrogen resistance
- Cushing syndrome from low doses of dexamethasone
- Insulin resistance
- Rhodopsin variants; dominant form of retinitis pigmentosa
- Vasopressin resistance (AVPR2)
- Increased metabolism—Atypical liver alcohol dehydrogenase (ADH)
- Defective receptor—Malignant hyperthermia / general anesthesia (Ca^{2+}-release channel ryanodine receptor) (RYR1, MHS1)

continues

TABLE 5-1 continued

Change in response due to altered enzyme induction
Porphyrias
Aryl hydrocarbon receptor (AHR) polymorphism (CYP1A1 and CYP1A2 inducibility polymorphism) correlated with cancer, immunosuppression, birth defects, chloracne, porphyria, and, possibly, eye toxicity and ovarian toxicity
Abnormal metal distribution
Iron (hemochromatosis, HFE), copper (Wilson disease, Menkes disease), and, possibly, lead, cadmium, and others
Disorders of unknown etiology (known to run in families)
Corticosteroid (eye drops)-induced glaucoma
Halothane-induced hepatitis
Chloramphenicol-induced aplastic anemia
Aminoglycoside antibiotic-induced deafness
Beryllium-induced lung disease
Hepatitis B vaccine resistance
Susceptibility to human immunodeficiency virus infection (polymorphism of CCR5 co-receptors)
Long-QT syndrome
Retinoic acid resistance and acute promyelocytic leukemia
Thombophilia (activated protein C resistance)
Lactose intolerance
Fructose intolerance
Beeturia; red urine after eating beets
Malodorous urine after eating asparagus
Reproductive disadvantage in F508 cystic fibrosis heterozygotes who smoke cigarettes (CFTR)
High risk of cerebral vein thrombosis in defective prothrombin (F2) heterozygotes
High risk of cerebral vein thrombosis in users of oral contraceptives

[a]Modified from Nebert (1999). See also refs. 11 and 18 of Nebert (1999). All of these are pharmacogenetic or ecogenetic in the sense that health risk correlates not only with the polymorphic state of the individual, but also exposure of that individual to a particular chemical (drug or environmental agent). Many of these are searchable in the online Mendelian Inheritance in Man (OMIM) database at http://www.ncbi.nlm.nih.gov/omim/. Not all of these are correlated with developmental defects.

defects resembling human defects (see examples in Chapter 6). The work on key developmental components in animals can greatly benefit the search for human variants of developmental components.

The final step, though, will be to evaluate how specific toxicants interact with those altered pathways to produce abnormal development. Relevant susceptibility genes of development can then be examined in human populations, and interactions between alleles of those genes and toxicant exposures can be identified. Whether allelic variants of genes controlling development, such as those encoding components of the major signal transduction pathways, will be more important, as important, or less important than those controlling DMEs remains to be determined and should be given high priority for future research.

SUMMARY

The sequencing of the human genome and a variety of animal genomes will provide fundamental information about genome organization, genome evolution, gene sequence variety, and genetic polymorphisms. Sequencing will also provide a platform for global systematic analysis of gene function and gene expression. Developmental toxicology and risk assessment are expected to benefit in major ways from the new methodologies and information, namely, in the analysis of gene-environment interactions in human development defects and in the analysis of toxicant action on developmental processes.

A quarter to a half of the human developmental defects are believed to be attributed to interactions of the genotype and environment—that is, the exposure of individuals of a particular genetic composition to particular environmental conditions to which they are more sensitive than are others. Complex gene-environment interactions present a great challenge to developmental toxicology. The best epidemiological methods, the most discriminating molecular assessments of exposure and effect, and the most detailed analysis of genetic differences will be needed to make progress in understanding gene-environment interactions. Recent improvements in high-throughput sequencing of the human genome and in the identification of polymorphic markers conveniently spaced along each chromosome increase the chances for progress in this direction. Within this new area of molecular epidemiology, recent insights into human differences in activity levels of various DMEs and the genetic basis for those differences, offer great promise. Other kinds of gene products that might be important in susceptibility but are less well known, include components of developmental processes, particularly the components of signal transduction pathways and genetic regulatory circuits. These components will be discussed in later chapters.

Methods now are available to describe patterns of simultaneous expression of thousands of genes of developing cells and tissues and, in principle, to describe the changes of expression in the embryos of normal experimental animals and those following testing with toxicants. The use of such methods is expected to improve the categorization and analysis of toxicant-induced developmental defects.

The amount of data generated by modern genomic methods is prodigious. For the full benefit of the data, departments or divisions of bioinformatics in universities and industries will be needed to keep track of the data and to analyze it with respect to questions about genome organization and function.

6

Recent Advances in Developmental Biology

The absence of an incisive understanding of the action of toxicants on development has been in large part attributable to the absence of understanding of development itself. Until a few years ago, there was no understanding of a "developmental mechanism" at the molecular level although there were explanations at the cellular and tissue levels, such as "gastrulation is the mechanism by which the organization of the egg is transformed into the organization of the embryo." Recent advances in developmental biology have been substantial enough for scientists to be confident for the first time that some aspects of development in some organisms are understood at the molecular level. Protein components are identified, their functions in developmental processes are known, and the time and place in the embryo of expression of the genes encoding them are known. This knowledge greatly benefits elucidating the mechanisms of developmental toxicity.

In this chapter, the committee, in response to its charge, evaluates the state of the science for elucidating mechanisms of developmental toxicity and presents insights of developmental biology. It will show the promise of the subject in the next decade for understanding the action of developmental toxicants.

A BRIEF HISTORY OF DEVELOPMENTAL BIOLOGY

Observations of embryos and embryonic stages were made and recorded in antiquity (e.g., Aristotle, fourth century BC) and with increasing attention in recent centuries (e.g., Malphigi in the 1600s, Wolff in the 1700s, and von Baer in the early 1800s). However, it was only in the late nineteenth century that scientists pursued a detailed description of the embryonic stages of a variety of vertebrates and invertebrates, aided by the then-recent improvements in light microscopy and in staining methods and stimulated by Darwin's proposals that the study of ontogeny (i.e., the animal's embryonic development) holds clues to phylogeny (i.e., its evolutionary origin). Among the highlights during the period of 1880-1940 were the detailed anatomical descriptions of developmental stages of embryos, including the first atlas of human embryos, reconstructed from microscopic sections, published by W. His, Sr., in 1880-1885. In vertebrate embryology, these descriptions revealed the organogenesis of the heart, kidney, limbs, central nervous system (CNS), and eyes. Developmental-fate mapping studies revealed the embryonic sites of the origin of cells of the organs and the rearrangements of groups of cells in morphogenesis. The stages of development were found to include, in reverse order, cytodifferentiation, organogenesis, morphogenesis (gastrulation and neurulation), rapid cleavage, fertilization, and gametogenesis. By the 1940s, anatomical descriptions of the embryos of related animals were integrated into coherent evolutionary schemes, taught in comparative embryology classes, revealing, for example, the modification of the gill slits of jawless fish to the jaw of jawed fish and further modification to the middle ear of mammals. Also, by this time, Haeckel's oversimplified scheme had been abandoned, namely, that ontogeny merely recapitulates phylogeny.

Experimental embryology also began in the late 1800s. In experimental studies, which mostly involved techniques of cell and tissue transplantation and removal, the central role of cytoplasmic localizations and cell-lineage-restricted developmental fates was recognized in the development of certain invertebrates by the early 1900s. In vertebrate development, the importance of inductions (also called tissue interactions) was recognized in the 1920s, following the stunning organizer transplantation experiments by Spemann and Mangold (1924) on newt embryos. By the 1950s, inductions had been found in every stage and place in the vertebrate embryo, for example, in all the kinds of organogenesis. Vertebrate development, including that of mammals, had become comprehensible as a branching succession of inductive interactions among neighboring members of an increasingly large number of different cell groups of the embryo.

Developmental mechanisms, as understood even in the 1970s, were descriptions of the movements and interactions of cells or groups of cells. They were cellular- or tissue-level mechanisms. The all-important "inducers" were materials of unknown composition released by one cell group and received by another group. Consequently, the recipient cells took a path of development different from the one that would have been taken if they were unexposed. The progression or momentum of development also was recognized: that the individual events of interactions and responses are time-critical, and that certain subsequent aspects of development never occur if one event is prevented.

Molecular mechanisms, however, were not understood at that time. Embryologists encountered the limits of the field in the 1940-1970 period, as they tried to discover the chemical nature of inducers and the responses of cells to them.

The basic information and methods of biochemistry, molecular biology, cell biology, and genetics were not yet available to analyze cell-cell signaling and transcriptional regulation in embryos. In light of discouraging results, some embryologists considered that the organizer concept was faulty and that inducers were an experimental artifact (see later discussion for recent successes in understanding inductions). Although Morgan and other early geneticists had proposed that inducers and cytoplasmic localizations elicit specific gene expression and that development was in large part a problem of ever-changing patterns of gene expression (Morgan 1934), the means were not at hand to pursue those insights. Roux, Spemann, and Harrison had outlined plausible lines of inquiry into determination and morphogenesis in the early part of the twentieth century; however, the means were also not available to pursue those questions at that time.

To many scientists in the 1940-1970 period, the study of development seemed messy and intractable. Researchers turned to more informative subjects such as the new molecular genetics of bacteria and phages (viruses that infect bacteria). From those inquiries came new insights in the 1950-1965 period on the nature of the gene and the code and the processes of replication, transcription, translation, enzyme induction, and enzyme repression. For example, it was only in 1961 that Monod and Jacob described gene regulation in bacteria in terms of promoters, operators, and repressor proteins (Monod and Jacob 1961). Those authors immediately saw the relevance to animal development. All of their insights made possible the invention of techniques for gene isolation and amplification, for in vitro expression of genes, for genome analysis, and, thereafter, for the new developmental biology.

With so little molecular information about developmental processes, there was scarcely any understanding of the action of developmental toxicants. For example, Wilson (1973) in his book *Environment and Birth Defects* could only raise the following possibilities for connections between inductions and developmental defects:

> It has long been accepted that cell interactions (induction) are an important part of normal embryogenesis, despite the fact that specific "inducer substances" have not been identified. [Failures] of normal interactions which may lead to deviations in development include, for example, lack of usual contact or proximity, as of optic vesicle with presumptive lens ectoderm; or the incompetence of target tissue to be activated in spite of its usual relationship with activator tissue, as in certain mutant limb defects; or the inappropriate timing of the interrelation, even though all parts are potentially competent. That the nature of cell-to-cell contacts and the manner of their adhesion are important determinants in both normal and abnormal development has been demonstrated.... Insufficient or inappropriate cellular interactions usually result in arrested or deviant development in the tissue ordinarily induced or activated by the interaction.

This committee will later argue that Wilson's insight was well directed and is now ready to be pursued.

ADVANCES IN DEVELOPMENTAL BIOLOGY

In the past 15 years, developmental biology has advanced remarkably, perhaps as at no other time in the field's history. It is now known that the trillions of cells of a large animal, such as a mammal, have the same genotype, which is the same as that of the single-celled zygote (the fertilized egg) from which the animal develops. That is to say, the genetic content of somatic cells does not change during the development of most animals. The recent clonings of Dolly the lamb (Wilmut et al. 1997), the Cumulina mouse family (Wakayama et al. 1998), and a nonhuman primate (Chan et al. 2000) reaffirm the fact that a specialized cell, such as a mammary or cumulus cell, carries the genes for all other kinds of cells of the animal. The scientific advances that led to these clonings were built on earlier nuclear transplantation successes in frogs, first by Briggs and King (1952), but particularly by Gurdon (1960), which had led to similar conclusions for a nonmammalian vertebrate. Despite the same genes, the cells within the individual organism differ greatly in their appearance and functions, meaning that they have the same genotype and different phenotypes. The cell types differ greatly in the ribonucleic acids (RNAs) and proteins contained within them. They differ in which subset of genes they express from their total genomic repertoire. At least 300 cell types are recognized in humans (e.g., red blood cells, Purkinje nerve cells, and smooth or striated muscle cells). The number of cell subtypes is much larger, perhaps numbering tens of thousands, when further differences are taken into account related to the cell's stage of development and location in the body, as has been discovered in recent years. Development can be viewed as evolution's crowning example of complex gene regulation. From the single genome, thousands of different gene combinations must be expressed at specific times and places in the developing organism, and from the developing egg the information for the selective use of combinations must be generated.

A major factor in this regulation is the transfer of chemical information (i.e., signals) between cells during development. From recent research, which has built on earlier findings, the following is now realized:

- Embryonic cells of arthropods and nematodes make many of their developmental decisions based on which chemical signals they receive from other cells just as vertebrate embryonic cells do. Later the embryonic cells of all these organisms will make further decisions based on other signals. The cycles of signaling and responding are repeated over and over as development progresses. With that in mind and the fact that one genotype supports hundreds or thousands of cellular phenotypes, development can be said to rely on "genotype-environment interactions," where the local environment of each cell is generated by neighboring groups of cells. The genotype and cell's previous developmental decisions determine its options for responses to the signals currently present (Wolpert 1969).
- The signaling pathways involved in this information transfer are known to be of 17 types (a few more may remain undiscovered). They are used repeatedly

at different times and places in the embryo, from the earliest stages through organogenesis and cytodifferentiation, and even in the various proliferating and renewing tissues of the juvenile and adult (see Appendix C).

- The signaling pathways are highly conserved across a wide range of phyla of animals (from chordates to arthropods to roundworms), presumably because they were present and already functional in the pre-Cambrian common ancestor of those animals.
- Many of the kinds of cell responses to signals also are conserved (e.g., responses of selective gene expression, secretion, cell proliferation, or cell migration). The response of developing cells to signals involves activation or repression of the expression of specific genes by transcription factors contained within genetic regulatory circuits. Signaling pathways frequently affect the activity of those factors. Many of the transcription factors and circuits are conserved across a wide range of phyla of animals.

Thus, an effective and general approach to the experimental analysis of developmental processes at all stages has been to inquire about the signaling pathways and transcriptional regulatory circuits that operate in the particular instance of development under study. Different organisms, which differ in aspects of their development, nonetheless use the same conserved signaling pathways and regulatory circuits, but in different combinations, times, and places, and have different genes as the targets of their transcriptional regulatory circuits. Processes of development, which seemed to confront scientists with infinite complexity and variety just a few years ago, now seem interpretable as composites of a small number of conserved elemental processes, namely, those of intercellular signaling, intracellular regulatory circuits, and a limited variety of targeted responses. These conclusions, which were reached by the analysis of development in animals as remote as mice, flies, and nematodes, give great validity to the use of model organisms in studying mammalian development, including that of humans, and in the future analysis of the action of developmental toxicants and in their detection.

Although the signal-response pathways are highly conserved, evolution has produced an increasing complexity of the "community" of pathways in vertebrates. This complexity is evident both in the increased number of closely related pathway components (diversifed protein family members) and in the increased possibilities for cross-talk among pathways. The redundant function of closely related components was made evident by existence of numerous targeted gene-knockout mutations in the mouse that produced little or no identifiable phenotypes—that is, the mice are normal or nearly normal under laboratory conditions (see Table 6-5, later in this chapter). It must be emphasized, however, that functional redundancy provides two advantages. It protects the organism by ensuring that a fundamental process can proceed even in the absence or reduced presence of a critical gene activity. On an evolutionary scale, the multiplicity of overlapping functions provides a basis for generating diversity without losing essential functionality.

The *Drosophila* Breakthrough

The recent molecular understanding of developmental processes and components was gained from the experimental analysis of a few model organisms such as *Drosophila melanogaster* (the fruit fly), *Caenorhabditis elegans* (a free-living nematode), *Danio rerio* (the zebrafish), *Xenopus laevis* (a frog), the chick, and the mouse (see Chapter 7 for proposals about their use in the assessment of developmental toxicities). *D. melanogaster* and *C. elegans* were chosen by researchers for their amenability to genetic analysis, afforded by their small size (hence, large populations) and short life cycle (hence, many generations). Nüsslein-Volhard and Wieschaus (1980) began a systematic search for developmental mutants of *Drosophila* in the mid-1970s. They submitted adults to high-frequency chemical mutagenesis and then inspected large populations of offspring for mutant individuals with strong and early developmental defects (before hatching) at discrete locations and discrete stages in the embryo. They discarded mutants with weak or pleiotropic effects as ones too difficult to analyze at their start. They examined mutagenized flies until the same kinds of mutants began to appear repeatedly in their collections. The recurrence was evidence that they had obtained all the different kinds of zygotic mutants (those affected in genes transcribed after fertilization) that mutagenized flies could yield under the conditions of inspection. This procedure is called "saturation mutagenesis," in which all the susceptible genes whose encoded products are important in development are thought to be revealed. Several laboratories, including those of Nüsslein-Volhard and Wieschaus, were also collecting maternal-effect mutants (those affected in genes transcribed in female germ cells before fertilization) and pursued this search to saturation.

The *Drosophila* mutants were categorized by phenotype and complementation behavior (putting two mutations together in a heterozygote to see whether they are alike or different) to establish the number of different genes whose mutations give the same phenotypic defect of development. Their categories included those embryos failing to develop the anterior or posterior end, odd or even segments, dorsal or ventral parts, mesoderm, endoderm, or nervous system. Further mutant combinations were made to establish epistasis (the interaction of different gene products, reflected in the dominance of one mutant defect over another) and to deduce plausible developmental pathways in which the actions of the encoded gene products could be related and ordered. By the late 1980s, a solid base of observations of *Drosophila* mutant phenotypes and gene locations had been built, and ordered pathways of function based on the mutant interactions had been proposed. This information served as the foundation for future molecular genetic analysis. The research was the first systematic and exhaustive approach to understanding an organism's development and to identifying components of developmental processes.

Synergy with Research Advances in Other Areas

Meanwhile, other researchers worldwide made advances in biochemistry, molecular biology, cell biology, and genetics. They learned an enormous amount about the function of proteins in replication, transcription, translation, secretion, uptake, membrane trafficking, cell motility, cell division, the cell cycle, cell adhesion, and apoptosis (programmed cell death), to mention but a few of the cellular processes. Researchers improved the methods to isolate genes, sequence them, manipulate sequences, make transcripts in vitro, detect messenger (m)RNAs in cells by in situ hybridization, translate RNAs to proteins in vitro, and make antibodies to proteins. In situ hybridization, which graphically revealed the time and place of expression of specific genes in the embryo, was to prove particularly important for connecting the new molecular analysis to the older developmental anatomy. Much of the work was initially done with single-celled organisms: bacteria, yeast, or animal cells in culture. Some insights and techniques came from the study of cancer cells in the search for oncogenes.

In the course of that work, many of the processes, protein functions, and protein sequences were found to be strongly conserved among organisms as diverse as yeast and humans or even bacteria and humans. Various proteins of different organisms, and also within the same organism, shared "sequence motifs" by which the protein could be recognized as a member of a protein family with a particular function and descended from a common sequence ancestor. Newly discovered proteins could be assigned a function from just their possession of a particular motif. As more motifs were found, it will be easier to categorize newly discovered proteins. For example, receptor tyrosine kinases were recognizable by their transmembrane hydrophobic motifs and adenosine triphosphate (ATP)-binding domains. G-protein-linked receptors could be distinguished by a seven-pass (serpentine) transmembrane motif. Transcription factors could be recognized by the sequence motifs of their deoxyribnucleic acid (DNA) binding domains (e.g., zinc finger, basic helix-loop-helix, homeodomain, or leucine zipper domains). Of the recently sequenced genomes of yeast and *C. elegans*, for example, about 40% of the open reading frames (ORFs) are recognizable by known motifs (Chervitz et al. 1998). Function can be assigned, at least preliminarily, to the products of those genes. Plans are afoot to define the function of the missing ORFs of yeast and make the functions of all proteins assignable from sequence. At the same time, there are plans to identify a large number of protein-binding sequences in the regulatory regions of genes to be able to predict better the conditions of expression of genes. These plans are among the aims of "functional genomics," as described in Chapter 5. All the information on sequences, motifs, and function is stored in databases available to researchers worldwide (e.g., the Basic Local Alignment Search Tool (BLAST) <http://www.ncbi.nlm.nih.gov/BLAST/>).

Drosophila Development at the Molecular Genetic Level

By the time the *Drosophila* mutants were characterized in the mid-1980s, techniques were well-suited for molecular genetic analysis of affected genes and gene products. This part of the work moved quickly, thanks to gene-cloning techniques, background information about gene sequence motifs and protein function, and databases available to researchers worldwide. The successful isolation of a gene responsible for a developmental phenotype (when the gene was mutated) could be validated by the rescue of the mutant phenotype by transformation with the wild-type gene (usually as DNA included in a P-element transposon). In situ hybridization, coupled with color stains, readily revealed the normal time and place of expression of the specific genes whose mutations had been isolated. Regarding the function of these developmental genes, many were found to encode proteins with familiar motifs, such as those for receptor tyrosine kinases or various transcription factors. In fact, a surprisingly large number turned out to be transcriptional regulators. Function could be rapidly concluded from sequence data. Other *Drosophila* genes encoded proteins whose specific functions were unknown, yet they were recognizable generally as secreted proteins by their signal sequences or as new transcription factors by the fact they accumulated in nuclei and could bind to DNA. In the course of this analysis, new intercellular signaling pathways were discovered, such as those involving the Decapentaplegic (DPP), Hedgehog (HH), Wingless (WG), and Notch/Delta ligands. (The whimsical names are those given by researchers to mutants based on the phenotypes.)

Hundreds of laboratories worldwide joined the work on *Drosophila* mutants, and the picture of early development took on a satisfying coherence and clarity, especially the steps of generation of segmentation and of the overall body organization in the anteroposterior and dorsoventral dimensions. These steps of early development are known collectively as "axis specification." The following is a brief summary of that picture to illustrate its completeness at the molecular level. The steps are stage-specific mechanisms of development. The mechanisms are now better understood in *Drosophila* than in any other organism. It is the kind of information scientists would like to have, but do not yet have, for mammalian development.

At the start of *Drosophila* development, the oocyte is provisioned with hundreds of maternal gene products that are uniformly distributed in the egg during oogenesis. Four gene products are spatially localized in the egg, however, and they provide the initial asymmetries on which the entire anteroposterior and dorsoventral organization of the embryo is built stepwise in development after fertilization. The four gene products include the following:

1. An mRNA located internally at the anterior end (encoding a transcription factor, named Bicoid).

2. An mRNA located internally at the posterior end (encoding an inhibitor of the translation of the mRNA for a transcription factor, named Nanos).

3. An external protein anchored to the egg shell at both ends of the egg (involved in the production of a ligand of a receptor tyrosine kinase in the egg-cell plasma membrane).

4. An external protein also anchored to the egg shell but at the prospective ventral side (involved in the production of a signal ligand of the Toll receptor in the egg-cell plasma membrane).

To exemplify the steps of use of those gene products, only one of the dimensions, the anteroposterior, will be described. The two mRNAs are initially at opposite ends of the egg. They are translated after fertilization, and the encoded proteins diffuse from the ends to form opposing gradients reaching to the middle of the egg. These proteins will act in concert to generate a gradient, high at the anterior end and low at the posterior end, of another transcription factor. The nuclear number increases rapidly in the uncleaved cytoplasm. The graded transcription factors, called members of the "coordinate class" or "egg-polarity class" of gene products, activate at least eight gap genes in nuclei along the egg's length at different positions, each position unique in terms of the local quantity of transcription factors of the coordinate class. (The terms "coordinate," "egg polarity," and "gap" also derive from mutant phenotypes.) The encoded gap proteins, which are all transcription factors themselves, accumulate in a pattern of eight broad and partially overlapping stripes along the egg's length. The proliferating nuclei are not yet separated by cell membranes—that comes later. These proteins in turn activate at least eight pair-rule genes, all of which also encode transcription factors. Complex *cis*-regulatory regions of the various pair-rule genes define their expression responses to the spatially distributed gap proteins. The pair-rule proteins then activate at least 12 segment-polarity genes, some of which encode transcription factors and some of which encode secreted protein signals. The pair-rule and gap proteins together also activate eight homeobox (*Hox*) genes to be expressed in broad stripes, as discussed in the next section. Thus, the early steps of development involve cascades of transcription factors distributed in space according to the initial gradients of a few agents and to the expression rules contained in the complex *cis*-regulatory regions of genes for yet other transcription factors. These key steps are accomplished in the first 3 hours of development, mostly before cell membranes are formed and gastrulation begins, although the final elaboration of the segment-polarity and *Hox* genes occurs after cells form.

Once the segment-polarity genes and *Hox* genes are activated, they maintain their expression in cells by an auto-activating circuitry, in some cases by the encoded transcription factor activating expression of its own gene. The coordinate, gap, and pair-rule proteins are then no longer needed. Their products disappear, and the genes are no longer expressed.

Similar conclusions apply to the development of the termini and the dorsoventral dimension, which also rely on initially asymmetric signals. The developmental mechanisms of the termini and dorsoventral dimension are of additional interest, because the signals bind to transmembrane receptors and activate signal transduction pathways, eventually leading to the activation of transcription factors and new gene expression. These inductions are the first to occur in the developing *Drosophila* egg. Approximately 100 genes and encoded gene products have been identified as necessary to establish the organization of the early gastrula. Hundreds more participate in the accomplishment of these events, but they are less well described at present. In most cases, these genes probably encode proteins required in numerous developmental processes and, hence, were not recovered under the conditions of the mutant inspections used here.

As shown in Figure 6-1A-D, a coherent scheme of early development was proposed and well supported by 1992, the first of such complexity and completeness at the molecular level for any organism.

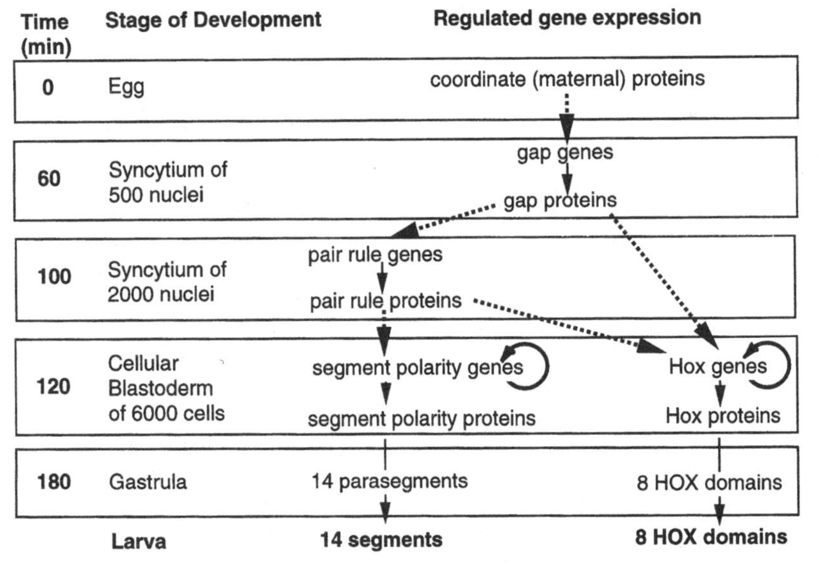

FIGURE 6-1A Outline of anteroposterior development in *Drosophila* and the steps of regulated gene expression (Ingham 1989). Heavy dashed arrows indicate the activation of specific gene expression by transcription factors. Thin solid arrows indicate transcription and translation. Note that *Hox* genes are activated by both pair-rule and gap proteins, whereas segment-polarity genes are activated by pair-rule proteins alone. In the anteroposterior dimension, segments and HOX domains are formed. Further explanation is given in Figure 6-1B.

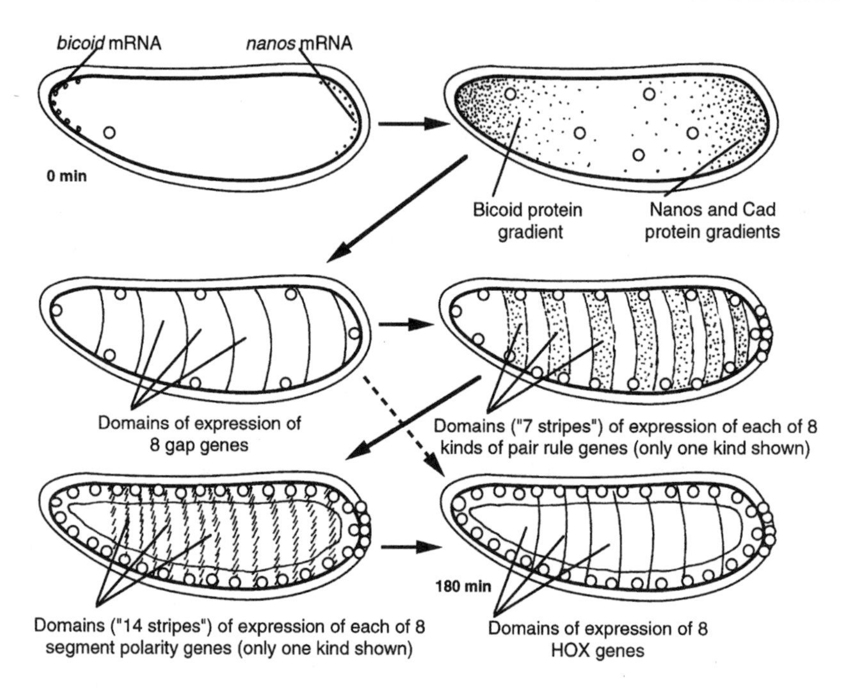

FIGURE 6-1B Anteroposterior development in *Drosophila* (Nüsslein-Volhard 1991). Figure 6-1B is shown diagrammatically here, for segment formation and HOX compartment formation. The coordinate proteins Bicoid, Nanos, and Cad are translated from mRNAs localized at the two poles of the egg during oogenesis. Translation generates gradients of proteins. Bicoid and Cad are transcription factors, whereas Nanos protein inhibits the translation of another translation factor (Hunchback) in the posterior half of the egg. The graded transcription factors activate eight gap genes, and different factor concentrations activate different gap genes. The gap proteins are also transcription factors. Each diffuses locally and inhibits other gap genes, setting up eight partially overlapping stripes of gap protein along the egg's length. The gap proteins activate eight pair-rule genes, each of which has a complex *cis*-regulatory region and is activated by seven combinations of gap proteins, each making seven evenly spaced stripes of protein. Thus, there are 8 × 7 or 56 stripes of pair rules along the egg's length, arranged in 7-fold repeats. The pair-rule proteins are all transcription factors. These activate eight segment-polarity genes, each of which has a complex *cis*-regulatory region activated by at least two combinations of pair-rule proteins, to give 14 stripes of expression each. Thus, there are 14 × 8 or 104 stripes of segment-polarity proteins. The 14-fold repeat is the basis for 14 segments of the posterior head, thorax, and abdomen. The pair-rule and gap proteins together activate *Hox* genes in eight domains in the posterior head, thorax, and abdomen. Cell outlines are not shown, but cells are present in the two lowest panels.

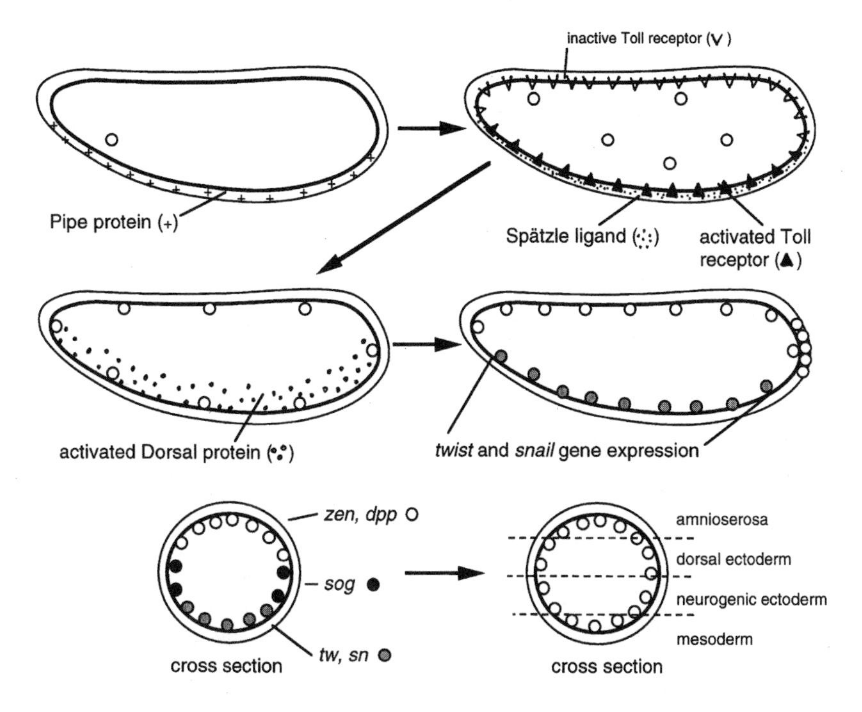

FIGURE 6-1C Dorsoventral development in *Drosophila* (Nüsslein-Volhard 1991). The egg shell contains Pipe protein on the future ventral side, deposited there during oogenesis. After fertilization, the egg secretes several proteins into the space between the egg shell and plasma membrane. Pipe activates one of the proteins, which then sets off others in a protease cascade, the last member of which cleaves the Spätzle protein, releasing a ligand that binds to the Toll transmembrane receptor, which is uniformly distributed over the egg surface but ligand-activated only on one side. The activated receptor, via several intracellular steps, activates the Dorsal protein, a transcription factor, which enters local nuclei and activates two genes, *Twist* and *Snail*, which also encode transcription factors. Those activate other genes for gastrulation and for mesoderm formation on the ventral side. Thus, the Pipe protein is involved in a kind of mesoderm induction. Active Dorsal protein also represses the *Zen* and *Dpp* genes on the ventral side. On the dorsal side, Dorsal protein remains inactive and the *Zen* and *Dpp* genes are expressed. Laterally, there is enough active Dorsal protein to repress *Zen* and *Dpp* but not enough to activate *Twist* and *Snail.* Here, the *Sog* gene is permissively expressed and not repressed, preparatory to neurogenic ectoderm formation. Thus, the dorsoventral dimension of the egg is divided into three domains of gene expression. Later, the Sog protein is secreted and diffuses to the *Zen, Dpp* region, inhibiting Dpp signaling and allowing the division of that region into two subregions the prospective amnioserosa and prospective dorsal ectoderm.

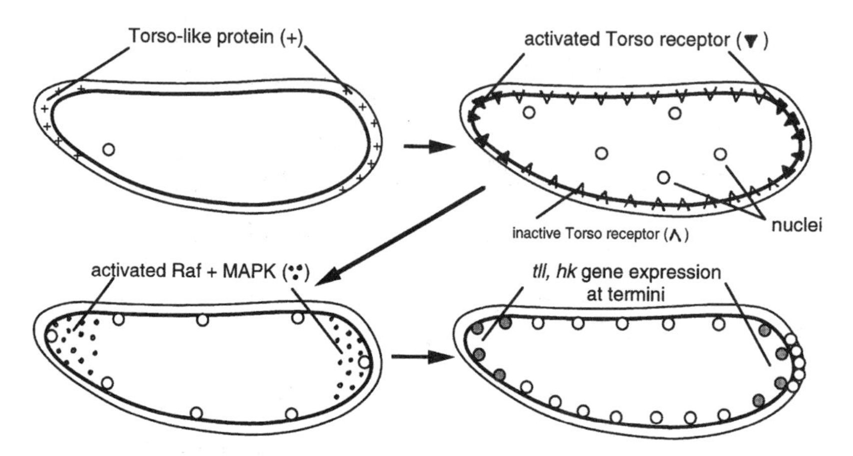

FIGURE 6-1D The development of termini in *Drosophila* (Nüsslein-Volhard 1991). The Torso-like protein is present in the egg shell at the two ends of the egg, deposited there during oogenesis. After fertilization, the egg secretes several proteins into the space between the plasma membrane and egg shell. The proteins include proteases that are locally activated at the end by the Torso-like protein and release a ligand that binds locally to the transmembrane Torso receptor, a member of the RTK signal transduction family. The activated receptor locally activates Raf and MAPK, which phosphorylate a transcription factor locally, inhibiting its repression of genes and allowing local expression of the *Tailless* (*Tll*) and *Huckebein* (*Hkb*) genes involved in formation of the endoderm, terminal ectoderm, and gut involution during gastrulation. Thus, the Torso-like protein is involved in endoderm induction.

Hox Genes and the *Drosophila* Connection to Vertebrate Development

Even though researchers in other areas widely appreciated the breakthroughs in *Drosophila* development, they questioned the relevance of the information to vertebrate development. Vertebrates, as chordates, were thought to have branched from arthropods long ago and last shared a very simple common ancestor in the pre-Cambrian era (about 540 million years ago). The two groups were thought to have evolved their segmentation and heads independently. One of the first significant similarities between vertebrate and fly development came from work on homeotic genes, now called *Hox* genes. As mentioned before, the *Hox* genes are expressed in eight broad bands or spatial compartments in the anteroposterior dimension of the body shortly after gastrulation but prior to organogenesis and cytodifferentiation. Their encoded products make each spatial compartment different from the others.

The study of the eight *Hox* genes of *Drosophila* was primarily pioneered by E. Lewis from 1940 to 1970. For his work in that area, he shared the Nobel Prize with Nüsslein-Volhard and Wieschaus in 1995. Lewis selected *Drosophila* mutants that exhibited mislocated body parts (e.g., wings in place of halteres (balancing organs) and legs in place of antennae). The term "homeotic" connotes such mislocation without distortion. In the homeotic mutant, the anteroposterior dimension of the animal has fewer anatomical differences along its length. For example, the *Ubx* mutant has an extra mesothorax located at the normal metathorax position but lacks a metathorax. It has four wings but no halteres, whereas normal *Drosophila* have two wings and two halteres. When the first two *Hox* genes (*Ubx* and *Antp*) were isolated, their sequences were compared (McGinnis et al. 1984a,b; Weiner et al. 1984), and a shared 60-base sequence was found, the homeobox. The sequence is the same in both genes except for a few bases. That sequence encodes the DNA-binding motif of the encoded proteins, which are members of a large and ancient family of transcription factors. The other six *Hox* genes were soon isolated from *Drosophila*, and those too had closely related homeobox sequences. Then the eight genes were shown to exist in a contiguous cluster (actually two subclusters in *D. melanogaster* but one in another arthropod, *Tribolium*), probably all tandemly duplicated and diverged from a few founder sequences in an ancestor of arthropods. Furthermore, the members are expressed in stripes in the anteroposterior dimension of the body, in an order identical to their gene order on the chromosome (a correspondence referred to as "colinearity" of gene order and expression).

In the mid-1980s frogs and mice were found to contain similar sequences, also arranged in contiguous gene clusters. Interestingly, their expression in mice showed the same anteroposterior colinearity as that in *Drosophila*. As an evolutionary explanation, the common ancestor of arthropods and chordates must have had a complex *Hox* cluster already functioning in its development. Vertebrates, however, differ from arthropods in having at least four multi-member clusters instead of one (Krumlauf 1994). A comparison of gene arrangements and domains of expression in *Drosophila* and mammalian (mouse) *Hox* clusters is shown in Figure 6-2.

Such genes are called selector genes because their encoded products, which are transcription factors, select which other genes will be expressed in that spatial compartment of the body. The thousands of target genes of a selector-gene product encode proteins involved in subsequent local development, including the many kinds of organogenesis of different parts of the body. *Hox* genes have a central role in development. Because of them, the coordinate, gap, and pair-rule proteins of early development do not have to directly activate those thousands of target genes in a region-specific way but activate only the *Hox* genes, whose encoded proteins then do the job of regulating sets of genes in their respective regions. Methods for the directed knockout of genes in mice were invented by the mid-1980s as a way to test gene function, and the *Hox* genes of mice were found to control aspects of local development in their compartments, especially in vertebrae, neural tube, and neural crest derivatives. Their selector role was similar to

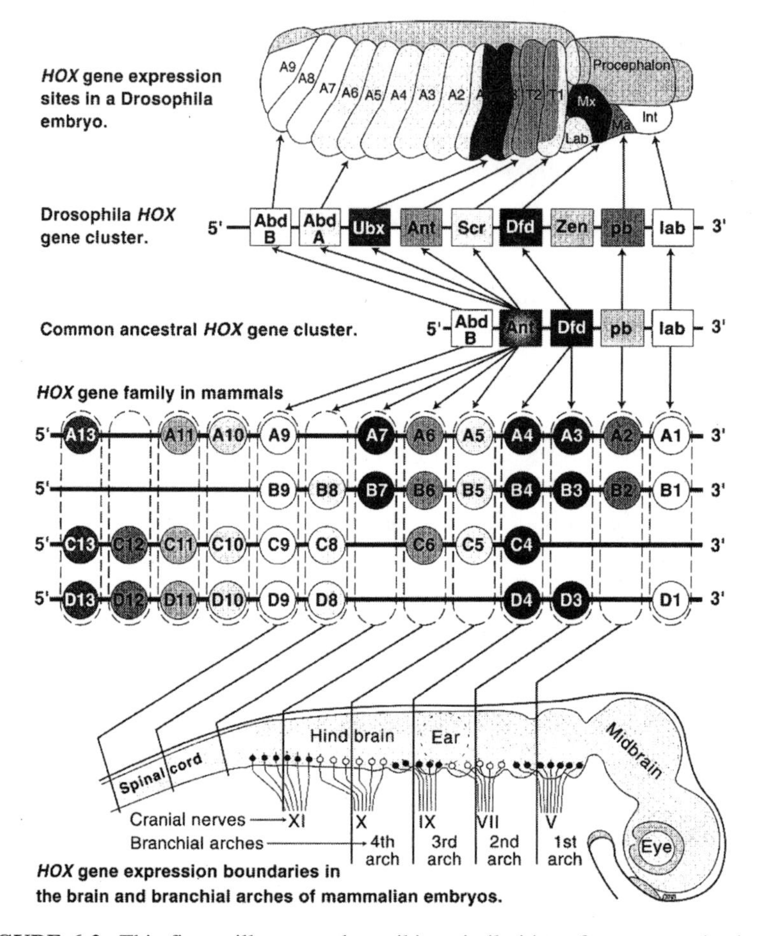

FIGURE 6-2 This figure illustrates the striking similarities of gene organization and expression of *Hox* clusters in *Drosophila* and mammalian (mouse) embryos. At the top is a 10-hour *Drosophila* embryo showing expression zones of individual *Hox* genes in thoracic (T1-3) and abdominal (A1-9) segments and parts of the head (Lab, labrum; Mx, maxillary; Ma, mandible; Int, intercalary segment). Note the colinearity of *Hox* gene expression sites along the anterior-posterior body axis to their 3′ to 5′ location along the chromosome. The greatly expanded vertebrate *Hox* gene family is shown in the middle. These genes are arranged in four clusters (labeled A, B, C, D), each on a separate chromosome. Having arisen by duplications early in chordate evolution, *Hox* genes in paralogous groups (e.g., A4, B4, C4, D4; shown enclosed in dashed boxes) are more closely related than are adjacent genes (e.g., B3 vs. B4 vs. B5). The four most 5′ paralogous groups have no close equivalent in arthropods; these are expressed in the tail and fins or limbs. Lines extending from each paralogous group to the schematic brain and cranial spinal cord show the rostral limits of expression of members on each group. Note, again, the colinearity between expression sites and relative chromosomal position of most *Hox* genes. The same is generally true for somites and, in the proximo-distal orientation, for limbs.

that in *Drosophila* (Behringer et al. 1993). However, many of the target genes of Hox proteins in mice and flies are clearly different.

The *Hox* clusters of *Drosophila* and chordates are under intense study. It is now known that genes of four mouse clusters are coordinated in an elaborate circuitry of auto- and cross-activation and repression, in which the genes near the 5′ end of the DNA sequence tend to repress genes near the 3′ end when both are initially expressed in same cell. Equivalent paralogs in different clusters tend to overlap in the target genes they activate and repress, but each has some unique targets, as shown by the phenotypes of single-*Hox* knockout mutants of the mouse. As a whole, the *Hox* genes operate as a complex genetic regulatory system rather than as independent members.

More recently, the *Hox*-like *Ems* and *Otd* genes have been discovered in *Drosophila* as expressed in the head in regions anterior to the expression compartments of the *Hox* genes. Homologs of these genes (called *Emx* and *Otx*) have been found expressed in the head of the frog and mouse anterior to the *Hox* gene domains of the posterior head, thorax, and trunk. This was a surprise, because evolutionary biologists had thought that the vertebrate head is unique to that group and has little in common with the head of a common ancestor of vertebrates and arthropods. However, even that complexity of body organization, like *HOX* compartments, must predate the branching of arthropods and chordates.

The Emergence of *Caenorhabditis elegans*

The free-living nematode *Caenorhabditis elegans* emerged as an important model system in the 1970s, as the result of pioneering work on its genetics by S. Brenner (1974). Chosen for its short life cycle (3 days) and general amenability for genetic analysis, small size (1-mm length), transparency, and simplicity (only 959 somatic cells), *C. elegans* quickly attracted a following among developmental biologists and geneticists. In particular, J. Sulston was primarily responsible for first describing the complete cell lineage from fertilization to adulthood (Sulston and Horvitz 1977; Sulston et al. 1983) and then spearheading the physical mapping and DNA sequencing of the genome. *C. elegans* recently became the first metazoan organism whose genome is completely sequenced (*C.elegans Sequencing Consortium* 1998). In the meantime, researchers from many laboratories isolated mutants and identified many important genes controlling development, the result being that *C. elegans* is now the most completely described and one of the best understood models for development (see Chapter 7). In some ways, the development of vertebrates is more similar to that of *C. elegans* than of *Drosophila* (e.g., having a cellular rather than a syncyctial early embryo), and in other ways less similar (e.g., having a highly invariant cell lineage and a fixed small number of cells, no Sonic Hedgehog signaling pathway, and few *HOX* genes). These two model animals complement each other usefully for research into fundamental mechanisms of metazoan development.

Conserved Developmental Processes

Researchers increasingly suspected similarities of development between fruit flies and mice and began to look systematically for homologs of *Drosophila* developmental genes in mice, frogs, and chicks. In the late 1980s, this was a new research approach. Its success has favored the impression that at a gross level, nematodes, flies, and mice are "all the same organism" and that what is learned about one will have relevance to the others. In a genetically tractable organism, such as *Drosophila* or *C. elegans*, a gene is isolated by using a screen for a particular kind of developmental failure, and then the role of its encoded product in development is efficiently deciphered in that organism. Homologs of "developmentally interesting" genes are then sought in vertebrates, such as mice or frogs, in which mutant searches are still daunting due to the comparatively small populations and slow development. The homolog's function is thereafter studied in the vertebrate, for which the *Drosophila* or *C. elegans* information is used as a guide. The mouse is attractive for such studies, because the homologous gene can be knocked out and the phenotype of the null mutant examined to learn about the function of the encoded product.

A surprising array of developmental components and processes is shared between *Drosophila* and vertebrates (i.e., between arthropods and chordates). In addition to the *EMX*, *OTX*, and *HOX* organization of the body plan, they share the compartments of the dorsoventral dimension (which are thought to be inverted in orientation in one group relative to the other); the presence and mode of organogenesis of limbs (appendages), eyes, heart, visceral mesoderm, and gut; the steps of cytodifferentiation during neurogenesis and myogenesis; and even segmentation. Although the anatomical structures themselves are very different between arthropods and chordates, a number of the underlying steps of development are the same. These are listed in more detail in Table 6-1. The last common ancestor of chordates and arthropods was, it seems, a pre-Cambrian animal of much greater complexity than previously realized. Divergent groups of metazoa (members of the animal kingdom) can be treated as "the same organism" in the experimental analysis of many fundamentals of development. From all of those similarities, the value of model systems for gaining an understanding of difficult basic problems in mammalian development, including that of humans, is undeniable. Humans, flies, and even roundworms are less different than widely thought just 10 years ago.

Signaling Pathways in Development

An important realization to come from the *Drosophila* research concerns the pervasive use of cell-cell signaling in most aspects of development, starting with the termini and dorsoventral dimension (see Figures 6-1A-D) and extending to organogenesis of many kinds. Inductive signaling was thought to be important in vertebrate development, as mentioned above, but insects and other invertebrates

TABLE 6-1 Similarities of Arthropods and Chordates

Developmental Process	Conserved Genes	Organisms That Share Process	Time Period Work Done
Anteroposterior organization	*Hox* gene complex: similar order of genes in the cluster and similar order of expression domains in the posterior head and trunk (thorax and abdomen)	*Drosophila* and mouse	1987-1992
Anterior head organization	*Ems-Otd (Emx-Otx)* selector genes: similar nesting expression domains in the anterior head	*Drosophila* and mouse	1992-1995
Dorsoventral organization	*Sog-Dpp-Tolloid (Chordin-BMP2,4-xolloid)*: similar gene expression domains, similar protein interactions in the neural versus epidermal regions; similar gene expression domains in the visceral mesoderm and heart. Was the chordate dorsoventral axis formed by inverting the axis of an arthropod ancestor?	*Drosophila* and *Xenopus*	1995-1997
Segmentation	*Engrailed* and *HH-SHH* expression domains are similar in posterior half of segment or somite; *Hairy* gene expression in alternate segments or somites	*Drosophila*, amphioxus, and zebrafish	1996-present
Appendage or limb patterning	Similar domains of WG-HH-DPP (WNT-SHH-BMP) signaling and expression of *En*, *Ap* (*En*, *Lmx*) selector genes	*Drosophila*, chick, and mouse	1994-1997
Eye specification	Similar domains of expression of *Eyeless-Pax6* and *Sine oculis-eye* selector genes	*Drosophila*, mouse, and human	1994-1997

Note: Although the organisms of these two phyla seem very different (e.g.,insects and crustaceans versus fish and mammals), they share many developmental processes at the level of their use of combinations of signaling pathways and genetic regulatory circuits. In italics are various similar conserved genes used in the conserved processes. These similarities serve as evidence that the pre-Cambrian common ancestor of chordates and arthropods was already complex in its anteroposterior and dorsoventral organization and perhaps segmented. Many aspects of cytodifferentiation are also similar (e.g., the use of MyoD in muscle and Achaete-Scute in nerve cells).

had been assumed to develop as composites of independent lineages of cells ("mosaic" development). This is not at all the case. Six signaling pathways are used repeatedly in early *Drosophila* development: the Hedgehog, Wingless-Int (Wnt), transforming growth factor β (TGFβ), Notch, receptor tyrosine kinase (RTK), and cytokine receptor (cytoplasmic tyrosine kinase) pathways. Comparative studies soon showed that these pathways exist in vertebrates as well, and most also exist in nematodes (except the Hedgehog pathway). Four other conserved pathways in addition to those six are used heavily in later development, mainly in organogenesis, and seven others come into use in the physiological functioning of the organism's differentiated cell types. The number of known pathways has now reached 17. Each pathway is distinguished by its unique set of transduction protein intermediates. The 17 pathways are listed in Table 6-2. Details of the components and steps of the individual pathways are given in Appendix C.

As a generalization, most of the pathways involve transmembrane receptor proteins that bind ligands at the extracellular face, as diagramed in Figure 6-3. Ligands arrive in some cases by free diffusion after secretion from distant neighbor cells. Others diffuse only short distances or remain attached to the surface of the cell of origin, reaching only the contacting cells. Activated receptors of the recipient cell activate the first intracellular component of a signal transduction pathway, and this then activates a subsequent component, and so on. Some pathways are long, with 7-10 intermediates. Others have one or two. The nuclear hormone receptor pathway is the shortest, having only one step. In this case, hydrophobic ligands penetrate the cell membrane on their own and bind to a receptor protein, which also functions as a transcription factor. In the longer pathways, a change of activity is passed along a series of on-off switches, which constitutes an information relay pathway, or signal transduction pathway. Ultimately, in some pathways, a protein kinase is activated at the end of the series, and that enzyme phosphorylates numerous target proteins, which change their activity (activated or inhibited) because of the phosphate addition. The target proteins are components of various basic cell processes, such as transcription, the cell cycle, motility, or secretion. Hence, these processes are turned on or off, and the change of function constitutes the cell's response to a signal. In many other pathways, a specific transcription factor is activated at the end of the pathway, and this factor is a pathway component. In development, the most frequent target of signaling pathways is indeed transcription. The pathways used in early development tend to have transcription as the only target. That is, particular transcription factors are phosphorylated or proteolyzed as a signal transduction step of the pathway, changing their activity in activating or repressing particular genes.

The pathways are used repeatedly at different times and places of development in *Drosophila*, nematode, and vertebrates, as listed in Table 6-3. *Drosophila* null mutants are usually lethal if they lack a step in any of those pathways. Lethality is an indication of the essentiality of those signaling functions. However, in the mouse (and probably all vertebrates), a null mutant for a step of a

TABLE 6-2 The 17 Intercellular Signaling Pathways

Period During Development	Signaling Pathway Used
Early development (before organogenesis and cytodifferentiation) and later (during growth and tissue renewal)	1. Wingless-Int pathway 2. Transforming Growth Factor β (receptor serine and threonine kinase) pathway 3. Hedgehog pathway 4. Receptor tyrosine kinase (small G proteins) pathway 5. Notch-Delta pathway 6. Cytokine receptor (cytoplasmic tyrosine kinases) pathway (STAT pathway)
Middle and late development (during organogenesis and cytodifferentiation) and later (during growth and tissue renewal)	7. Interleukin-1-Toll Nuclear Factor-Kappa B pathway 8. Nuclear hormone receptor pathway 9. Apoptosis pathway 10. Receptor phosphotyrosine phosphatase pathway
Larval and adult physiology (after cell types have differentiated)	11. Receptor guanylate cyclase pathway 12. Nitric oxide receptor pathway 13. G-protein coupled receptor (large G proteins) pathway 14. Integrin pathway 15. Cadherin pathway 16. Gap junction pathway 17. Ligand-gated cation channel pathway

Note: The pathways are shared by most animals (metazoa). Note the six pathways used heavily in the early development of most animals (i.e., at stages before organogenesis and cytodifferentiation begin). Along with those pathways, four more are used in later development in the periods of organogenesis and cytodifferentiation, growth, and tissue renewal. Seven pathways are used mostly in the physiological function of differentiated cells; much of that function also involves signaling. Each pathway is identified by the particular transduction intermediates it contains. Most pathways are unique to metazoa, although components of each are often found in single-celled eukaryotes in other signaling roles, such as pathways of checkpoint control, stress response, infection response, mating, and feeding.

pathway is often not lethal. The mutant is born with a limited abnormality of anatomy and sometimes of behavior. The genetic basis for some of these mouse developmental defects is fully known. In some cases, the mutant mice live to adulthood and reproduce.

Are the signaling pathways less important in vertebrate development than in *Drosophila*? No, the nonlethality in vertebrates reflects the fact that the pathways have a substantial redundancy of signaling components. It is postulated that early in vertebrate evolution (as jawless fish arose), the genome underwent a quadruplication (Holland et al. 1994). In addition, some genes underwent tandem dupli-

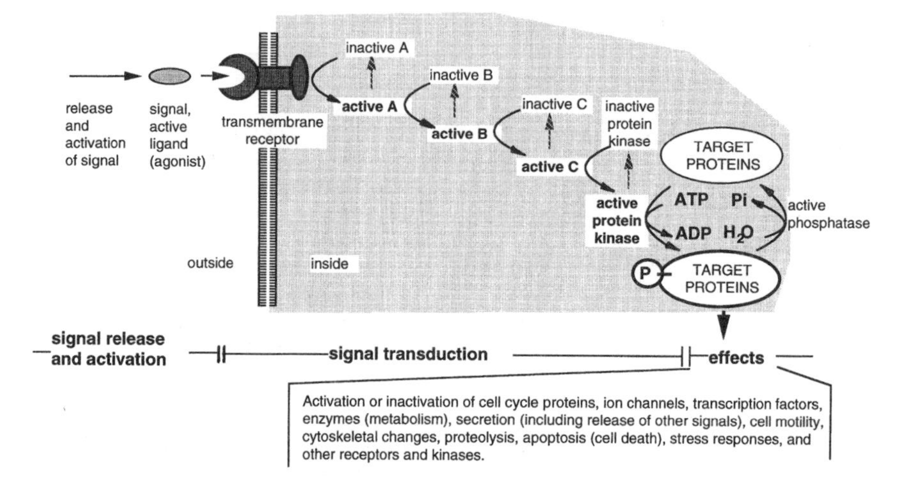

FIGURE 6-3 A generalized signal transduction pathway (information transfer via a series of on-off switches. The active ligand is shown on the left, approaching the transmembrane receptor protein. Inside the cell is a multistep signal transduction pathway composed of switch-like intermediates, A, B, and C, that can exist in active and inactive states. In the absence of signal, the intermediates are inactive. During binding, the receptor becomes active and activates one intermediate that activates the next, and so on in series, until eventually a protein kinase is activated. This pathway transfers information, not energy or materials. The kinase enzyme is specific for transferring phosphate from adesine triphosphate to a serine, threonine, or tyrosine residue of select protein targets within the cell. Phosphorylated target proteins change activity, becoming either active or inactive, and therefore the basic cellular process of which they are a part, changes activity. In this way, the signal has effects (i.e., it triggers cellular responses). Target proteins might be components of the processes of transcription, translation, the cell cycle, cell movement, differentiation, or other signaling pathways. Transcription is a particularly frequent target. Fourteen of the 17 involve transmembrane receptors; the other two involve intracellular receptors and the ligands pass through the plasma membrane readily.

cations. Many of the initially identical genes later diversified their sequences, leading either to diversified functions of encoded products or to diversified *cis*-regulatory regions setting different conditions of expression. Despite this diversification, extensive redundancy still remains. For example, there are more than 24 *TGFβ* ligand genes and 11 *Wnt* ligand genes in mouse, but only 3-5 *TGFβ* and 1-3 *Wnt* genes in *Drosophila*. In mouse development, the various genes for a single step of a single kind of signaling pathway are expressed at different times and places in the embryo. When one gene is knocked out, the defect in development is limited in scope to a few times and places where no related gene is expressed to provide overlapping function.

TABLE 6-3 Various Uses of Signaling Pathways in Vertebrate Development

Pathway	Use
Wnt pathway	Dorsalization of the body (fish, frogs) induction of the organizer or node; posteriorization of the neural plate, midbrain development; dermamyotome induction, somite dorsoventral organization; dorsalization of the fin or limb; female reproductive development; kidney development; dorsoventral differences of limb, hematopoiesis
Transforming growth factor β pathway (BMP,TGFβ, GDF, VEGR, Nodal, Activin, Dorsalin)	Mesoderm induction; induction of organizer, left-right asymmetry; ventralization of mesoderm and ectoderm; neural-crest development to neurons; chondrogenesis of limb, bone development; digit-web spacing, tooth, heart; notochord induction of floor plate of neural tube; notochord and floor-plate induction of sclerotome of somite; prechordal mesoderm induction of prosencephalon; left-right asymmetry, somitogenesis, lung; fin or limb development; zone-of-polarizing-activity (ZPA) induction of anteroposterior axis; gut and visceral mesoderm, hair follicle, skin, and tooth development; spermatogenesis
Hedgehog pathway (Sonic, Indian, Desert)	Notochord induction of floor plate of neural tube; notochord and floor-plate induction of sclerotome of somite and the dorsoventral organization of neural tube; prechordal mesoderm induction of prosencephalon; inhibit cyclopia; ZPA induction of anteroposterior axis of fin or limb development; gut and visceral mesoderm development; hair development
Receptor tyrosine kinase pathway (EGF, FGF, PDGF, EPH)	Mesoderm maintenance; limb (apical ectodermal ridge), vasculogenesis; hair follicle, inner ear, retinotectal projection; astrocyte differentiation, branchial arch signal to neural crest; heart, lung, and tooth development
Notch-Delta pathway (Delta, Serrate, Jagged)	Several steps of neurogenesis; oligodendrocyte differentiation, retina development; somitogenesis, inner-ear development; feather-bud development; blood-cell development (e.g., thymocytes)

Note: These pathways are used repeatedly in early development. The particular uses in developmental processes are well conserved across vertebrates.

As shown in Table 6-4 and Table 6-5, many single-null mutants of the mouse are born live and have minimal defects that can be scored as developmental defects (e.g., skeletal abnormalities). Some have motor coordination defects, such as the WNT1 null mutant, which lacks the entire cerebellum. Some live for a few days, and others reach adulthood and are fertile. In Table 6-4, five pathways are surveyed partially (one or two components eliminated per pathway). In Table 6-5, one pathway is surveyed exhaustively (all components eliminated one at a time). Each kind of mutant is defective for one component of one signaling

TABLE 6-4 Survey of Phenotypes of Mouse Mutants Lacking Components of Any of Several Signaling Pathways

Signaling Component	Viability of Null Mutant	Phenotype of Null Mutant	References
Wnt pathway			
WNT-1	Adulthood	No midbrain, cerebellum, and rhombomere 1; behavioral deficits	McMahon et al. 1992
Axin	Early lethal (day 8-10)	Twinning	Vasicek et al. 1997; Zeng et al. 1997
Transforming growth factor β pathway			
TGFβ 1	Adulthood	Immune defects, inflammation	McCartney-Francis et al. 1997
TGFβ 2	Perinatal death	Defects of heart, lung, spine, limb, and craniofacial and spinal regions	Sanford et al. 1997
GDF5	Adulthood	Fused skeletal elements in limbs, one-third of joints missing; like brachypodism mutant.	Storm and Kingsley 1996
BMP5	Adulthood	Thin axial bones; abnormal lung, liver, ureter, and bladder; like a short ear mutant	Mikic et al. 1996
BMP7	Adulthood	Defects in eye and kidney; skeletal abnormalities; polydactyly of hindlimbs	Dudley and Robertson 1997
Noggin	Juvenile	Bone hyperplasia, joints not formed; neural tube and somite defects	Brunet et al. 1998
Hedgehog pathway			
Sonic Hedgehog (SHH)	Perinatal death	Cyclopia, defects of spinal cord, axial skeleton, and limbs	Chiang et al 1996
Patched receptor	Homozygotes, early lethality	Open neural tube	Goodrich et al. 1997
	Heterozygotes, adulthood	Rhabdomysarcomas, hindlimb defects, large size. Like Gorlin syndrome in humans.	Hahn et al. 1998
Notch pathway			
Notch 1	Perinatal death	Disordered somites. Like Danforth short tail mutation?	Swiatek et al. 1994; Conlon et al. 1995
Delta (DII1)	Perinatal death	Disordered somites	Hrabe De Angelis et al. 1997

continues

TABLE 6-4 continued

Signaling Component	Viability of Null Mutant	Phenotype of Null Mutant	References
Cytokine pathway			
STAT1	Adulthood	Defects in interferon viral resistance	Durbin et al. 1996; Meraz et al. 1996
STAT3	Early embryonic lethality		Takeda et al. 1997
Nuclear receptor pathway			
Progesterone receptor	Adulthood	Males normal; females: no ovulation; no mammary glands, uterine hyperplasia	Lydon et al. 1995, 1996; Chappell et al. 1997
Retinoic acid receptor β	Adulthood	Fertile, small size, transformations of cervical vertebrae	Ghyselinck et al. 1997
RARα and RARβ	Perinatal death	Visceral abnormalities, reduced thymus, spleen	Ghyselinck et al. 1997

Note: Null mutants are shown for components of signaling pathways emphasized in the text. A null mutant is defined as producing no protein function for that encoded gene product. Note the differing viability. References are minimally indicated but fully enough to find the citation in widely available databases, such as Medline, by using the author and signaling component.

TABLE 6-5 Phenotypes of Mouse Mutants Lacking Components of the Receptor Tyrosine Kinase Pathway

Signaling Component	Time of Death	Affected Tissues and Organs	Phenotype	Human Syndrome or Disease[a]	References
Signals					
FGF2	Viable	Brain, skin	Delayed wound healing, lower neuronal density		Ortega et al. 1998
FGF3	Neonatal	Tail, inner ear	Developmental defects		Mansour 1994
FGF4	Early post-implantation	Inner cell mass	Growth failure after implantation		Feldman et al. 1995
FGF5	Viable	Hair follicle	Continuous hair growth	Hypertrichosis universalis[b]	Hebert et al. 1994
FGF6	Viable	Muscle	Defective muscle regeneration		Floss et al. 1997
FGF7	Viable	Keratinocyte and hair follicle	Rough coat		Guo et al. 1996
FGF8	Gastrulation	Mesoderm	No heart and somites		Meyers et al. 1998
GDNF	Neonatal	Kidney, neural crest	Renal agensis, enteric nervous system defects, neoplasia	Hirschprung's disease, multiple endocrine	Pichel et al. 1996; Sanchez et al. 1996; Moore et al. 1996
HGF	Mid-gestation	Skeletal muscle, liver	No muscles, small liver		Schmidt et al. 1995
IGF1	Reduced, some neonatal death[c]	Organs, muscle, bone	Severe growth deficiency	Dwarfism; neoplasia[b]	Powell-Braxton et al. 1993; Liu et al. 1993
IGFII	Viable	Fetus and placenta	Severe growth deficiency	Beckwith-Wiedemann syndrome	DeChiara et al. 1990
NGF	Neonatal	Nervous system	Lack of sensory and sympathetic neurons		Crowley et al. 1994
PDGFA	Mid-gestation or neonatal	Lung, alveolar, myofibroblasts	Emphysema		Bostrom et al. 1996

Receptors					
EGFR	Perimplantation, mid-gestation, or postnatal [c]	Inner cell mass, placenta, skin, kidney, brain, liver, lungs	Multiple abnormalities, lung immaturity, wasting	Cancer	Threadgill et al. 1995; Miettinen et al. 1997; Sibilia and Wagner 1995; Downward et al. 1984
EPHA2	Viable	None	No phenotype		Chen et al. 1996
EPHA8	Viable	Neurons	Altered axonal projections		Park et al. 1997
FIT1	Viable	Vascular endothelium	Abnormal vascular channels		Fong et al. 1995
FLK1	Mid-gestation	Vascular endothelium	No blood or blood vessels		
FLK2	Viable	Hematopoietic stem cells	Deficient B cells		
FGFR1	Mid-gestation	Mesoderm	Abnormal mesoderm patterning	Pfeiffer syndrome, Acrocephalopoly-syndactyly type ll; stem cell leukemia/lymphoma syndrome[b]	Deng et al. 1997; Yamaguchi et al. 1994
FGFR2	Periimplantation	Inner cell mass	Growth failure	Apert, Crouzon, Pfeiffer, Jackson-Weiss, Cutis gyrata syndrome[b]	Arman et al. 1998
FGFR3	Viable	Skeleton, inner ear	Skeletal overgrowth	Achondroplasia; hypochandroplasia; thanatophoric dwarfism; deafness; Muenke syndrome	Colvin et al. 1996
FGFR4	Viable	None	Slight growth retardation		Weinstein et al. 1998
GCSFR	Viable	Immune system	Immune deficiency	Kostmann disease	Liu et al. 1996
HGFR	Mid-gestation	Skeletal muscle, liver	No muscle, small liver		Bladt et al. 1995
IGF2R	Perinatal	Fetus	Overgrowth, cardiac defects		Lau et al. 1994; Wang et al. 1994
IR	Neonatal	Fetus	Diabetic ketoacidosis, growth retardation	Leprechanism	Joshi et al. 1996

continues

TABLE 6-5 continued

Signaling Component	Time of Death	Affected Tissues and Organs	Phenotype	Human Syndrome or Disease[a]	References
LIN12	Mid-gestation	Somites	Somitogenesis delayed/ disorganized		Conlon et al. 1995
NGFR	Viable	Nervous system	Defective sensory neurons		Lee et al. 1992
NTRK1	Neonatal	Central nervous system (CNS)	Neuropathies	Cancer, congenital neuropathy with anhidrosis	Smeyne et al. 1994
PDGFRA	Mid-gestation	Skeleton, neural crest	Spina bifida oculta and other skeletal abnormalities, increased neural crest apoptosis		Soriano 1997; Payne et al. 1997
RET	Perinatal	Kidney, neural crest	Renal agenesis, enteric nervous system defects	Hirschprung's disease, multiple endocrine neoplasia	Schuchardt et al. 1994, 1996
Intracellular					
GAP	Mid-gestation	Undefined	Postimplantation arrest and death		Henkemeyer et al. 1995
PLCY1	Mid-gestation	Unknown	Growth retarded		Ji et al. 1997
PRKM4/ MAPKK4	Mid-gestation	Undefined	Death		Yang et al. 1997b
PTPN11/SHP2	Mid-gestation	Undefined	Failure to gastrulate		Saxton et al. 1997
SAPK/ERK1	Viable	T cells	Altered susceptibility to apoptosis		Nishina et al. 1997
SERK2/SAPK/ ERK/JNK3	Viable	CNS	Altered response to kainic acid damage		Yang et al. 1997a
H-RAS	Viable	None	None	Cancer	Johnson et al. 1997
K-RAS 2	Mid-gestation	CNS	Increased cell death	Cancer	Koera et al. 1997
N-RAS	Viable	None	None	Tumors	Umanoff et al. 1995

RAS/GAP	Mid-gestation	CNS, heart, allantois	Disorganized, organogenesis defects		Henkemeyer et al. 1995
RAS-GRF	Viable	CNS	Impaired long-term memory		Brambilla et al. 1997
VAV	Viable	T cells	Defective selection of thymocytes		Turner et al. 1997
Target Proteins					
C-FOS	Viable	Osteoclasts	Osteopetrosis, toothless		Johnson et al. 1992; Wang et al. 1992
FOSB	Viable	Mammary gland	Failure to nurture offspring		Gruda et al. 1996; Brown et al. 1996
C-JUN	Mid-gestation	Fibroblasts and others	Altered response to mitogens		Hilberg et al. 1993; Johnson et al. 1993

[a] Sources: Online Mendelian Inheritance in Man (<ww3.ncbi.nlm.nih.gov/Omim/>) and reference no. 2.
[b] Possible association between gene and syndrome.
[c] Variable phenotype depending on genetic background.

Note: Null phenotypes of mice produced by gene targeting for some genes involved in one signaling pathway, the receptor tyrosine kinase pathway, and human syndromes or diseases associated with mutations or deletions in these genes. In addition to the entries in this table, many of the mouse mutations have been combined to make double or triple mutants. Additional follow-up studies, some using hypomorphic alleles and chimeras, have been reported (see Mouse Knockout Database: <http://tbase.jax.org>).

pathway. Of relevance to this report, some mutant phenotypes resemble those of animals treated in embryogenesis with various developmental toxicants. Said otherwise, some toxicant treatments produce phenocopies of mutants. For example, SHH mutants have cyclopia, as do normal embryos treated with cyclopamine (Beachy et al. 1997). Researchers of human developmental defects have benefitted by analysis of these mouse developmental defects for which the genetic defects are accurately known.

Antagonists of Signaling

Although only a small number of signaling pathways operate in early development, these have several regulatory features of relevance to developmental toxicity. One is that many of the pathways engage in "cross-talk" with one another, so that activation of one enhances or suppresses the activation of another. Cross-talk occurs at all levels, from effects on ligand availability to effects on target function. For example, the active RTK pathway can lead to phosphorylation and activation of GSK3 in the Wnt pathway, or phosphorylation and inactivation of a SMAD protein of the TGFβ pathway. Signaling via one pathway, such as the SHH pathway, can lead to repression of gene expression of the components of another pathway, such as the Nodal TGFβ pathway. Signaling via the RTK or Wnt pathways can negatively affect signaling via the Notch pathway.

A second regulatory feature is the production by cells of antagonists of signaling, some of which are listed in Table 6-6. For example, embryonic cells can produce proteins, such as Chordin or FRZB, that bind directly to the BMP4 or WNT8 ligands, respectively, and block their capacity to bind to their receptors. Hence, signaling is prevented even though the signal is present. The Chordin protein is further regulated by a protease that degrades it, an antiantagonist. Thus, the pattern of activity of a signaling pathway is subject to extensive modulation, both positive and negative. Several kinds of adjacent cells can affect the outcome of signaling. It is plausible that some of these antagonist proteins are targets of toxicants. Disrupting the activity balance of agonists and antagonists in a region of the developing embryo would be expected to disrupt development, leading to the over- or under-development of an organ. That could occur without a change in level of the agonist or antagonist.

Finally, in many pathways, the operation of the pathway leads to the generation of self-inhibiting components in the cells receiving signals, a feedback mechanism, which is thought to have importance in the controlled spatial responses of cells to diffusible signals. Toxicants upsetting these feedbacks would upset development.

Molecular-Stress Pathways and Checkpoint Pathways

Molecular-stress pathways and checkpoint pathways are not pathways of intercellular signaling but of *intra*cellular signaling, reflecting an individual cell's

Table 6-6 Natural Antagonists of Signaling Pathways

	TGF-β pathway Antagonists	Wnt Pathway Antagonists	RTK Pathway Antagonists	Hedgehog Pathway Antagonists	Notch/Delta Pathway Antagonists
Ligand	Noggin, Chordin, WIF1, Cerberus, Dickkopf, Xnr3	Frzb, Cerberus, Sizzled	Inactivity of Rhomboid		Fringe
Receptor site for ligand binding (plasma membrane)	Inactivity of one-eyed pinhead		Kekkon 1, Argos	HIP, Patched	Wnts, EGF
Receptor tail	Smad 7, 8				Dishevelled, Inactivity of Presinillin
Activated intermediate 1	MAPK?	Axin	Sprouty, Puckered, Erp phosphatase	PKA	
Activated intermediate 2		Nemo-like kinase			
Actived transcription factor in nucleus (nuclear membrane)	Brinker				Hairless

Note: Cell-cell signaling in metazoa involves antagonists as well as agonists, making signaling more complex and conditional. Antagonists can act at any of several steps in a pathway. Those acting at the ligand or receptor site levels are secreted and affect many nearby cells. Those acting at intracellular steps affect only the cell in which they are produced. Source: Adapted from Artavanis-Tsakonas et al. (1999) and Perrimon and McMahon (1999).

sensing of disruptions of normal cell function and development due to either physical or chemical agents of the environment (the molecular-stress pathways) or the cell's own internal imbalance or errors in its synthetic activities (the checkpoint pathways). These pathways are widely present in single-celled eukaryotes (e.g., yeast) and even prokaryotic cells, as well as in animals. Several molecular-stress and checkpoint pathways are listed in Table 6-7 and illustrated in Appendix C.

Molecular-stress pathways are activated when the cell suffers some chemical alteration, such as damage to DNA (e.g., by X-ray, UV, or alkylating agents) or denaturation of proteins (e.g., by hyperosmotic conditions, oxidation, heat, or alcohol). The cell's signaled response is one of repair and homeostatic counteraction. In the case of the cytosolic unfolded protein pathway (previously called the "heat shock response"), chaperone proteins, such as Hsp90, help to refold denatured proteins, restoring their activity. These same chaperones play a folding role in the normal synthesis and deployment of intrinsically unstable proteins, such as cell-surface receptors. Recent experiments have suggested that if Hsp90 is partially disabled by mutation or overloaded by stress, variant proteins in some members of a population might be unable to fold correctly, resulting in developmental defects (Rutherford and Lindquist 1998). Another example would be the multidrug transport proteins (P-glycoprotein of mammals) that are induced in the presence of high drug levels and serve to export a wide variety of drugs from the cell.

In checkpoint pathways, the cell's response is one of delaying certain synthetic processes until other processes are complete. These controls are important in coordinating the timing and extent of cellular processes, such as ensuring the completion of DNA synthesis before mitosis begins or ensuring the attachment of chromosomes to the spindle before anaphase begins. An example of relevance to developmental toxicology is that when colchicine (a Vinca alkaloid) inhibits microtubule formation in a mitotic cell, the cell is prevented from initiating anaphase because of a checkpoint control pathway, which assesses the attachment of kinetochores to spindle microtubules (Rudner and Murray 1996). While chromosomes remain unattached, anaphase is not initiated. When the inhibitor is removed, the cells assemble a spindle and proceed with anaphase. Mutant cells have been isolated that lack components of the control (such as the MAD2 protein), and these cells initiate anaphase in the presence of colchicine, without a spindle. They suffer extensive aneuploidy.

Checkpoint and molecular-stress pathways work together. For example, damaged DNA inside the cell triggers various stress pathways, leading to DNA repair. During the repair, the checkpoint pathways delay DNA synthesis or mitosis until repair is complete. In the context of this committee's evaluation, there are two relevant points about those widely distributed pathways:

- They offer possibilties for the detection and analysis of developmental toxicants, because they indicate the cell's state of stress in the presence of toxi-

Table 6-7 Molecular-Stress Response and Checkpoint Pathways (see Appendix C for illustrations)

Checkpoint Pathways	Function and Cell Response
G1/S checkpoint	Monitors nutritional state, biosynthetic capacity, and cell adhesion and imposes G1 arrest until cell is prepared for S phase.
G2/M checkpoint	Monitors completion of DNA synthesis (S phase) and imposes G2 arrest until cell is prepared for M phase.
Metaphase/anaphase checkpoint	Monitors attachment of chromosomes to the spindle and imposes metaphase arrest until cell is ready for anaphase.

Molecular-Stress Response Pathways	Function and Cell Response
DNA damage (genotoxic stress)	Kinases are activated at DNA damage site by single stranded DNA and 5′, 3′ ends, leading to p53 activation and transcription of genes encoding p21 inhibitors of cyclin-dependent kinases, and hence G1/S or G2/M arrest until repair is complete.
Cytosolic unfolded protein pathway	Activated by heat, alcohol, anaerobiosis, and amino acid analogs, leading to activated transcription of genes encoding chaperone proteins until protein refolding is complete.
Endoplasmic reticulum (ER) unfolded protein pathway	Unfolded proteins activate receptor thre/seri kinase, leading to release of a nuclease that degrades some mRNAs and reduces translation (G1 arrest), and splices some mRNAs leading to transcription of genes encoding chaperone proteins until protein refolding is complete.
Apoptosis (cell death)	Triggered by intracellular damage or extracellular signals, leading to caspase protease activation and cell destruction.
Ultraviolet, hyperosmotic shock, free-radical oxidation pathways	Mediated by MAP kinases, leading to transcription, until damage is reversed.

Note: Many of these are found in single-celled eukaryotes as well as in most or all cells of animals.

cants. Broadly acting toxicants are likely to show up as triggers of stress responses.

- The intercellular signaling pathways of metazoa probably arose in evolution as elaborations and reworkings of the more ancient molecular-stress and checkpoint pathways of single-celled eukaryotic ancestors. This is surmised because a number of the intermediates (e.g., protein kinases) of the molecular-stress and checkpoint pathways are also used in the metazoan signaling pathways.

One more kind of molecular-stress and checkpoint pathway should be noted: the apoptosis pathway, which is also a signaling pathway (see Table 6-2). It can be activated by either extracellular or intracellular signals and leads to the "programmed" death and destruction of a cell. It is a tightly controlled process in which a cell is destroyed but neighboring cells are unaffected. Apoptosis is not found in single-celled organisms. It is an invention of metazoa and is used in normal embryonic development as well as in recovery attempts of teratogen-damaged embryos. In normal development, where apoptosis is also known as programmed cell death, it is important in the shaping of tissues and organs (e.g., the elimination of cells from the interdigital spaces of the human hand). Cells undergoing apoptosis are found throughout most embryonic mesenchymal tissues, presumably reflecting the elimination of cells that have not been able to successfully integrate the signals impinging upon them. Some mouse mutants, such as Hammertoe, fail to initiate the normal amount of programmed cell death in normal limb development, and an abnormal limb results (Zakeri et al. 1994).

Apoptosis is also the ultimate molecular-stress and checkpoint pathway, for it eliminates cells too damaged to be restored to a normal state by the various repair and checkpoint pathways. For example, if DNA repair is incomplete and the cell attempts to divide, it is killed and autolyzed. It has been proposed that cell death is less detrimental to the multicellular organism than having live cells with highly modified DNA, perhaps proliferating uncontrollably and interacting aberrantly with other cells. Apoptotic cell death is an early response of embryos to many if not all teratogens (Scott 1977; Knudsen 1997). Often, teratogen-induced cell death occurs in the areas of normal programmed cell death but in an expanded area (Alles and Sulik 1989). If cell death is not too extensive, embryos are thought to recover by compensatory cell proliferation (Sugrue and DeSesso 1982). Excessive teratogen-induced cell death, however, is directly linked to abnormal development. For example, eye defects induced by 2-chloro-2'-deoxyadenosine are associated with excessive teratogen-induced cell death (Wubah et al. 1996).

The intracellular signals of apoptosis are not yet known. Key components in the execution phase of the apoptotic pathway are the intracellular cysteinyl-aspartate proteases known as caspases, particularly caspase-3 (Colussi and Kumar 1999). These enzymes are normally present in all cells as inactive precursors that become activated by cleavage at specific internal motifs, in response to cytochrome c leaked by mitochondria into the cell's cytoplasm. Once activated, these caspases function to degrade specific target substrates such as poly(ADP-ribose)-polymerase (PARP), DNA-PKs, and lamins. Thereafter, chromosomal DNA is broken down. Treatment of cells with such developmental toxicants as hyperthermia, cyclophosphamide (an alkylating agent), and sodium arsenite (thiol oxidant) leads to the activation of caspase-3, cleavage of PARP, fragmentation of DNA, and cell death (Mirkes and Little 1998). It is not known how cells in the embryo recognize exposure to a developmental toxicant and initiate the apoptotic response, but perturbation of the redox status of the cell and oxidative stress are often, if not always, involved. As in other pathways, the apoptotic pathway engages in cross-talk, for example, with the nuclear factor-kappaB (NF-kB) and INF and FAS pathways. A recent report demonstrates that heat shock (43°C) can rapidly activate the stress-activated protein kinase pathways mediated by c-JUN terminal kinase (JNK) and p38 (Wilson et al. 1999).

As noted for the drug-metabolizing enzymes discussed in Chapter 5, these molecular-stress and checkpoint pathways deserve attention as elements of the organism's defense against physical and chemical interventions. It remains to be learned whether polymorphisms of defense components exist in humans, compromising their responses to environmental agents. The extent to which the germ line, gametes, and early embryos operate these molecular-stress and checkpoint pathways is also poorly understood.

Developmental Differences

Although *Drosophila* and mouse development share more similarities than anyone thought 15 years ago, significant differences do exist. Mice share more aspects of development with other chordates (the chordate phylum includes vertebrates, cephalochordates, and urochordates) than they do with *Drosophila*, and they share still more aspects with other mammals. There appear to be "nested similarities" of development (i.e., the more recent the common ancestor of two groups, the more shared features of their development). Regarding *HOX* genes, for example, chordates have four more kinds of genes (*HOX 10-13*) than do arthropods. These differ slightly in sequence from the others and are located at the 5′ end of each cluster. They are expressed in the postanal tail, which is a chordate structure not shared by arthropods, and also in the developing vertebrate limb. Still, the difference between chordates and arthropods is a modification of a shared feature, namely, the use of *HOX* genes to divide the anteroposterior dimension of the animal into nonequivalent spatial compartments.

Chordates, but not arthropods, share the development of a dorsal hollow nerve cord, a notochord, and a segmentally arranged pharyngo-branchial apparatus, in addition to a postanal tail. They also share a kind of development involving a centralized "organizer" group of cells, the Spemann organizer, which releases inducers important in the placement, orientation, and scaling of later development by surrounding cells. The inducers secreted by the organizer have now been identified. Several inducers are secreted protein antagonists of the TGFβ and WNT signals and are used by surrounding cells to maintain their ventral posterior paths of development. The inducer antagonists disinhibit and hence release the inherent capacity of the surrounding cells to undertake dorsal anterior kinds of development (e.g., to form the neural tube rather than epidermis) (Harland and Gerhart 1997; Smith and Schoenwolf 1998; Weinstein and Hemmati-Brivanlou 1999). Few researchers would have guessed a few years ago that this subtle depression of neural development is the organizer's function in all chordates. Nonetheless, it should be noted that Holtfreter (1947), building on a discovery of Barth (1939), suggested that neural inducers provide little information except to release the inherent capacity of ectoderm cells to develop as neural tissue. This suggestion came from Barth and Holtfreter's findings that ectoderm would develop neural tissue if merely shocked briefly by ion imbalances or pH extremes.

Even though the organizer mode of development is distinctive to chordates, the components of the process are common to a wide range of other animals. For example, one antagonist, the Chordin protein, exhibits significant homology with the SOG protein of *Drosophila*. The SOG protein antagonizes a TGFβ inductive signal (called Screw) in *Drosophila* as part of the development of regions of neural versus epidermal development (Neul and Ferguson 1998). Furthermore, in both *Drosophila* and frogs, there is a specific metalloproteinase that degrades the signal-antagonist complex, releasing the signal. The chordate and *Drosophila* inductive processes have deep similarities, though differing in details of time, place, and circumstances of use.

As a final example of differences, the dorsoventral dimension of arthropods looks quite different from that of a mouse, but recent analysis has shown that a number of similar genes are expressed in the nerve cords, hearts, body muscle, visceral mesoderm, and gut of both. It is currently accepted that these organs were present in primordial form in a common ancestor, but the arrangement of the organs in chordates is the inverse of that in arthropods. That is, the nerve cord is dorsal in chordates and ventral in arthropods, and the heart is ventral in chordates and dorsal in arthropods. The inversion of the dorsoventral axis is thought to have occurred in the chordate line after hemichordates split off (Nübler-Jung and Arendt 1996).

Recognizing the fact that *Drosophila* does not share all details of early development and organogenesis with vertebrates, researchers have begun a systematic collection of developmental mutants of the zebrafish, a small vertebrate with a short life cycle (see Chapter 7), suitable for the production of a large mutant collection. The organs of embryonic zebrafish, more than the organs of *Drosophila*, resemble those of mammalian embryos in structure and function. In light of the extensive conservation of developmental processes found thus far, it is expected that in most cases what is true for fish development, as learned from those mutants, will be true for mammalian development, down to the level of molecular details of components and processes. That is not meant to deny differences among organisms (e.g., mammals undergo placental development with extensive extra-embryonic tissues not found in a zebrafish), nor to dismiss the possibility that developmental biologists might be misled in some instances by the study of model organisms. The greater part of mammalian development can be understood, however, by the study of other organisms' development. Ultimately, mammalian development will have to be understood in all the details of its differences, but even this pursuit will benefit from the context of knowledge of the processes shared with other organisms. For example, unique mammalian processes, such as extra-embryonic tissue formation or more extensive forebrain development, are still expected to entail many of the same signal transduction pathways and genetic regulatory circuits as used elsewhere in development.

The Evolutionary Perspective

In light of the availability of base sequences for a variety of kinds of genes in a variety of organisms, the place of metazoa (the multicellular animals) among the kingdoms of living organisms has been recently re-evaluated. It now appears that animals share a common ancestor more closely with plants (especially fungi) than with protozoa such as ciliates or amoebae. These three multicellular kingdoms arose from a common eukaryotic ancestor (probably single celled) perhaps 1.2 billion years ago, whereas eukaryotic single cells go back 2.2 to 2.7 billion years and prokaryotic life goes back perhaps 3.5 billion years (Feng et al. 1997; Pace 1997). The conservation of basic biochemical, genetic, and cell biological functions has been surprisingly extensive in that long lineage. At least 3 billion years ago, ancient prokaryotes originated the processes of replication, transcription, translation, energy metabolism, and biosynthesis, and those processes have been carried forward to this day with little change in all life forms, including animals. The comparisons of the whole genomes of bacteria, yeast, and now the nematode, *C. elegans,* show clearly the conservation of the protein-coding sequences of genes. At least 2 billion years ago, single-celled eukaryotes originated the basic cell biological processes of mitosis, meiosis, a cdk-cyclin-based cell cycle, an actin-based cytoskeleton and myosin-based movements, a tubulin-based cytoskeleton and kinesin-dynein-based movements, membrane-trafficking, and membrane-bounded organelles. These processes and structures have been carried forward by the single-celled eukaryotes and animals with little change to this day.

In light of this conservation of ancient processes, what have metazoa added in the past 1.2 billion years? Their innovations include abundant cell-cell signaling, extracellular matrix, cell junctions, and a wide range of responses to intercellular signals based on complex genetic regulatory circuits and protein phosphorylation. The *C. elegans* genome shows that, compared with yeast, metazoa have greatly expanded the number of genes encoding proteins of signal transduction, the cytoskeleton, and transcriptional regulation and have greatly increased the size and complexity of the *cis*-regulatory regions of genes. Metazoa seem to have evolved in a regulatory or informational direction, that of determining the time, location, and circumstances within a multicellular population for activating and inhibiting the many conserved biochemical and cell biological processes brought forward from their single-celled ancestors. All these ancient processes have been made contingent on cell-cell signaling.

As mentioned above, the importance for developmental toxicology of the discovery of extensive conservation of components and processes among seemingly disparate animals is the conclusion that the study or testing of toxicants in model animals can provide relevant information about humans, as long as the extrapolation is done within conserved responses, of which there are many.

Organogenesis

Organs are usually defined as containing two or more tissues, each tissue containing differing cell types and cell functions, coordinated in a higher level of organization and function than the independent tissues. Second to the organism's overall body organization, organs are the most complex level of organization of cells. Organogenesis is the organ-forming phase of embryonic development. It begins once the basic anteroposterior and dorsoventral organization of the embryo is established by gastrulation and neurulation. During organogenesis, cytodifferentiation takes place, and then the organ begins to function.

A fundamental question about organogenesis concerns the means by which the different parts of the organ are brought into complex alignment and integrated function. In the first half of the twentieth century, organ formation was described in detail by light microscopy, and the inductive interactions of different cell groups involved in organ formation were revealed by experimental analysis. In general, the different tissues of the organ were not found to form independently and then come together in perfect apposition. Rather, tissues that are nearby as a result of extensive movement during gastrulation and neurulation interact with each other and also with surrounding tissues. Combinations of signals establish positional identity and initiate the progressive delineation of organ-specific gene activations. Thus, it is not necessary that all participants in early organogenesis have position and cell-type specific information. Cell signaling operates throughout organogenesis. Recently, the local signals and responses have been identified in several kinds of organogenesis, the responses often being experimentally proven by using "marker" or "reporter" genes activated at various stages of the process and visualized by staining specific mRNAs by in situ hybridization. Extensive molecular descriptions and cellular and genetic analyses have defined key regulatory pathways that facilitate the development of many vertebrate systems, including the following:

- Neural tube: regionalization of forebrain, midbrain, hindbrain, and spinal cord (for reviews, see Wassef and Joyner 1997; Brewster and Dahmane 1999; Dasen and Rosenfeld 1999; Veraksa et al. 2000).
- Neural tube: dorso-ventral organization of brain and spinal cord (for reviews, see Edlund and Jessell 1999; Lee and Jessell 1999).
- Sensory systems: optic vesicle and eye, otic vesicle and inner ear, and olfactory epithelium (for reviews, see Fekete 1999; Holme and Steel 1999; Kraus and Lufkin 1999; McAvoy et al. 1999).

- Neural crest: autonomic and sensory ganglia and glia and melanocytes (for reviews, see Francis and Landis 1999; Gershon 1999a,b; LaBonne and Bronner-Fraser 1999).
- Neural crest: midfacial and branchial connective tissues and teeth (for reviews, see Francis-West et al. 1998; Peters and Balling 1999; Schneider et al. 1999; Tucker and Sharpe 1999; Vaglia and Hall 1999).
- Paraxial mesoderm: somites, skeletal muscle, vertebrae, and ribs (for reviews, see Brand-Saberi and Christ 1999; Relaix and Buckingham 1999; Burke 2000; Rawls et al. 2000; Summerbell and Rigby 2000).
- Intermediate mesoderm: kidneys, gonads, reproductive ducts, and sex determination (for reviews, see Sariola and Sainio 1998; Horster et al. 1999; Parker et al. 1999; Swain and Lovell-Badge 1999).
- Cardiovascular system: heart, angiogenesis, and hematopoiesis (for reviews, see Baldwin and Artman 1998; Mercola 1999; Morales-Alcelay et al. 1998; Tallquist et al. 1999).
- Limb: growth and specification of axes (for reviews, see Martin 1998; Ng et al. 1999; Vogt and Duboule 1999).
- Pharyngeal endoderm: thyroid and thymus (for reviews see Bodey et al. 1999; Missero et al. 1998).
- Gut tube: lungs, liver, pancreas, stomach, and intestines (for reviews, see Gretchen 1999; St-Onge et al. 1999; Warburton and Lee 1999).

The familiar conserved signaling pathways are used over and over in many different contexts in organogenesis and other steps of development, as listed in Table 6-3. For example, the Sonic Hedgehog (SHH) signaling pathway is involved in establishing asymmetry in the early gastrula, inducing floor plate and motor neurons, separating the single eye field into paired optic primordia, maintaining proliferation in migrating neural crest cells, establishing patterning of the medial and lateral nasal prominences and tooth induction, inducing sclerotome segregation and epaxial muscle formation in somites, development of the prostate gland, determining left-right asymmetry, establishing the anteroposterior (rostrocaudal) axis of the limbs, delineating the tracheo-esophageal diverticulum, and establishing sites of formation and branching patterns in lung and pancreatic epithelia. Comparable matches could be made for members of the fibroblast growth factor (FGF), TGFβ, BMP, and WNT signaling families. Although the signaling pathways involve the same or closely related signaling molecules, the responses made by cells are distinct because of the genes and gene products they express prior to and in response to the many different combinations of these signaling factors.

The Vertebrate Limb: The Best Known Organogenesis Model

The limb is by far the most studied organ rudiment of vertebrates, supported by over 50 years of experimental embryology and intensive recent molecular genetic analysis. Cell-cell interactions, based on known signaling pathways, are well understood, as are the patterns of cell proliferation within the developing limb. Surprisingly, many homologous genes and signaling pathways are conserved between the vertebrate limb and the developing leg or wing of *Drosophila*. The researchers who followed the seemingly remote leads from the early *Drosophila* work on appendages have made rapid progress on vertebrate limb development in the past few years.

Limb development is discussed here to show (1) the importance of precise temporal and spatial signaling for organizing a complex organ rudiment, and (2) the interaction of multiple signaling pathways to establish the organ's three-dimensional morphology.

The developing limb contains three axes: proximodistal, anteroposterior (rostrocaudal), and dorsoventral. Each axis has its own secreted signals and these are integrated in the limb bud, as shown in Figure 6-4 and as summarized by Johnston and Tabin (1997) and Ferretti and Tickle (1997). The bud originates as a locus of rapidly dividing cells in the somatic mesoderm and overlying epidermal ectoderm of the flank of the trunk. Outgrowth of the paired anterior and posterior limbs involves the maintenance of a high rate of cell division within the bud. Local trunk structures, such as the mesonephros, somites, and notochord, secrete FGF10 onto the flank tissues, and this signal initially keeps the bud proliferating. As outgrowth begins, the ectoderm overlying the bud locally thickens to form a ridgelike structure called the apical ectodermal ridge (AER), which then secretes several FGFs, notably FGF8, and takes over for the flank in maintaining cell proliferation in the adjacent bud mesoderm. The bud is then self-sufficient for this signaling and the flank stops serving as an FGF source. (The bud can be transplanted to a remote site, such as the yolk sac, at this stage and will develop autonomously.) The area of rapid division in the bud is called the progress zone (PZ). It lies just beneath the AER. As cells proliferate in it, some are displaced away from the AER and are no longer exposed to FGF. They slow their division rate and stop changing their specification (i.e., their capacity to develop as one

---

FIGURE 6-4 Development of the limb bud in tetrapods (the four-legged vertebrates). (Panel A) Cross section of mouse embryo after ventral closure, at the level of the forelimbs. The limb buds emerge as small protrusions from the left and right flank. They consist of a surface layer of epidermal cells (ectoderm) and an internal mass of mesenchymal cells (mesoderm). The latter engage in rapid proliferation. (Panel B) Close-up view of a bud, cross section, dorsal, ventral. The apical ectodermal ridge (AER) secretes FGF onto the underlying mesenchyme. The zone of polarizing activity (ZPA) of the mesenchyme secretes SHH onto the ridge. A reciprocal activation circuit is completed in which the AER and ZPA keep each other active. In the mesenchyme close to the ridge is the progress zone (PZ), a population of rapidly dividing cells. Their division is kept going by FGF and SHH. Dividing cells keep changing their option for a future developmental path,

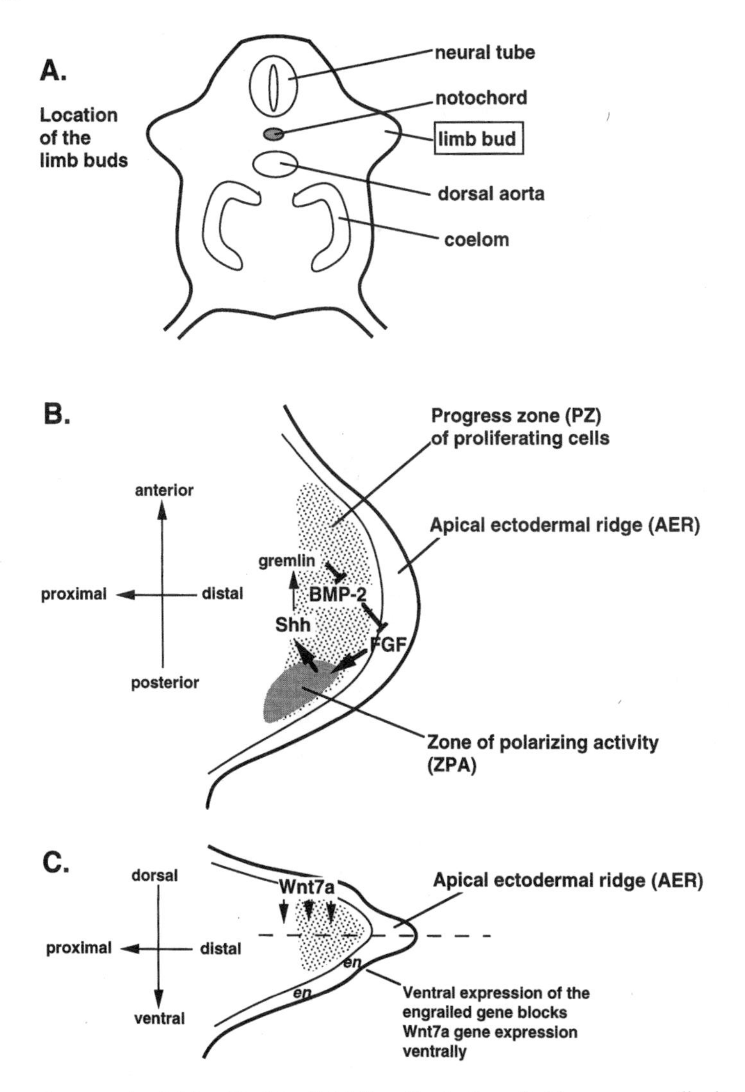

in a progression, proximal (shoulder) to distal (hand). As they divide, some are displaced toward the flank, out of the PZ, too far away to receive FGF signals. Their division slows, and they adopt the developmental path of the step of the progression at which they were when they were last in the PZ. The SHH from the ZPA spreads in a gradient toward the anterior edge of the bud. The anteroposterior differences of the limb are signaled by this gradient, perhaps through a BMP2,4 signaling coupled to SHH signaling. (Panel C) The limb bud in cross section, at right angles to panel B, to show the dorsoventral plane. The dorsal epidermis secretes WNT7A onto the mesenchyme. That signals it to develop dorsally. The ventral epidermis does not secrete WNT7A, because the cells express the *En* gene encoding a transcription factor inhibiting the *Wnt7a* gene from expression. See text for further information and references.

part or another of the future limb). As a consequence, cells leave the PZ in a proximal to distal order of specification (i.e., the first to leave have the capacity to form the upper limb, the next to form the lower limb, and the last to form the hand or foot). When cells leave the PZ, they start to differentiate into cartilage, bone, and connective tissue of their specified limb level. Precursor cells of limb muscles migrate into the bud from the adjacent somitic myotomes, and nerves extend in from spinal ganglia and the spinal cord. Interactions between the AER and PZ are reciprocal (i.e., the AER maintains mitosis within the PZ by way of FGF, and the PZ maintains the thickened AER by way of a yet unknown signal).

The second axis of the limb, the anteroposterior axis, is also established through cell-cell interactions and secreted signals. The posterior portion of the limb bud contains a specialized region called the zone of polarizing activity (ZPA). This zone was originally recognized as a signaling center, because when it was transplanted to the anterior side of another limb bud, the bud develops a mirror-image duplicated limb. The ZPA secretes SHH, which diffuses across the anteroposterior dimension of the bud, establishing a gradient and setting off the formation of a second gradient of BMP2 and 4, two kinds of TGFβ signals. SHH is both sufficient and necessary to establish the anteroposterior pattern of the limb. Retinoic acid, acting via a nuclear receptor, might also have a role.

The third axis of the limb, the dorsoventral axis, is established through interactions between the nonridge dorsal ectoderm of the limb and the underlying bud mesoderm. That was initially shown by experiments in which the dorsal and ventral regions of the ectoderm were rotated with respect to the mesoderm. The dorsal ectoderm releases the WNT7A signal, which induces the expression of the *Lmx-1* gene in the underlying mesoderm and which suppresses expression of the *engrailed-1* gene, thereby restricting its expression to the ventral ectoderm. Mutants defective in WNT7A signaling develop double ventral limbs. Mutants defective in *engrailed-1* expression develop double dorsal limbs with double sets of fingernails.

It remains to be learned how the signaling pathways of each axial dimension are coordinated with those of the other dimensions, and how the integration of these pathways leads to the formation of unique skeletal structures in precise locations within the limb. The limb exemplifies the advanced understanding of vertebrate organogenesis at a molecular genetic level (i.e., of the signal pathways and the genetic regulatory circuits involved in changing transcription and regulating cell proliferation). This kind of understanding is prerequisite to understanding the action of toxicants on embryogenesis. On the basis of new information, mechanisms of action of toxicants have been recently proposed, although these have yet to be tested. For example, thalidomide leads to a failure to form proximal parts of the limb (the upper and middle parts of the limb), but the hand or foot is usually formed. That developmental outcome is paradoxical, because other treatments, such as the removal or inactivation of the AER, lead to truncation of the limb in the reverse order; the upper and middle parts are present but the hand or foot is missing. Tabin (1998) recently proposed that thalidomide reduces cell proliferation in the PZ, by means yet unknown, and the few cells that remain there in prolonged contact with FGF8 secreted by the AER are specified as hand or foot cells, the last normally to emerge from the zone. This is an example of an incisive prediction about a long-known toxicant made on the basis of recent knowledge. Yet, the chemical basis for thalidomide's specific effect on cell proliferation in the PZ escapes even a proposal at this time, and so the hypothesis is incomplete. Stephens and Fillmore (2000) have suggested that thalidomide interferes with integrin gene expression in limb bud mesenchyme cells and, thereby, with their ability to stimulate angiogenesis at the level needed for continued rapid proliferation.

The capacity of the limb bud to develop normally after injury has been studied. Large numbers of cells can be removed at early stages, and as long as representatives remain of the AER, ZPA, and PZ, development will be normal. Immigrating muscle cells from any myotome will enter and adapt to the limb bud, and nerves from any spinal cord level will enter and make neuromuscular connections, although the CNS circuitry of that level may not be appropriate for normal limb movement. The robustness of limb development, like that of other parts of the body, is substantial, because each of the interactive cell groups is much larger than minimally necessary and is capable of proliferation. Robustness is not unlimited, however, and total removal of a key signaling or responding group is deleterious. Regenerating limbs, such as those of newts, have surmounted even that limit, but they are the exception among vertebrates.

SUMMARY

This committee has been asked to evaluate the state-of the-science for elucidating mechanisms of developmental toxicity. It seems self-evident that the knowledge about the basic processes of development provides developmental biologists with an understanding of normal development not even thought possible a decade ago, and also provides developmental toxicologists with improved tools to understand the mechanisms by which chemicals cause abnormal development.

In the last decade, great advances have been made in the understanding of developmental processes on a molecular level in model organisms, such as *Drosophila* and *C. elegans*, and in several vertebrates, including the mouse. For the first time, molecular components and their functional interactions have been identified. Developmental processes can be described for the first time as organized networks of these components and their functions. The examples examined so far primarily concern early development before organogenesis but also organogenesis in a few cases. Cell-cell interactions by way of intercellular signals are pervasively and repeatedly used. In all aspects of development, including organogenesis and cytodifferentiation, signaling is expected to be of central importance. A small number of signal transduction pathways are used in these interactions. There are approximately 10 kinds in early development and organogenesis. (Seven more are used by differentiated cell types.) The number of allelic variants that exist in these human genes remains to be studied. These pathways are conserved among animal phyla, as are many of the genetic regulatory circuits involved in the responses of cells to signals. In addition, many of the basic cell processes, such as proliferation, secretion, motility, and adhesion, are also highly conserved among animals. This extensive conservation gives strong justification to the use of model animals, including *Drosophila*, *C. elegans*, zebrafish, and mouse, to learn about basic aspects of mammalian, even human, development.

The understanding of development is far from complete. Although a number of main components have been identified for early processes, their interactions and use in combinations introduce substantial complexity to an inclusive understanding of development. Few of the many types of mammalian organogenesis have been analyzed, and only a few of the 300 types of cytodifferentiation have been studied. As more components and interactions are revealed, it will become important to establish readily accessible databases containing all the information about development.

7

Using Model Animals to Assess and Understand Developmental Toxicity

The recent advances in developmental biology described in Chapter 6 have established the central importance of a small number of highly conserved signal transduction pathways that mediate cell interactions crucial for animal physiology, reproduction, and development. It seems likely that many developmental toxicants might affect development by acting on those pathways. Application of the methods that have been so successful in elucidating them should now allow scientists to investigate that possibility and to determine the mechanisms by which developmental toxicants act. This chapter reviews the experimental approaches primarily responsible for the recent advances in knowledge about animal development and discusses how those approaches might be applied to developmental toxicology. Chapter 8 discusses how those approaches might lead to improved qualitative and quantitative risk assessment.

MODEL ORGANISMS AND THE GENETIC APPROACH

Single-Cell Organisms

Model organisms have been important throughout the study of modern biology. In the 1940s and 1950s, biochemical analysis of bacteria was important in working out the enzymatic pathways of metabolism. In the 1960s and 1970s, bacteria, especially *Escherichia coli* and its viruses (called phages), provided models for the new science of molecular biology and the elucidation of basic mechanisms for deoxyribonucleic acid (DNA) replication, transcription, and translation in prokaryotes. Since then, the budding yeast *Saccharomyces cerevisiae* and more recently the fission yeast *Schizosaccharomyces pombe* have served as models for intensively investigating the molecular mechanisms of these and other functions unique to eukaryotic cells, such as the cdk-cyclin-based cell cycle, mitosis, meiosis, ribonucleic acid (RNA) splicing, regulation of chromatin structure, secretion, dynamics of the cytoskeleton, stress pathways, checkpoint pathways, and, to some degree, intercellular signaling and differentiation, the last two associated with yeast mating. Most of these cellular functions have been highly conserved during eukaryotic evolution, so that knowledge gained from yeast research is directly applicable to understanding human cell processes. However, understanding the interactions of cells and tissues in development and physiology of higher eukaryotes requires study of metazoans (i.e., multicellular animals). It should be appreciated, though, that as the processes are understood in metazoa, the components of each process can be introduced into yeast and the individual processes reconstituted there for further detailed study. For example, it has been found that a number of human cell-cycle proteins function well in the yeast cell cycle, when replacing the yeast cell's components.

Utility of Model Animals

Much has been learned about human development and physiology through the study of model animals, a small set of diverse metazoans that have particular advantages for laboratory research. There are several reasons for their utility. Research on humans and other primates is expensive and limited by ethical considerations. The most commonly studied model animals are relatively inexpensive to maintain and are well suited for experimental manipulation. Most important, as outlined in Chapter 6, recent research has shown that there is a remarkable degree of similarity in the developmental mechanisms of all animals. Not only individual genes and proteins but also entire pathways of signaling and response and their functions in developing embryos appear highly conserved throughout evolution. This means that, although the embryology of simpler animals might appear superficially very different from that of humans, knowledge gained from those models can often be applied directly to understanding human developmental mechanisms.

On the other hand, there are important developmental and physiological attributes that can be investigated only in vertebrates, such as the adaptive immune system, or in mammals, such as placentation and lactation. Therefore, it is useful to study a representative range of model animals—from invertebrates that are only distantly related to humans but have particular experimental advantages, to rodents and other mammals that are less convenient but more closely related to humans.

Model Animals for Study of Development

For study of development, the currently most intensively investigated model animals, in order of increasing complexity, are the free-living soil roundworm (nematode) *Caenorhabditis elegans*, the fruit fly *Drosophila melanogaster*, the frog *Xenopus laevis*, the zebrafish *Danio rerio*, the chick, and the laboratory mouse. Also particularly useful for certain investigations are sea urchin, sea slug (*Aplysia*), puffer fish, and a few mammals, including the rat. This set of model animals is somewhat different from those most widely used in the 1950s. Why have these species been chosen for recent intensive study? For four of them, the principal answer is genetics.

The genetic approach has become established in the last three decades as one of the most powerful tools for elucidating biological mechanisms. It allows researchers to compare wild type with a mutant phenotype and to identify new genes involved in controlling a biological process and to determine their functions in the organism. Genes that control important functions are identified by mutations that cause defects in those functions. These genes are then mapped, cloned, and identified at the molecular level so that the proteins they encode can be studied using methods of biochemistry and cell biology. This approach has proved to be extremely powerful, not only for basic research in model organisms but also for medical research on heritable human diseases. The approach was followed, for example, in the mapping, cloning, and subsequent study of the cystic fibrosis gene, the breast cancer susceptibility gene, and many others.

The four model animals chosen primarily on the basis of their convenience for genetic analysis are *C. elegans*, *Drosophila*, zebrafish, and mice. All are relatively small, easy to maintain in large populations in the laboratory, and have short generation times, which allow for rapid analysis of breeding experiments. The remaining animals are not well suited for classical genetic analysis, primarily because of much longer generation times, but have compensating advantages of convenience and manipulability or simplicity. Sea urchins, because of their reproductive properties, have been particularly valuable in studies of fertilization and gene regulation in early embryos. *Aplysia* are used in nerve growth and development studies. Puffer fish are useful for genomics because of their remarkably small genome size (400 megabases (Mb)) compared with most other vertebrates (about 3,500 Mb, including humans). The frog *Xenopus* has eggs and embryos that can be obtained in quantity and are relatively large (about 1 millimeter (mm) in diameter). The eggs and embryos are convenient for biochemical analysis as well as microsurgery and can easily be microinjected with cloned genes, RNAs, proteins, drugs, and so forth to study the developmental effects of those molecules. The embryos have been used in toxicant tests, such as the frog embryo teratogenesis assay–*Xenopus* (FETAX). FETAX is currently under consideration for validation (Bantle et al. 1996; NIEHS 1998). Chick embryos, more closely related to mammalian embryos, are readily accessible for observation and microsurgery (unlike those of mice, which develop in the uterus) and are convenient for tissue transplantation experiments. Putative developmental toxicants can be added directly to the embryo, thereby bypassing the modifying effects of maternal metabolism and selective transfer by the placenta. Rat, rabbit, and guinea pig have long been standard systems for physiological and toxicological investigation. However, because of the power of genetic analysis, the four genetically tractable model animals (*C. elegans*, *Drosophila*, zebrafish, and mouse) have become mainstays of recent research in developmental biology and, for the same reason, are also likely to be particularly valuable in emerging approaches to developmental toxicology. These systems are described in more detail below, following a brief review of methods in genetic analysis.

Rationale and Strategy of the Genetic Approach

Genetic analysis has a powerful advantage in that it can "dissect" functionally and define the important components of any biological process without knowing anything about the process in advance—simply by isolating mutations that affect it, using those mutations to define the genes that control the process, and then cloning and characterizing those genes and their gene products, thereby revealing molecular mechanisms. Over the past two decades this approach has been successfully applied to many aspects of animal development, as indicated in Chapter 6. It can also be applied to elucidating the mechanisms of action of developmental toxicants. The general steps in the standard genetic approach, described below, are sometimes referred to as "forward genetics" (going from the mutant phenotype to the gene) in contrast to the more recently developed methods of "reverse genetics" (going from the gene back to a phenotype) made possible by molecular biology and genomics (see Chapter 5 for some of the genomic methods). Although the terms forward and reverse genetics are now generally accepted, it should be noted that the term "reverse genetics" has had a history of use in earlier medical genetics literature to describe the progression from mapping of a heritable disease state to cloning of the responsible gene (called "forward genetics" elsewhere).

Forward Genetics

The steps in this approach are as follows:

1. Choose a *defective phenotype* of interest (e.g., failure to develop a particular structure or increased sensitivity to a toxicant) that is specific and selectable or easily recognizable.

2. Using *mutagenized* populations, carry out a *saturation screen* for mutants with the defective phenotype (i.e., a screen large enough so that mutations are likely to be found in every gene required in development of the normal phenotype).

3. Use classical genetic analysis of these mutations to define the genes they represent by *genetic mapping* and *complementation tests* and to determine their *null phenotypes* (i.e., the effects of complete loss of gene function). The incisive-

ness of studying null mutants is worth mentioning in the context of developmental toxicology. Their phenotypes match the toxicologist's ideal of what the "perfect" toxicant would generate for observation if it completely inhibited just one target component of the organism.

4. If possible, establish the order of function of the identified genes by constructing double mutants to determine which of two distinguishable phenotypes takes precedence (*epistasis test*).

5. Identify additional modifier genes by using *suppressor* and *enhancer* screens in a sensitized genetic background for secondary mutations that make the defective phenotype of an existing mutant less or more severe.

6. Using fine-structure genetic mapping and *positional cloning*, obtain genomic clones of each gene for molecular analysis and verify their identities by demonstrating that each gene in the corresponding mutant animal carries a DNA sequence alteration.

7. From suitable *complementary (c) DNA libraries* of cloned cDNA copies of the animal's messenger (m) RNA population, isolate cDNAs corresponding to each gene, sequence them to determine the predicted amino acid sequence of each encoded protein, and carry out a *similarity search*, comparing those sequences with the sequences available in databases, which often can be used to discern motifs and reveal the functional class to which a protein belongs. (Function was initially deduced for the class from other kinds of studies—biochemical, cellular, developmental, and physiological.)

8. Determine when and where the mRNA and the protein encoded by each gene are found during development by using, respectively, nucleic acid probes and antibodies made to fusion proteins. A faster but sometimes less reliable alternative is to make *reporter constructs*, which carry the promoter region of the cloned gene fused to a gene encoding a reporter protein that can be detected by its activity (e.g., the *E. coli* β-galactosidase gene *lacZ*) or fluorescence (e.g., green fluorescent protein (GFP)). Embryos into which such a construct has been introduced (by DNA transformation) can be observed at various stages to determine when and in what cells and tissues the promoter is active. Generally (but not always), an active provider will reflect the expression pattern of the normal gene.

9. Supplement that information with *genetic mosaic analysis*, by producing animals in which only certain cells or tissues are mutant, to discover where a gene must normally function and whether its functions are cell autonomous (i.e., intracellular) or cell nonautonomous (i.e., intercellular).

10. Isolate and biochemically analyze proteins encoded by the mutationally identified genes to study further the function of the proteins.

All these steps are not always carried out. The most important and difficult step, once mutants have been obtained, has been positional cloning of the gene. However, shortcuts are becoming available with the accumulation of genomic mapping and sequence information and the development of new technologies (see following sections). For example, if a mutation defining a gene of interest has been mapped to a region of the genome for which the entire DNA sequence is known, the *"candidate gene"* approach can be used to identify it. Computer analysis of the genomic sequence can predict which sequences in the region represent coding sequences and open-reading frames (ORFs) of genes and what proteins these DNA sequences encode. It is then often possible to guess one or a few most likely candidate genes and confirm that one of these is correct by sequencing one (or preferably more) mutant allele and finding the responsible sequence alteration(s) or by expressing the candidate gene to see if its encoded product reverts the mutant phenotype back to wild type.

A new method called genomic mismatch scanning (GMS), using DNA microchip technology, will allow more rapid identification of the candidate gene and the mutational lesion in one step. Oligonucleotides representing the entire sequences of all candidate genes in the region to be tested, as well as all possible single base-change mutational variants of each sequence, are synthesized and fixed in an indexed array on a microchip (see description of the method in Chapter 5). The chip is then annealed to differently labeled probes from nonmutant and mutant forms of the cloned gene. By comparing these patterns, both the correct candidate gene and the nature of the mutational lesion can be determined.

Reverse Genetics

With the increasing availability of genomic sequence information, the following somewhat different approach is becoming more useful for studying biological processes, especially in organisms such as mammals, for which the forward genetic approach is difficult. It is called reverse genetics, because it starts with a cloned gene of potential interest. The cloned gene is then used to obtain animals with defects in the gene or its expression for functional analysis. The steps in this approach are as follows:

1. Identify a gene of interest from its sequence (e.g., the mouse homolog of a developmentally important gene in *Drosophila*) and obtain a clone of the gene by standard methods based on sequence similarity (such as screening a mouse library (collection) of genomic DNA clones with the cloned *Drosophila* gene).
2. Determine its expression pattern (as described above) for clues to its function.
3. Inactivate the gene (often referred to as "knocking out," "targeted inactivation," or "homologous recombination" of the gene) and observe the phenotypic consequences for more definitive information on function. This can be done either transiently, by injection of an antisense or double-stranded mRNA that specifically prevents gene expression, or permanently (preferable, but requiring considerably more effort), by generating animals that carry a null mutation in the gene. In the nematode *C. elegans*, the double-stranded mRNA method works

particularly well (described in more detail below). In the mouse, mutations can be obtained efficiently by targeted recombination of mutant DNA constructs introduced into the germ line (described in more detail below). In flies (*Drosophila*) and nematodes (*C. elegans*), the desired mutant individual can be screened from a large population after random transposon insertion or chemical mutagenesis.

4. Once mutations are obtained, they can be subjected to any of the genetic analyses described above.

Again, emerging technologies, such as microchips carrying ordered arrays of cDNAs to allow rapid analysis of how a mutation affects mRNA populations, will accelerate and enhance the above approaches.

Extrapolation to Humans

For many genes identified by forward or reverse genetics in model animals such as the mouse, and particularly for genes relevant to human disease states, the next step is to isolate and characterize the corresponding (*orthologous*) gene in humans. Several recent developments have simplified the task of cloning human homologs for molecular analysis. Extensive and detailed maps of molecular markers are now available for many areas of the human genome, and rapid progress is being made on the remainder in connection with the Human Genome Project. Comparison of mouse and human maps demonstrate extensive linkage conservation (*synteny*) between the two genomes (i.e., the arrangement of orthologous genes has been conserved over large regions from the last common ancestor). Considerable linkage conservation is found even between fish and mammals. As ancestral species diverged hundreds of millions of years ago and evolved into present-day species, local gene order in most instances has been maintained while large blocks of contiguous genes have been rearranged. For example, genes *A-B-C-D-E* found on mouse chromosome 12 might be found as *A-B-C-D-E* or, in reverse order, as *E-D-C-B-A* on human chromosome 7. As a result, if the chromosomal location of a gene responsible for a trait in the mouse is known, it is now possible to predict quite accurately the chromosomal location of its ortholog in humans (see Web site at http://www.informatics.jax.org, under mammalian homology and comparative maps). This approach will also be useful in defining human genes that affect responses to developmental toxicants (e.g., the genes for various enzymes that metabolize exogenous chemicals).

There also are large libraries (expressed sequence tag (EST) libraries) of sequences representing pieces of mRNAs transcribed from genes at various times and tissues in the human (see description of EST methods in Chapter 5). Transcripts from almost 90% of all human genes are estimated to be sequenced and present in these libraries. The transcripts are of great value for isolating the human homologs of genes and gene products that have been well characterized in other organisms.

THE MAJOR MODEL ANIMALS FOR GENETIC ANALYSIS

The four genetically tractable model animals, *C. elegans*, *Drosophila*, zebrafish, and mouse, are useful for somewhat different reasons. Relevant characteristics of each are described briefly below, along with some of their experimental advantages and disadvantages (see also Tables 7-1 and 7-3). The potential utility of each animal for identifying and investigating mechanisms of developmental toxicants is discussed later in this chapter.

The Nematode *Caenorhabditis elegans*

History, Biology, and Genetics

Caenorhabditis elegans is a roundworm found commonly in soils all over the world. It has become widely exploited as a model animal largely because of the early efforts of Brenner (1974), who recognized its experimental advantages and pioneered its genetic analysis. The adult is about 1-mm long, just visible to the naked eye. It feeds on bacteria, such as the common bacterium *E. coli*, and is easy to grow and breed on agar plates in the laboratory. *C. elegans* is one of the simplest animals known, with a small fixed number of somatic cells: 959 in the adult hermaphrodite and 1,031 in the adult male. It is transparent throughout the life cycle, so that its entire development can be analyzed in living animals with the light microscope. Its generation time is only 3 days, and development is rapid (Figure 7-1). Embryogenesis is complete by 14 hours after fertilization. The first-stage (L1) larva hatches from the egg, growing and molting through three larval stages (L2, L3, and L4) as its reproductive system develops before the final molt to adulthood. Adult males make sperm and can mate with hermaphrodites, making genetic crosses possible. The hermaphrodites are essentially females but produce some sperm during late larval development and can self-fertilize, which simplifies genetic analysis. *C. elegans* has a genome size of about 100 megabases (Mb) packaged into six small chromosomes, including five autosomes and a sex (X) chromosome (hermaphrodites have two and males one). Extensive genetic and physical maps have been constructed, and its genome has recently become the first in a metazoan to be completely sequenced under the auspices of the Human Genome Model Organisms Project (*C. elegans* Sequencing Consortium 1998). The genome includes about 19,000 genes.

Because of its transparency and the invariance of cell-division patterns throughout *C. elegans* development, it has been possible to describe embryonic and larval development completely at the cellular level. By observation of developing animals using Nomarski microscopy, Sulston and coworkers were able to define all the larval cell lineages (Sulston and Horvitz 1977) and later the entire embryonic cell lineage (Sulston et al. 1983), so that the ancestry of every cell in the adult organism is now known. Perturbation of normal development by laser

TABLE 7-1 Comparison of Four Model Animals for Genetic Analysis and Humans As a Reference

Animal	Adult Size (cm)	Genome Size (Mb)	Period of Organogenesis (d)	Generation Time (wk)	Experimental Advantages
Nematode (*Caenorhabdits elegans*)	0.1	97	0.2-0.4	0.4	Convenient forward and reverse genetics, complete genome sequence known, complete description of development available, simplicity, transparency
Fruit fly (*Drosophila melanogaster*)	0.4	180	0.5-1	2	Most convenient forward genetics, many genetically defined signaling pathways known, extensive knowledge of development
Zebrafish (*Danio rerio*)	3	1,700	1-4	12	Vertebrate, good forward genetics, transparency, external, well-studied development, accessible to test chemicals in water
Mouse (*Mus musculus*)	6	3,000	6-15	10	Placental mammal, closest model to humans, good forward and reverse genetics, well-studied development
For comparison:					
Human	170	3,500	14-60	27 yr (1,400 wk)	

Abbreviations: cm, centimeter; d, day; Mb, megabase; wk, week; yr, year.

ablation of specific cells has provided information on inductive cell interactions during embryogenesis and larval growth. This knowledge has been extremely useful in analyzing the genetic control of cell-fate determination and the roles of cell signaling pathways by using genetic approaches, as described further below. For more comprehensive reviews on current knowledge of *C. elegans*, see Wood et al. (1988) and Riddle et al. (1997).

Transgenic Technologies

DNA Transformation. Cloned genes can be reintroduced into the *C. elegans* genome by injection of DNA into the syncytial region of the hermaphrodite gonad (Mello et al. 1991). The injected DNA recombines to form large replicating extrachromosomal arrays, which become incorporated into developing oocytes

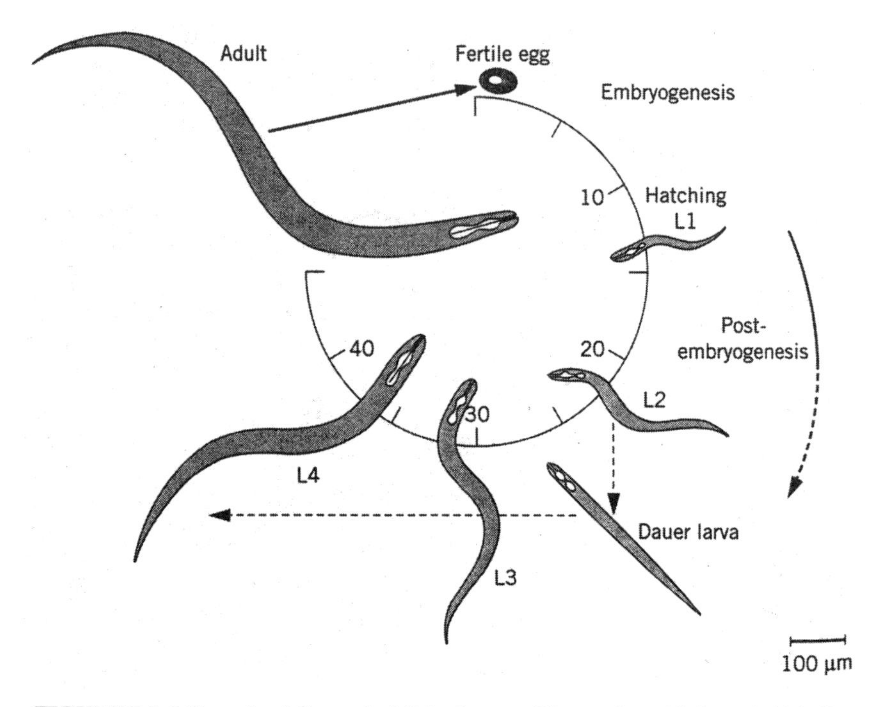

FIGURE 7-1 Life cycle of *Caenorhabditis elegans.* The numbers 10 through 40 indicate hours after fertilization of the egg. L1 through L4 indicate larval stages, each ending in a molt, a shedding of the tough cuticle. The dauer larva is a diapause stage entered when food (usually bacteria) is in short supply. Source: Wood (1999). Reprinted with permission from *Encyclopedia of Molecular Biology;* copyright 1999, John Wiley & Sons.

and then embryos. These arrays can be transmitted to most cells of the resulting animals and through the germ line to their progeny. Although the genes on such arrays are present in high and somewhat variable copy number, they are efficiently expressed and can be useful for many types of investigations, such as transformation rescue in positional cloning and analysis of a cloned gene's expression patterns using *lacZ* or GFP reporter-gene constructs. From transmitting lines, more stable integrated lines can be obtained in which the array has inserted randomly into a chromosomal locus, allowing various gene-trapping technologies for identifying loci with tissue-specific expression patterns. Targeted insertion of transgenes by homologous recombination has not yet been achieved. Reverse genetics using targeted gene disruption is therefore difficult but can be accomplished by random transposon insertion (Plasterk 1995) or deletion mutagenesis followed by appropriate screens, or it can be accomplished by RNA-mediated gene interference (RNAi), as discussed next.

RNAi. A powerful tool for reverse genetic analysis has been provided by the discovery that introduction of double-stranded mRNA for a particular gene into *C. elegans* will specifically inactivate that gene, resulting in loss-of-function phenotypes that generally mimic the gene's null phenotype for at least a generation or two (Fire et al. 1998). Although the mechanism of this inactivation, referred to as RNAi, is not yet understood and gene expression in some tissues is more susceptible to inactivation than expression in other tissues (Montgomery et al. 1999), it is clear that RNAi will be extremely useful for rapid functional tests of genes identified by genome sequencing as potentially important, for example, in development or in responses to environmental toxicants. Moreover, recent results indicate that the technique is applicable to *Drosophila* (Kennerdell and Carthew 1998) and perhaps to other organisms as well.

Signaling Pathways in Development

Most of the progress in understanding *C. elegans* development has come from application of forward genetics as described above, combined with laser ablation experiments to identify required cell interactions. A variety of inductive events, which in *C. elegans* can be analyzed at the single-cell level, are mediated by signaling pathways that are still under investigation. However, it is already clear that nematode development uses most of the pathways described in Chapter 6, often in developmental contexts similar to those found in more complex metazoans. Two exceptions are the Hedgehog and cytokine signaling pathways, which *C. elegans* appears to lack (Ruvkun and Hobert 1998).

The Fruit Fly *Drosophila*

History, Biology, and Genetics

Drosophila melanogaster is the common fruit fly found worldwide in orchards, where adult flies lay eggs on rotting fruit. Since the beginning of this century, fruit flies have been cultured in the laboratory in half-pint milk bottles and more recently in shell vials and plastic tubes by using a solid food, typically composed of agar, cornmeal, dried yeast, and molasses. At 25°C the life cycle takes approximately 2 weeks. Embryogenesis and the first two larval stages require 1 day each; the third larval stage, 2 days; and the pupal stage, 4-5 days (Figure 7-2). Two-day-old adults begin to lay eggs. Because of the short life cycle, ease of rearing in large populations, and the many diverse phenotypes readily visible under a simple dissecting microscope, many mutations have been accumulated in the organism since its initial use by T. H. Morgan and his associates at Columbia University in the 1920s. The study of fly genetics has been instrumental in many classic discoveries in eukaryotic genetics, such as linkage, gene mapping, recombination frequency, and chromosomal aberrations. Discov-

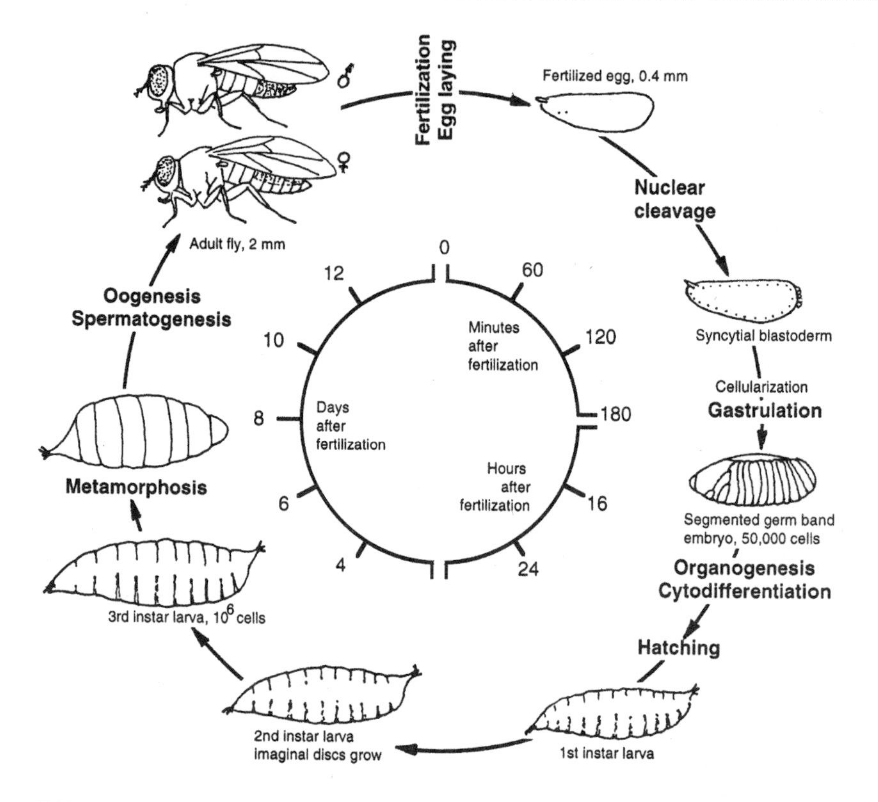

FIGURE 7-2 Life cycle of *Drosophila melanogaster*. The larva hatches 1 day after the egg is fertilized. First, second, and third instar are larval stages, each ending with a molt. During pupation most of the larval tissues are destroyed and replaced by adult tissues derived from the imaginal discs that were growing in the larva. Times are given for the life cycle at 25°C. Source: Adapted from Wolpert et al. (1998).

ery of the giant polytene salivary gland chromosomes in the 1930s provided a cytological basis for those genetic theorems and thus made *Drosophila* a key organism for genetic analysis.

As discussed in Chapter 6, the use of fruit flies for developmental studies awaited the saturation screens for lethal and female sterile mutations. These screens were conducted in the late 1970s and 1980s and led to the discovery of cascades of gene functions responsible for the organization of the egg and early pattern formation in the embryo. The advent of recombinant DNA and cloning quickly led to the isolation and sequencing of key genes, which affect the regional specification of body parts. Such genes were defined by the homeotic mutations studied by E. B. Lewis. These studies led to the startling discovery in 1983 that sequences of amino acids coded for by homeotic genes (the homeobox

DNA sequence, and the homeodomain protein sequence) were conserved not only in different homeotic genes in flies but also across the whole animal kingdom, including humans. This conservation of genetic structure and function has become the cornerstone of modern developmental biology, forming the basis for the usefulness of model organisms in understanding human developmental mechanisms.

D. melanogaster has a genome size of approximately 180 Mb, a third of which is centric heterochromatin (regions rich in simple sequence repeats that remain condensed during interphase). The 120 Mb of euchromatin (unique sequence, decondensed during interphase) are located on two large autosomes, one dot chromosome, and paired XY sex chromosomes. The 120 Mb of euchromatic DNA have now been sequenced (Adams et al. 2000), and are estimated to encode approximately 13,600 genes, somewhat fewer than the C. elegans genome but with comparable functional diversity. The polytene chromosomes of *Drosophila* provide a cytogenetic map of the euchromatic portion of the genome, and by means of *in situ* hybridization to those large chromosomes, molecular markers have been identified within most subdivisions of the map.

Transgenic Technologies

One of the principal tools for *Drosophila* research is the availability of a transposon, called the P-element, which can be used as a vehicle for introducing nearly any genetic construct into the *Drosophila* genome at high efficiency via a relatively easy process of transformation. In addition, flies possessing single P-elements, containing dominant markers, such as the bacterial β-galactosidase gene, can be used as mutagens to disrupt coding sequences and as markers of the disrupted gene for cloning. Similarly, P-elements lacking a strong promoter for the expression of β-galactosidase can insert adjacent to enhancer elements that activate the enzyme in an enhancer-specific manner and thus identify potential new genes for study. By a combination of saturation screens and newer insertional mutagenesis experiments, it is possible to accumulate large sets of mutations of known genes and of related gene functions. Among reverse genetic technologies, for determining the function of a gene identified only by its nucleotide sequence, is insertional mutagenesis using transposable elements or RNAi, which has recently been shown to be effective in *Drosophila* as well as *C. elegans*.

Signaling Pathways in Development

One of the most useful outcomes of the genetic analysis of *D. melanogaster* development has been the identification of developmental pathways that are conserved in most organisms. Initially, these sets of genes were identified because of their similar phenotypes. As indicated above, using epistasis relationships, genes could be put into developmental sequence. More recently, the innovative use of a sensitized genetic background has led to the identification of additional components in those pathways. These new components have been difficult to discover, because they are either stored in the egg and thus show a maternal perdurance or they are used in multiple pathways, the consequence being that their mutant phenotype diverges from that seen in any one pathway or is generally lethal. In this section, the committee will describe sensitized mutants in some detail, because it believes that they will be useful in the future for toxicant assays.

The screen used by Simon et al. (1991) has set the pattern for many subsequent screens (and the committee will later draw attention to the potential use of this strategy for assaying toxicants). The *Sevenless* gene encodes a tyrosine kinase receptor that is required specifically for the formation of rhabdomere 7, one of eight light-receptor cells in each ommatidium of the *Drosophila* compound eye. The authors conducted a genetic screen on flies bearing weak, temperature-sensitive mutations in the *Sevenless* gene. They screened for genes for which inactivation of one copy caused the Sevenless phenotype. Hence, even if such genes are needed in multiple tissues, the activity of the remaining intact gene would suffice in all but the eye. Increases or decreases in gene activity in the tyrosine kinase pathway would affect the intermediate eye phenotype of the fly strain used. The screen detected a series of genes functioning downstream of the receptor, including homologs of *Ras*, *Raf1*, and a guanine nucleotide exchange factor (Figure 7-3). The dose-dependence of the Ras mutation suggested that a small (two-fold) reduction downstream should modify the rough-eye phenotype imparted by the mutant Ras and do so only in the eye. The suppressors so discovered are the *Drosophila* homologs of components of the MAP kinase pathway. These are used by many tissues, so their mutations would be recessive lethal. This study provided the basis for integrating the downstream effector pathway for many tyrosine kinase receptor functions.

Since this initial screen, screens using sensitized genetic backgrounds have become commonplace in developmental genetics of flies. For example, the initial set of downstream functions of the decapentaplegic pathway (a TGFβ signaling pathway), now called SMADs, was discovered in a screen for mutations that enhanced the phenotype of a weak *Dpp* allele (Raftery et al. 1995). Mutations in the Notch pathway have identified additional components functioning downstream of the receptor (Xu and Artavanis-Tsakonas 1990).

A major difficulty in identifying components of these critical signaling pathways is that they play essential roles in many developmental processes; hence, mutations in the genes involved are lethal. Genetic approaches have been developed in *Drosophila* to circumvent that phenomenon. The site-specific recombinase system from yeast (FRT-FRP system, Golic and Lindquist 1989) has been particularly useful (Figure 7-4). In the case of essential genes that are expressed during oogenesis, germ-line clones that are homozygous for a lethal mutation are made in the background of an ovary expressing a dominant female sterile mutation. These clones, lacking the dominant female sterile mutation, are

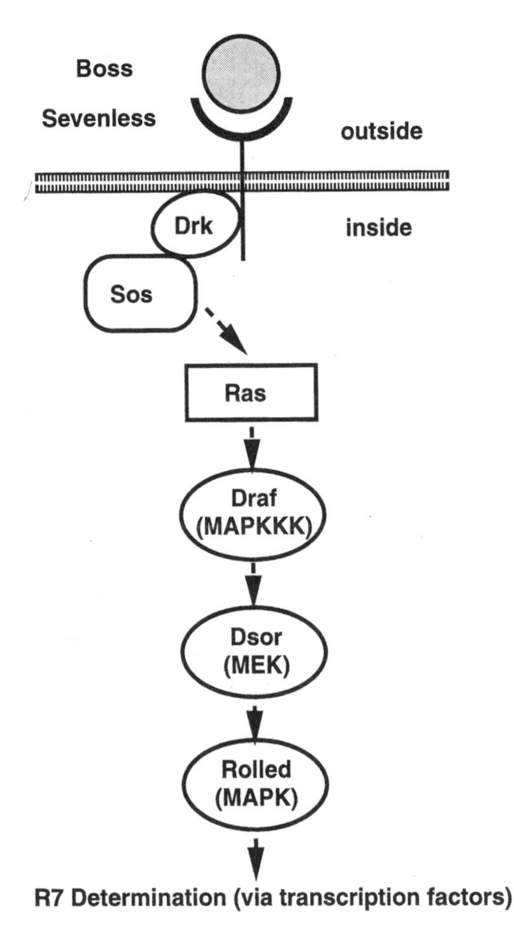

FIGURE 7-3 The signal transduction pathway by which the development of the R7 photoreceptor cell is induced in the eye of *Drosophila*. This pathway is a receptor tyrosine kinase pathway. Many of the components of the pathway were first discovered in *Drosophila* by doing selections in mutants in which the pathway was operating close to threshold due to a component of reduced activity. These "sensitized strains" revealed secondary mutations easily, and might be useful for detecting toxicant effects on development. BOSS, the ligand presented on the surface of neighboring cells R2 and R5. SEV, the transmembrane receptor that binds the BOSS ligand protein from an adjacent cell. DRK, an adapter protein binding to the phosphorylated cytoplasmic tail of the receptor. SOS, a GTP:GDP exchange factor protein activated when binding to DRK. RAS, a small G protein active when GTP is bound. DRAF, a protein kinase that phosphorylates DSOR, which in turn phosphorylates ROLLED, which in turn phosphorylates a transcription factor, which then activates specific gene expression involved in R7 determination. MARPKKK, mitogen-activated protein kinase kinase kinase; MEK, mitogen-activated protein kinase/extracellular signal-related kinase kinase; MAPK, mitogen-activated protein kinase. See text for details. Source: Adapted from Simon et al. (1991).

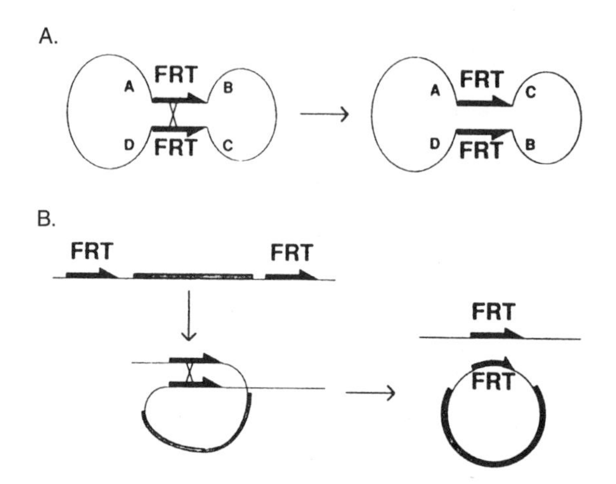

FIGURE 7-4 The FLP recombination method for removing DNA sequences from the genome at specific times and places in the animal to inactivate or activate specific genes. This method is now widely used in *Drosophila* and in mice. (Panel A): A circular piece of DNA is shown, containing two FRT sites in opposite 5′-3′ orientation. When crossover occurs, the gene order ABCD changes to ACBD. (Panel B): Two FRT sites are in the genome in the same orientation. When cross-over occurs, a circle of DNA is looped out carrying the equivalent of one FRT site. The other remains in the genome. See text for details. Source: Golic and Lindquist (1989). Reprinted with permission from Cell Press, copyright 1989.

able in many instances to produce mature eggs lacking the gene product (Perrimon et al. 1989; Chou and Perrimon 1996). This provides the opportunity to determine the phenotype of embryos lacking the function of the gene in question. Similarly, somatic clones can be produced in embryos or larvae heterozygous for various mutations. These somatic clones, induced at high frequency by the FRT-FRP system, will be homozygous for the new mutations and show a phenotype (Xu and Rubin 1993). Thanks to these powerful methods, components of several of the signal transduction pathways have been identified first in *Drosophila* and *C. elegans* and later confirmed in vertebrate cell systems.

The Zebrafish

History, Biology, and Genetics

The zebrafish is a relative newcomer to the list of model animals. It has become important as the first vertebrate to be subjected to large-scale genetic screens, which were shown to be feasible by C. Nüsslein-Volhard and others in the 1980s (Haffter and Nüsslein-Volhard 1996). Such screens are possible because, unlike other vertebrate model systems, adult zebrafish can be reared in large numbers (each is 5-cm long) at reasonable expense (although zebrafish are more expensive than *C. elegans* and *Drosophila*), and they are fecund, laying hundreds of eggs at regular intervals. The embryos are permeable to, and in fact, bioconcentrate many chemicals added exogenously in the water. The effects on development might be assayed simply and visually, although such tests have not been done systematically. Most important for assaying effects, the embryo is transparent and develops rapidly (128 cells develop 3-4 hours after fertilization), so all organs are visible and established during a few days. The organs, including heart, vessels, kidney, and liver, are nearly identical to those in the early human embryo. The zebrafish generation time is about 12 weeks. Its genome, carried on 26 chromosomes, is small for a vertebrate, 1,700 Mb (about half the size of the human genome).

Large-scale mutagenesis screens in zebrafish identified genes involved in development of vertebrate-specific body plans and tissues. One such chordate feature is the transient embryonic "backbone," the notochord. The notochord generates signals (e.g., proteins of the Hedgehog family) that pattern embryonic development of adjacent tissues, including the nervous system and muscles. The neural crest is found only in vertebrates. The migratory population of neural-crest cells emanates from the neural tube and disperses widely, contributing to neural ganglia, pigmentation, jaw structures, and the major blood vessels from the heart. Other important vertebrate organ systems, without close cognates in invertebrate model genetic organisms, include the bony skeleton, an endothelial lined vascular system, a chambered heart, and gut derivatives such as the pancreas and liver, and the kidneys.

Organogenesis in Development

Individual mutants from large-scale screens of zebrafish were found to have perturbations in organogenesis and in other aspects of development that are highly informative. For example, the notochord is ablated entirely in some mutants, and in others, the notochord is structurally present but particular notochord signals are absent. Neural-crest derivatives, such as neurons, craniofacial structures, and melanocytes (pigment cells), are affected by mutations. Different modules of organ form or function are selectively removed by individual mutations. For example, the *Pandora* mutation eliminates the heart ventricle; *Slo Mo* causes a slow heart rate. Those mutations, therefore, provide an entrance point to specific pathways of organogenesis. In some cases, the phenotypes resemble congenital disorders, such as aortic coarctation (as noted in Chapter 2, cardiac defects are the most common of live-born human developmental defects). Others have phenotypes that are common in the adult, such as heart failure. Current screens are pinpointing even more subtle phenotypes, in part by inclusion of molecular probes to reveal particular cell populations. For example, a large number of mutants have been isolated as ones affected in the projection of retinal nerves to the tectum of the brain.

Large-scale efforts are under way in many laboratories to clone the mutant genes. Genetic and physical maps, crucial for positional cloning, are being constructed. Additionally, a large-scale EST project is under way, along with development of methodologies to map the ESTs. Because there is extensive synteny (conservation of chromosomal gene order over short distances) even between zebrafish and mammals, the dense EST maps of mouse and human should suggest candidate genes once the map position of an EST is established. The large insert libraries needed for positional cloning are now available. One important advantage of zebrafish for positional cloning, compared with the mouse, is the ease and thrift of scoring thousands of embryos in mapping crosses. The large numbers greatly enhance genetic resolution, thereby delimiting the chromosomal region in which to search for a mutant gene. With current maps, and some guesswork, more than 30 zebrafish mutant genes have been located and identified as of late 1999. One mutant gene had not been previously described in any other species and is critical to normal endoderm (gut) differentiation. Other mutant genes are related to known genes in other organisms and have refined the understanding of how signaling pathways pattern the early vertebrate embryo.

The Mouse

Biology and Genetics

Although lower vertebrate and nonvertebrate organisms are valid model systems for studying many aspects of cell and molecular processes that are shared by widely disparate organisms, certain characteristics are restricted to mammals. These characteristics include placentation, intrauterine development, lactation, and aspects of immunology and carcinogenesis. To study these characteristics, only a mammalian model is ultimately appropriate. The laboratory mouse provides a small, tractable, and genetically well-characterized model, in spite of minor differences among mammals in the details of development and metabolism.

Many different mammalian models have been used for different aspects of biomedical research, but among mammals, the mouse is perhaps the most versatile and best studied. Among the advantages, mice are among the smallest mammals and have a short generation time of around 10 weeks. They are prolific breeders and their reproductive cycles are easily monitored for the timing of pregnancies. They have been bred in captivity for biomedical research for nearly a century and are docile animals. All these features add up to a great practical benefit in cost efficiency when large numbers of animals are required for research (e.g., in genetic and toxicological studies). As many as 3,000 pups (and up to 5 generations per year) can be raised per year per square meter of area in an approved animal facility, assuming cage racks are 5-6 shelves high (Silver 1995).

Recent advances in genomic research in humans and mice have reinforced the mouse as a model genetic system. They have closely related genomes of approximately the same size (about 3,000 Mb) and probably diverged from a common ancestor 80 million years ago. Counterparts of most human genes can be found in the mouse. In terms of the genomic structure, large segments of chromosomes containing the rank order of hundreds to thousands of genes have been preserved virtually intact (synteny) between the mouse and human, facilitating the application of forward and reverse genetic techniques. Laboratory mice present a vast resource of defined genetic strains, including inbred and recombinant inbred strains with characterized allelic differences that can serve as models for human genetic polymorphisms. There is a growing resource of naturally occurring and induced mutants (including a large variety of knock-out null mutants) that are commercially available, easily obtained, and easily maintained. Furthermore, the embryos of mice are accessible to embryological and genetic manipulation and have been widely used in the development of transgenic technologies.

Transgenic Technologies

Manipulating the mammalian genome has become a commonplace experimental procedure during the past two decades, and transgenic animals have been widely used in many research areas. "Transgenic" is a term that was originally coined to describe animals that had a foreign or "trans" gene inserted at random into their genome by experimental means. Its use has been broadened as more sophisticated techniques for altering the genome have been developed, and it can now be used to include any animal whose genome has been altered by addition of genetic material or by alteration of existing genes by gene targeting. Transgenic techniques, which were first devised in the 1975-1985 period, have been applied to a variety of experimental animals and agricultural animals, although by far the most common mammalian subject of gene manipulation remains the laboratory mouse.

The transgenic approach, whereby genes can be isolated, altered, and then returned to the animal, has provided a new means to investigate experimentally the function of genes and their regulation in different tissues and at different times during development. As it became clear that foreign genes could indeed function after insertion into the genome of a host, and that transgene expression was, to some extent, under experimental control, the practical uses of transgenic animals began to emerge, including uses in toxicology.

Although the first successful transgenics were made using a viral vector to deliver DNA to mammalian embryos through viral infection, the direct microinjection of cloned genes into the pronucleus (haploid nucleus) of a fertilized egg has proved to be the more versatile method and has been used widely. The principle is simple: a gene of interest is cloned, with or without regulatory elements, and is microinjected into the pronucleus. One or more copies of the gene will integrate into a chromosome at random, either immediately or after one or two cleavage divisions. Because integration is a rare event, individual eggs rarely contain more than one integration site. If the transgene enters the germ line, it subsequently behaves as a Mendelian gene in meiosis.

Table 7-2 summarizes transgenic technologies commonly used in experimental mice, and those technologies are described in detail below.

Overexpression and Misexpression. A common use of transgenics is to overexpress a given gene either in the tissue where it is normally expressed or at ectopic sites (where it is not normally expressed). A transgene is constructed that includes the coding region of the gene and its own regulatory elements or those of another gene that will drive expression constitutively, inducibly, or ectopically. Some of the earliest transgenic experiments were ones in which transgenic mice were produced with a human-growth-hormone gene driven by the inducible metallothionein promoter. The use of a human gene provided a means of distinguishing the expression of the endogenous mouse-growth-hormone gene from the expression of the transgene.

Promoter and Enhancer Analysis. To identify the regulatory elements of a given gene, a series of transgenes can be constructed containing ever increasing amounts of the 5′ regulatory region linked to an unrelated gene with an assayable gene product (a reporter gene) or the gene itself. Then, a series of transgenic animals is produced with those constructs and assayed for the expression of the reporter gene or gene product. In this way, regulation of levels of expression and tissue specificity of expression can be assigned to specific positions in the regulatory region of the gene under study. This has been widely done, for example, in the analysis of the expression of the 39 *Hox* genes of the mouse.

Antisense Transgenes. Transgenes can be constructed to contain antisense sequences. If expression of an antisense transgene is directed to the tissues where the endogenous gene of complementary coding sequence is being expressed, antisense RNA transcribed from the transgene can hybridize to the endogenous mRNA and reduce its translation. This is useful for assessing the function of the endogenous protein.

Gene Trapping. A serendipitous means of finding new genes includes the phenomenon of insertional mutagenesis, where, by chance, a transgene inserts into a chromosomal gene causing a mutation that results in an unexpected mutant phenotype. A similar procedure was discussed above for P-element insertion in *Drosophila*. Another means of screening for new genes in the mouse is to make transgenics using a promoterless reporter gene. Successful expression of the reporter transgene will indicate its intergration near an endogenous promoter, and the expression pattern can provide information about the endogenous gene that

TABLE 7-2 Transgenic Technologies Commonly Used in Laboratory Mice

Method	Purpose	DNA Construct	Desired Effect	Variations
DNA microinjection into zygotic pronucleus	Overexpression	Gene of interest with strong promoter	Higher than normal expression	Tissue specific, ubiquitous or inducible promoter
	Misexpression	Gene of interest with ectopic promoter	Ectopic expression	Temporal or spatial misexpression
	Promoter and enhancer analysis	Regulatory regions driving reporter gene expression	Mapping of regulatory elements	Various reporter genes (e.g., *lacZ*), green fluorescent protein, luciferase
	Antisense	Endogenous promoter with antisense gene	Reduced expression of endogenous gene	
	Insertional mutagenesis	Any	Phenotype associated with serendipitous disruption of unknown gene	
	Gene trapping	Promoterless reporter gene	Expression of reporter gene indicating expression of endogenous gene	Various reporter genes (e.g., *lacZ*), green fluorescent protein, luciferase
Targeted mutagenesis via homologous recombination in embryonic stem cells				
DNA electroporation into embryonic stem cells	Mutations	Homologous genomic regions flanking selectable marker	Mutation of endogenous gene	Insertion into or deletion of endogenous genomic sequences to produce null (knockout), hypomorph, or dominant negative mutations

continues

TABLE 7-2 Continued

Method	Purpose	DNA Construct	Desired Effect	Variations
	Expression reporting	Homologous genomic regions flanking in-frame reporter gene and selectable marker	Disruption of endogenous gene and expression of reporter gene under control of the endogenous gene's promoter	Various reporter genes (e.g., *lacZ*), green fluorescent protein, luciferase
Targeted mutagenesis via homologous recombination in embryonic stem cells				
DNA electroporation into embryonic stem cells	Point mutations	Homologous genomic regions with desired point mutation flanking lox-P-flanked selectable marker	Replacement of homologous region and removal of selectable marker	Different recombinase systems possible
	Conditional mutations	Homologous genomic regions containing lox-P site in intron, flanking lox-P-flanked selectable marker	Produce mutation by removing regions flanked by lox-P sites at desired time	Mutation produced at different times or in specific tissues
	Conditional restoration	Homologous genomic regions with lox-P-flanked selectable marker	Restore mutated gene function by removing selectable marker at desired time	Function restored at different times or in specific tissues
	Knockin	Homologous genomic regions of gene 1 flanking coding regions of gene 2	Replacement of coding region of gene 1 with that of gene 2 to misexpress gene 2 in expression domain of gene 1	

has been "trapped" by the inserted transgene. Cloning of the insertion site then leads to the identification of the chromosomal gene.

Embryonic Stem Cell; Mediated Gene Targeting. A major limitation of the DNA microinjection method for making transgenic animals is that it does not allow the targeted alteration of endogenous genes. A technically more complicated method of gene targeting in embryonic stem (ES) cells provides the means for accomplishing that alteration, particularly for the directed inactivation of genes. All the knock-out mutants for signal-transduction components, which were discussed in Chapter 6, were produced by this method. The method, worked out in 1980, is simple in principle. A gene-targeting construct is made that incorporates a selectable marker flanked by cloned sequences of the gene to be targeted. It is introduced, usually by electroporation, into ES cells in vitro. ES cells are obtained from the inner cell mass of a mouse embryo at the blastocyst stage. The cells can be cultured in a Petri dish in artificial medium, where they proliferate for tens to hundreds of generations. Random integration of the targeting construct DNA into a chromosomal site of the ES cell will occur as a rare event (1 in 1,000 cells), and even more rarely (1 in 1 million cells), a double-crossover event of recombination will occur between the homologous DNA of the construct and the endogenous gene, the result being that the selectable marker is inserted into the endogenous allele. Cells that have incorporated the transgene are selected using the selectable marker. They are cloned, and the cell clones that have undergone homologous recombination are distinguished from random integrants by analysis of the DNA.

Cells with the desired genetic change are then introduced into early embryos to produce chimeric pups (i.e., ones containing both normal cells and genetically altered cells). The gene alteration is recovered in the subsequent generation if, and only if, the ES cells, which would be heterozygous, contribute to the pup's germ cells. With further mating, homozygotes can be obtained for study. Although technically demanding, the flexibility and precision of this method ensures its widespread application. A glance at recent databases indicates almost exponential growth in the number of reported mutations produced this way during the past 10 years (more than 1,000 published at last count). The majority of these were intended as loss-of-function mutations (null or knockout mutations or targeted deletions), although gene targeting is increasingly being used to produce other types of mutations (such as more subtle mutational alterations of a resident gene, sometimes referred to as "knockin" mutations). Some of the possible types of mutations are listed below.

Knockout or Null Mutations. Gene targeting has been most commonly used to disrupt the function of an endogenous gene. The simplest means of accomplishing this is to make a targeting construct with the selectable marker, such as neomycin resistance (the selection cassette), in a position that will disrupt gene transcription or translation. This can be accomplished either by inserting the cassette in a critical region or by replacing a portion of the endogenous gene with the selection cassette producing a deletion. The selection cassette usually has its own promoter and polyadenylation sequences. Note that such a mutation cannot properly be called a knockout, or assumed to be a null mutation, until evidence has been obtained that no gene product is produced in the mutant. Expression of a partial gene product might result in a phenotype other than the null phenotype.

Expression Reporting. A variation on a simple knockout construct can provide information on the expression pattern of the targeted gene. For this purpose, a reporter gene (e.g., the *E. coli* gene *lacZ* or the jellyfish gene for GFP) is used without a promoter. The targeting construct is made in such a way that the reporter gene is in frame and is driven off the targeted gene's promoter following homologous recombination. In this way, the reporter-gene product will be produced in all cells that would normally express the targeted gene. Usually, the reporter gene is studied in a heterozygote so that the other allele provides a functional gene product.

Point Mutations. Various schemes have been devised to produce mutations that are more subtle than complete loss of function. These usually involve a targeting construct that replaces the endogenous sequence with a homologous sequence containing a point (or other type) mutation. The trick is then to remove the selectable marker. This is most commonly done by using the bacterial Cre recombinase system, which is similar to the FLP recombinase system described in Figure 7-4. In the targeting construct, the selectable marker is made to be flanked by lox-P sites, short DNA sequences that are the substrate for Cre recombinase. Correctly targeted cell clones are then transiently transfected with the *Cre* gene, whose encoded product causes recombination between the lox-P sites, popping out the intervening DNA. The targeted allele is then left with a point mutation and a single lox-P site, which, if located in an intron, has no effect on the targeted gene.

Conditional Mutations. The Cre recombinase system (similar to the FLP recombinase system) has proved to be extremely versatile in gene-targeting schemes. The greatest potential is perhaps in the production of so-called conditional and tissue-specific mutations. The targeting construct is similar to that described above for making point mutations. The difference, however, is that a third lox-P site is introduced into a neutral position in the endogenous gene, in addition to the lox-P-flanked selectable marker. Then, following transient *Cre* expression, some cells will be recovered in which the lox-P-flanked selectable marker has been removed but two lox-P sites still remain. Any further expression of *Cre* will result in the removal of the intervening DNA, producing a deletion that can be planned to result in a null mutation. Provided the construct has been engineered so that the two remaining lox-P sites do not interfere with endogenous gene expression, normal mice can then be made with this ES cell line following removal of the lox-P-flanked selectable marker. The final step is the removal of the lox-P-flanked DNA in specific tissues of the whole animal, which is accomplished by breeding the gene-targeted mice with transgenic mice expressing *Cre*, as a transgene, in specific tissues.

A variation on this scheme can be used to restore normal expression of a mutated gene. In this case, a lox-P-flanked selectable marker is inserted into a gene, by gene targeting, to disrupt its function. Function can be restored in specific tissues in the resulting mutant mice by mating with transgenics expressing tissue-specific *Cre* in order to remove the inserted deleterious DNA, leaving behind a single lox-P site in a noncritical position.

Limitations and Pitfalls of Transgenic Technologies

Variations in the design of any transgene or gene-targeting construct, as well as local features of the integration or target site, will affect the outcome of transgenic experiments. Integration of a transgene is random. Thus, its expression might be affected unpredictably by other promoters or enhancers at its integration site, or the transgene might be entirely silenced by integration into a transcriptionally inactive chromosomal region. This has been termed the "neighborhood effect." Expression might also be influenced by epigenetic phenomena, such as methylation. Another chance event that will alter the intended experimental outcome, but which can be exploited, is insertional mutagenesis, described above. If the transgene happens to integrate in a position that causes the disruption of an endogenous gene, the experiment might be more informative about the endogenous gene than the transgene.

With gene targeting, a possible complication can arise if the selectable marker, which is essentially a foreign transgene, remains in the genome. Either this gene or its promoter could potentially affect expression of the targeted gene or neighboring genes. In gene-targeting experiments that involve deletions, it is possible to unknowingly remove cryptic regulatory regions located within introns and thus, potentially, to affect expression of nearby genes.

Phenotypic effects observed following any mutational change, whether through a transgene, gene targeting, or a naturally occurring mutation, are subject to what are called "genetic background effects." The mutant phenotype might vary in different animals depending on what other genes that animal possesses (its genetic background). For example, a mutation might show a different phenotype in two inbred strains if those strains carry different alleles in other genes that directly or indirectly modify the phenotype of the mutant gene. These background effects will be largely unpredictable, but this situation provides material to identify and isolate modifier genes and thus to increase understanding of genetic pathways. The judicious use of inbred strains, in which the mice are theoretically 98% genetically identical, allows these background effects to be studied (e.g., see Lander and Schork 1994).

Animal-Cloning Technology

Recently, it has become possible to produce small numbers of genetically identical mice by the procedure of fusing individual cumulus cells from an adult female into individual enucleated eggs, thereby providing each egg with a diploid nucleus. This Cumulina family is in its third generation of transfers (Wakayama et al. 1998). Recently, mice have been cloned from fibroblasts derived from adult tail snips (Wakayama and Yanagimachi 1999a,b; Wakayama et al. 1999). In testing situations in which genetic variability is a problem, such clones could provide a uniform population.

POSSIBLE APPLICATIONS OF MODEL ANIMAL RESEARCH TO DEVELOPMENTAL TOXICOLOGY

Are Simple Toxicological Tests Possible?

Using the New Knowledge

The new knowledge gained from model animal research should be applicable to developmental toxicology in at least three important ways:

1. In developing more effective assays to test for environmental toxicants.
2. In assessing the risks of known toxicants.
3. In investigating toxicological mechanisms, the understanding of which will allow development of new therapeutic approaches to toxicant-induced defects.

In keeping with the third charge to the committee to evaluate how this information might be used to improve qualitative and quantitative risk assessment, this section deals primarily with possible new model organism approaches to toxicant detection and to the analysis of the mechanism of action of toxicants on developmental processes. The committee will draw upon these new approaches in Chapters 8 and 9 in proposing a multilevel, multidisciplinary strategy to improve developmental toxicity assessment.

Learning from the Ames Test

Ideally, scientists would like to have an inexpensive test system analogous to the Ames test, which is used, and sometimes misused, for detecting potential carcinogens. The Ames test was based on the assumption that many carcinogens are mutagens and that most mutagens are carcinogens. This test uses sensitized bacteria to measure the mutagenic activity of test samples. It is inexpensive, rapid, and suitable for testing many compounds under many conditions.

Is a similar test possible for developmental toxicants? Probably not. Whereas carcinogens act in a limited number of ways, primarily by inducing mutations in the DNA of somatic cells, developmental toxicants probably act by a large variety of mechanisms involving many aspects of development. Attempts to use very simple metazoans, such as hydra, to test for general effects on development have proved to be unsuccessful, because the results obtained were not interpretable as predictive of mammalian responses. On the other hand, the rodent assays, which are now considered the most predictive of human developmental responses, are expensive and slow and hence suitable for assaying only a small number of compounds. As currently performed, they also might detect only gross effects.

From the knowledge now being gained about developmental mechanisms, it seems possible that many developmental toxicants (those that defy the drug-metabolizing defenses of the animal) will prove to act by perturbing the signaling pathways involved in the many inductive interactions between cells and tissues. (As previously emphasized, signaling pathways appear to be highly conserved among most animal phyla.) However, this hypothesis remains largely untested. Do known developmental toxicants affect signaling pathways, and if so, is this how they cause developmental defects? Pursuit of these questions is a search of mechanisms of developmental toxicity. Using the simple and relatively inexpensive animal model systems amenable to genetics, scientists should be able to answer these questions. If the answers are yes, as is the committee's hypothesis, it should be possible to design evaluation approaches for potential developmental toxicants with the use of animals, having sensitized genetic backgrounds and reporter-gene outputs, to detect effects on specific signaling pathways, as described in further detail below. Results will have to be used with caution, so that false positives are not overinterpreted. However, a judiciously applied battery of such tests could represent a major advance in developmental toxicity testing.

Sensitized Genetic Models for Testing of Specific Pathways

Some mutations essentially shut down a pathway by completely inactivating a component. Others that produce less inactivation cause no visible phenotype, although they bring the pathway close to a threshold of function and, therefore, render it sensitive to changes of activity of other components of the pathway—changes that by themselves might be asymptomatic. In signaling pathways, such sensitization can be accomplished either by raising or by diminishing the level of activity of a particular component, depending on its activating or inhibiting contribution. Change in the activity of a second component of this pathway due to mutation would cause the threshold to be crossed to altered function and phenotypic consequences. Hence, depending on how the assay is established, a phenotype might be enhanced or suppressed by perturbing a second element. In attempting to define a pathway, genetic screens for new dominant mutations that enhance or suppress the pathway-defective phenotype in a mutagenized, genetically sensitized strain of the test animal can be used to identify genes for new pathway components. Because the sensitization can often be designed to affect the pathway only in a particular tissue, this approach can succeed, even if a pathway is used in many places and many stages of development and null mutations cause phenotypes too pleiotropic to be interpretable.

Such sensitized strains should also be useful for identifying toxicants that modify the activity of a pathway component. Advantages of testing on animals having a tissue-specific sensitized pathway include the following:

- The chemical's effect can be assigned to a pathway without knowing the particular target protein, or even all elements, of the pathway.
- Biologically relevant thresholds of effect can be sought, because low doses should suffice.
- Phenotypes are more readily and reliably assessed, because they are revealed in a tissue-specific manner.

A variety of sensitized models and other approaches to assaying effects of known and potential developmental toxicants on specific pathways should be possible in the test animals considered here. All the model animal systems provide opportunities for developing methods of toxicity assessment and for investigating toxicological mechanisms (these opportunities are discussed in greater detail in Chapter 8). (Questions will be considered below on extrapolation to humans and the differences between animals and humans in uptake and metabolism of toxicants and in developmental processes.) Although the readouts of the assays involve scoring the development of various invertebrate organs, these readouts are chosen because they are likely to reveal effects of toxicants on conserved signaling pathways, and not because the organs are like mammalian organs. A relevant point is that human polymorphisms of signaling components might sensitize certain individuals to the detrimental effects of environmental agents.

Test animals for which genetic manipulations are difficult are not mentioned in the following section. For example, the FETAX test makes use of the frog *Xenopus laevis*. Although much has been learned about the development of *Xenopus* by mRNA injections into the egg, assays of cDNA libraries, and in situ hybridization, the organism is not yet amenable to easy genetic manipulation, and transgenesis procedures are in the early stages of use (Kroll and Amaya 1996). Thus, the committee does not believe that it equals the genetic model organisms for use in sensitive and ultimately informative assays of toxicants.

Caenorhabditis elegans

Suitability for Developmental Toxicology

Advantages of *C. elegans* include its low maintenance cost in the laboratory and properties mentioned above: its facility of genetic analysis, including rapid reverse genetics using RNAi, its anatomical and developmental simplicity, and its transparency throughout the life cycle, allowing visualization of internal phenotypes at the cellular level and expression of fluorescent reporters, such as GFP, in living specimens. One possible disadvantage is that, as a soil organism, it might have evolved resistances to some chemicals that can act as developmental toxicants in higher animals. These differences could in principle be characterized and genetically modified. Another possible disadvantage is that its collagenous cuticle might be impermeable to many test compounds. However, since larvae and adults constantly ingest materials from their environment, compounds that are not rapidly degraded should enter the animal through the gut.

Assays with Sensitized Pathways

Many of the signaling pathways described in Chapter 6 have now been demonstrated to function in *C. elegans* development. Of particular potential utility for toxicological applications are pathways important for postembryonic development but not essential for viability up to that point, so they can be assayed in living animals. A few examples of such pathways follow. For some, sensitized strains are already available; for the others, they can be easily constructed.

Receptor Tyrosine Kinase (RTK) Pathways. One of the best-studied signaling pathways is not essential for viability in *C. elegans* but does mediate the induction of the hermaphrodite vulva in the hypodermis. The gonadal anchor cell releases an epidermal-growth-factor-like signal to nearby hypodermal cells that receive it via an appropriate RTK and the downstream components of a typical Ras signaling pathway. Defects in this pathway lead to an easily visible lack of a vulva (and, hence, inability to lay eggs) or to multiple vulva-like structures (Sternberg and Horvitz 1991). This organogenesis operates in the last larval stage, when the animal feeds actively. Sensitized strains for screens for enhancer and suppressor mutations are already available and could be used to test for effects of toxicants on pathway function.

Transforming-Growth-Factor (TGF) β Pathways. Also nonessential for larval viability are two distinct pathways responding to different TGFβ-superfamily ligands, which interact via receptor serine and threonine kinases with typical downstream Smad protein components. One is involved in controlling development of *C. elegans* larvae into a diapause form (the dauer larva) under adverse conditions, and the other is in control of body size and patterning of the tail of the male (Riddle and Albert 1997; Padgett et al. 1998; Suzuki et al. 1999).

Notch and Delta Pathways. Pathways involving the Notch-like receptor LIN-12 and a Delta-like ligand, also nonessential for larval viability, affect postembryonic gonadal and vulval development. Another pathway involving the Notch homolog GLP-1 and a Delta-like ligand controls germ-line proliferation in the hermaphrodite. However, since GLP-1 signaling is also essential for early inductions in the embryo, further genetic modification would have to be carried out in order to use the GLP-1 variant of the pathway.

Wnt Pathways. Several recently discovered pathways involving WNT-like ligands and the homologous receptors appear to be important for establishing cell polarities and resulting patterning processes throughout *C. elegans* development; however, some of these pathways appear to be required only postembryonically and are nonessential for viability (Wood 1998).

Stress Pathways. GFP reporters have been made for various heat-shock proteins of the cytosolic unfolded protein pathway. Because *C. elegans* is transparent, transgenic animals carrying such reporters could provide a convenient readout of stress-pathway activation in response to toxicants.

Apoptosis Pathway. Much of the current fundamental knowledge of cell-death control was worked out in *C. elegans*. The pathway of interacting gene products that controls apoptosis during normal *C. elegans* development is well defined (Metzstein et al. 1998). It is considerably simpler than the pathways that are emerging in mammals, as is true for the simple model organisms in general; therefore, all aspects of the mammalian mechanisms are not represented in *C. elegans*. For example, there is only one cysteine protease, CED-3, in *C. elegans*, and there are at least 10 in humans (Salvesen 1999). Nevertheless, the control pathways are fundamentally similar. Sensitized pathways that can be produced by mutations in *C. elegans* should prove useful in testing for toxicants that cause developmental defects by way of apoptosis. As noted in Chapter 6, a variety of toxicants increase apoptosis in affected rodent embryos.

Behavioral Development Pathways. Although the behavioral repertoire of *C. elegans* is limited, the neural and molecular bases for several behaviors are well understood in the context of the completely mapped connectivity of its simple nervous system, which includes only 302 neurons. Because these behaviors are easily scored in the laboratory, assays for abnormal development or function of neuronal signaling pathways could provide simple, inexpensive, and useful screens for neurotoxins and other toxicants affecting development.

Genetic analyses have identified over 100 genes required for development of animals with normal movement. The "uncoordinated" (UNC) phenotypes resulting from mutations in these genes can be the consequence of either neuronal or muscular defects (Moerman and Fire 1997; Ruvkun 1997). Assays for toxicant effects on movement could therefore detect interference with the normal development and function of both muscles and the neurons that control them. Moreover, a skilled observer can distinguish many different UNC phenotypes associated with specific defects that are genetically and often physiologically understood. Therefore, comparison of toxicant-induced abnormal movement to behavior of known mutants could rapidly provide initial evidence on the point of action of the toxicant.

More specialized assays could be developed based on the animal's chemosensory abilities. *C. elegans* normally senses and responds appropriately to a variety of ionic and volatile chemoattractants (e.g., Na^+, K^+, ethanol, and ketones) and chemorepellants (e.g., acid pH, Cu^{2+}, octanol, and benzaldehyde) in their environment (Bargmann and Mori 1997). These responses are mediated in the head by chemosensory neurons, which send information for control of movement via the major anterior ganglion, called the nerve ring, and the ventral and dorsal nerve cords. Chemoreception involves a large family of 7-membrane-pass cell-surface receptors, which are divergent from chemoreceptors in vertebrates. However, the neurotransmitter receptors (both ligand-gated channels and G-protein linked), neurotransmitter synthesis and release pathways, and G-protein-linked second-messenger pathways involved in chemoresponses are highly conserved between *C. elegans* and mammals (Bargmann 1998). Extensive genetic and neuron ablation studies have shown that defects in the development of these signaling pathways lead to abnormal chemosensory responses, which are easily and inexpensively assayed on agar plates in the laboratory.

Another potentially informative assay system could be based on the *C. elegans* dauer pathway. The dauer (enduring) larva is a state of diapause that allows a population to survive periods of limited food and overcrowding. Under such conditions, the animals produce a pheromone that can be detected by sensory neurons in the head. Presence of the pheromone induces molting of L2 larvae (see Figure 7-1) to an alternative form of the L3 larva called the dauer larva or simply dauer, which is resistant to dessication, does not feed, and has a low metabolic rate and an increased lifespan of several months compared with the normal lifespan of about 2 weeks. This response is mediated by a signaling pathway that is clearly homologous to the metabolically crucial pathway of insulin signaling in mammals. Genetic analysis has identified components of the pathway that are required for dauer formation under starvation conditions, as well as components required for preventing dauer formation when food is available (Riddle and Albert 1997). Defective function of either class of components is easily assayed in the laboratory.

Not only could such assays serve as the basis for testing toxicant effects on nervous system development and function, but also the extensive knowledge of the cells and molecules involved should make it possible to rapidly identify the components affected by a toxicant that causes behavioral defects.

Testing for Pathway Readouts Using Functional Genomics Technology

A limitation of the above approach is that only genetically well-defined pathways can be tested for toxicological responses, and only one pathway can be tested at a time. More global tests will soon be possible that take advantage of the complete genomic sequence of the nematode. (As discussed in Chapter 5, functional genomics refers to the profiling of all transcription from the genome.) The targets of most signaling pathways in development are transcription factors; thus, the end result of the signal is a change in the pattern of transcripts synthesized by responding cells. DNA microchips carrying ordered arrays of cDNAs representing all the *C. elegans* expressed genes will soon be available, allowing rapid comparisons of mRNA populations from different stages of development and from wild-type and mutant animals at the same stage. As data from such tests accumulate, it should be possible to define "fingerprints" of the mRNA changes that result from perturbation of specific signaling pathways. Once these fingerprints are defined, comparison of animals treated versus not treated with a developmental toxicant could indicate whether a particular pathway or pathways are being affected. Such DNA microchips with arrays for detecting 6,000 mRNAs have been successfully used to detect changes of gene expression in yeast cells in mating circumstances or not (Spellman et al. 1998) and in cultured human cells exposed or unexposed to serum (Iyer et al. 1999).

RNAi Testing of Candidate Genes Affecting Toxicant Responses

Several genes known to encode proteins affecting toxicant responses in mammals have homologs in *C. elegans*. Examples are the genes for the aryl hydrocarbon receptor (AHR) and the AHR nuclear translocator ARNT (Powell-Coffman et al. 1998), whose normal functions are not yet fully understood in any animal. As an approach toward understanding the role of such gene products, their normal functions can be conveniently investigated in *C. elegans* by reverse genetics using the RNAi technique described above.

Drosophila

Suitability for Developmental Toxicology

Both the low cost of rearing flies and the ease of constructing special stocks are special virtues of *Drosophila* for the testing of potentially dangerous chemicals. One difficulty, however, will be to determine the effective dosage of the chemical being tested and the most effective route of delivery. Development of the eye and wing takes place during pupal development, although most of the growth of the eye and wing primordia (imaginal discs) occurs in the larva. The pupal case is relatively impermeable to anything except very lipophilic compounds. These compounds can be dissolved in acetone and applied directly to the pupal case. For example, juvenile hormone, an isoprenoid, readily passes through the cuticle. However, water-soluble test chemicals would have to be included in the food fed to 3- and 4-day-old larvae (third instar) or injected into pupae or flies (see later discussion of successful cases).

Assays with Sensitized Pathways

Because *Drosophila* has already shown the utility of sensitized screens for detecting new components in developmental pathways, there is little doubt that appropriate genetic stocks can be developed that would be useful for identifying potential toxic compounds. Such compounds would formally act like secondary mutations in reducing or increasing the activity of some other pathway component. In cases in which the sensitized strains are vigorous, as was the case for the *sevenless* mutants in the Simon et al. (1991) screen, mutant flies can be used without further genetic manipulations. In many instances, however, the sensitized strains are weak due to the fact that the signaling pathway is used for multiple functions, resulting in reduced viability. In these cases, the FRT-FRP system for producing clones of cells (see Figure 7-4), homozygous for sensitizing mutations, will be extremely useful. Careful construction of strains will also provide for "twin clones," one for the homozygous mutation and the other for clones homozygous for the wild-type genes. This twin clone will provide a record of the loss of the mutant clone in these instances where the homozygous mutant cells become inviable due to chemical treatment.

The compound eye and the wing are two excellent organs for the analysis of potential toxic chemicals. Both are nonvital organs under laboratory conditions, and indeed eyeless and wingless flies are viable and fertile. Both organs are easily scored for developmental effects. Both have a relatively large precursor population so that the frequency of clones is relatively high, if the FRT-FRP system is used. Finally, many of the major signaling pathways function in establishing the regular pattern. A few examples will illustrate these features.

Ras Signaling in Eye Development. Earlier in this chapter, the committee described an example of a screen that uses a sensitized genetic background to identify genes in the receptor tyrosine kinase pathway involved in *Drosophilia* eye development (Simon et al. 1991). The use of sensitized systems to reveal chemical modifiers has in fact been well demonstrated for the *Drosophila* eye. Peptidomimetics that block isoprenylation interfere with RAS membrane binding and activity and, as shown in Figure 7-5, when injected into flies, can abrogate the abnormal phenotype of the activating RAS val12 mutation (Kauffmann et al. 1995). The same agents block activated Ras-induced tumors in mice, confirming cross-species relevance of the assay used as a chemical and genetic indicator.

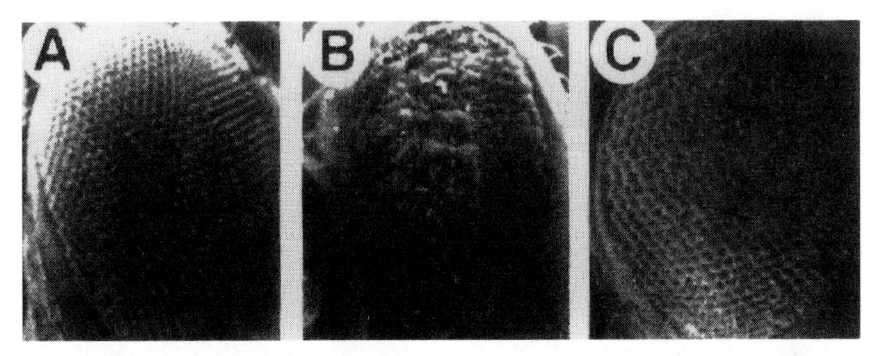

FIGURE 7-5 Effects of peptidomimetics on eye development in *Drosophila.* (Panel A) Normal compound eye with 800 ommatidia, each a bundle of seven photoreceptor cells and a bristle. (Panel B) The eye of a RAS val12 mutant in which the receptor tyrosine kinase (the sevenless pathway) is overactive. Note the disarrangement of ommatidia. (Panel C) A pupa was injected with a peptidomimetic that blocks isoprenylation, hence interfering with Ras membrane binding and activity. This reduces the activity of the pathway toward normal. The eye is more normal in development. (Kauffmann et al. 1995).

Transforming-Growth-Factor (TGF) β Pathways. The *Dpp* gene in *Drosophila* has been instrumental for identifying downstream components of the mammalian TGFβ cascade, such as the SMAD genes (Chanut and Heberlein 1997). *Dpp* is required in both the eye and the wing for normal development, and there are specific alleles of *Dpp* that strongly affect each of those organs. In addition, stronger alleles could be used in combination with a clonal analysis.

Hedgehog Pathways. Although *Hh* functions in the eye, its effect is not easily seen in small clones, presumably because of the diffusible nature of the HH protein (Heberlein et al. 1995). Effects might be more visible in wing clones, or it might be useful to develop sensitized systems for one of the downstream functions of the Hedgehog pathway, such as protein kinase A or the transcription factor *cubitus interruptus* (equivalent to the *GLI* oncogene in mammals).

Notch Pathways. Notch is required both for eye development and formation of the sensory components along the wing edge (Baker and Yu 1997). A wide variety of mutations specifically affecting the wing are available; therefore, clonal analysis should be useful.

Wnt Pathways. The Wingless ligand is required for establishing pattern in the wing and legs, and for limiting the eye domain. Clonal analysis of sensitized mutations should be efficient in detecting components of this widely used pathway.

Fibroblast-Growth-Factor (FGF) Pathways. The FGF pathway, which is a subset of the RTK pathways, is not used in either the eye or the wing in *Drosophila*, but it is required for proper tracheal branching, a process that is readily visible in living larvae by a variety of techniques. Consequently, specific interference with this pathway might also be detectable (Klambt et al. 1992). The tracheae are branched airways of the respiratory system of *Drosophila*. Their development has interesting similarities to angiogenesis in vertebrates, which involves FGF signaling.

Stress Pathways. Much of the original characterization of these pathways (e.g., the heat-shock response) was done in *Drosophila*. For convenient scoring of chemical effects, a variety of transgeneic strains have been constructed with reporter genes that are activated under conditions of stress.

Cell-Cycle Control. *Drosophila* might also be useful for the analysis of compounds that interfere with control of the cell cycle or that activate various checkpoint control pathways. Compounds stored in the egg during oogenesis support nearly all of the critical cell divisions required to achieve the free-living larval stage. During the larval stages, growth of the larva occurs by polyploidization and cell enlargement rather than by cell division. The only dividing cells in the larva are the precursors of the adult structures, and they increase in number prior to metamorphosis in the pupal stage. Most loss-of-function mutations in genes that are required for the cell cycle result in larval lethality, often at the larval and pupal boundary (metamorphosis), when the absence of adult precursor cells becomes critical. Consequently, it should be possible to test various chemicals for their ability to thwart the growth of adult precursor cells.

It could be argued that toxicant effects of the cell cycle are better studied in defined cultured conditions with, for example, culture-adapted differentiated cells. Indeed, cell lines are available, or could be prepared, with one or the other of the 17 intercellular signaling pathways functioning in culture, and these would be valuable for assessments of toxicant effects (e.g., effects on activin's action as an erythroid differentiation factor in human erythroleukemic cell lines). The advantage of a proliferating developing system, such as the *Drosophila* imaginal discs described above, is that several signaling pathways simultaneously influence proliferation, in their full complexity of signal release via the Golgi and signal modification in the extracellular space, in addition to signal reception and transduction. Thus, the initial net cast to find toxicant effects might be wider than that in a cell culture system. Still, the latter could be very useful in focused identification and analysis of toxicant effects.

Possible Assays of Behavioral Development in Drosophila

Because of complex courtship behavior, feeding habits, and biological rhythms, flies have become an excellent organism for identifying genes involved directly in behavior. Some of the major discoveries of genes controlling specific behavior traits have been accomplished in *Drosophila*. For example, the recent identification of genes controlling circadean rhythms in mammals (Takahashi 1995) owes its origin to the genetic screens that revealed the *per* locus in *Drosophila* (Konopka and Benzer 1971). Subsequent *Drosophila* screens have disclosed other components of the molecular clock—such as *Timeless*, *Cycle*, and *Clock*—which are also conserved in other organisms for photo-period regulation. Similarly, genetic screens in *Drosophila* have identified the mutants, such as *Dunce*, *Turnip*, *Cabbage*, *Amnesiac*, and *Rutabaga* that affect associative learning (Dubnau and Tully 1998). These genes have been shown to be part of the adenyl cyclase pathway.

Possibly the most elaborate behavioral patterns of *Drosophila* involve courtship behavior, which includes a series of species-specific activities essential for successful mating (Hall 1998). Starting from these successful pioneering studies, many of which were initiated in the laboratory of S. Benzer, *Drosophila* has become a useful organism for exploring the genetic basis for alcoholism (Bellen 1998), susceptibility to drugs (McClung and Hirsh 1999), aging (Lin et al. 1998), and other neurobehavioral traits. Because of the simplicity of its life cycle and rearing needs, the fly will continue to be an important first step in the identification of genes controlling a wide variety of behaviors. It should also be extremely useful for detecting deleterious effect of toxicants on specific behaviors, and as described above, the tests could be done on sensitized strains.

Zebrafish

The zebrafish is an appropriate model organism in which to test potentially toxic chemicals for several reasons. First, the zebrafish is a vertebrate and therefore of more direct relevance to the role of pathways in development of vertebrate-specific tissues and organs, such as the neural crest (Kelsh et al.1996; Schilling et al. 1996) and parts of the heart (Stainier et al. 1996), which are affected in many congenital anomalies. Second, chemicals can be added directly to the water (Stainier and Fishman 1992). Third, the zebrafish embryo is transparent, so tissue and organ development can readily be assayed. Viable, preferably dominant, mutations would be needed as the sensitized target to make such screens achievable on a large scale. Some viable pigmentation mutations have different heterozygous and homozygous phenotypes, indicating a dose responsiveness to the defect (Haffter et al.1996), and therefore are candidates, but these mutations have not been cloned, so the affected pathways remain to be determined. The potential is great for devising assays for chemical effects on development.

The zebrafish embryo has been used for toxicological assays (Ensenbach and Nagel 1995, 1997; Henry et al. 1997; Mizell and Romig 1997); however, it has not been developed for sensitized assays. Such assays are now feasible and would be advantageous. Also, there are now a number of mutants for components of signaling pathways (e.g., BMP2, Nodal, and Cripto).

One of the original hopes for the study of zebrafish was that it would permit the genetic dissection of behaviors, including learning. In principle, the transparency of the embryo offers the opportunity for simultaneous assay of neuronal activity. Several behaviors are established early during development. For example, the embryo becomes motile and will respond to touch within the first day of postfertilization (Saint-Amant and Drapeau 1998), and eyes follow a striped drum (termed the optokinetic response) by 3 days (Brockerhoff et al. 1995). Screens have already identified genes that modify these activities (Brockerhoff et al. 1995; Granato et al. 1996). Mutations that perturb locomotory behavior have been shown to affect a variety of sites, including receptors, the CNS, or muscle (Granato et al. 1996). Rhythmic activity, such as circadian rhythms (Cahill et al. 1998) and cardiac pacemaking (Baker et al. 1997), are embryonic in their time of onset and can be dissected by genetics (Baker et al. 1997). Whether these behaviors may be modified is not known, but modification would provide a means to garner genes for learning, addiction, and memory. The effects of chemicals on the development of these behaviors remain to be examined, but the availability of specific behavioral assays is at hand.

A radiation hybrid map of the zebrafish genome has been completed, the map coverage being 81.9% of the genome. The map is based on a panel of 94 radiation hybrids (Geisler et al. 1999). A large-scale insertional mutagenesis screen in the zebrafish, with the goal of isolating about 1,000 embryonic mutations, is under way (Amsterdam et al. 1999). This approach is similar to the Nobel Prize-winning approach of isolating a large number of *Drosophilia* mutants—described in an earlier chapter.

Mouse

Transgenic Animals and Developmental Toxicology

Mice and rats have long been the mammals of choice for toxicological tests. Their advantage over the previously discussed model organisms is their similarity, as mammals, to humans. An advantage of mice over rats is the advanced state of the procedures for genetic manipulation and the large number of mutant and inbred strains already available. The disadvantage of mice and rats, compared with the other model organisms presented above, is their expense. To use them to test tens of thousands of compounds under a range of exposure conditions or in combinations is not feasible. Thus far, little use has been made of sensitized strains and reporter strains to improve toxicant detection and to learn more about mechanisms of toxicity.

Furthermore, little analysis has been done on the differences between mice and humans that lower the validity of cross-species extrapolations. There are likely to be two main kinds of differences: (1) differences in the steps by which the active or activated toxicant is introduced to the embryo or fetus, and (2) differences in the components of developmental processes, making the toxicant have more effect or less effect on developmental function and resulting in a more severe or less severe developmental defect. Enough is now known about absorption, distribution, metabolism, and excretion, especially regarding drug-metabolizing enzymes (DMEs), to make systematic comparisons between humans and rodents possible. Many components of signaling pathways and genetic regulation, which are central to development, are also now known in test animals and humans, and those components could be compared systematically as well.

As summarized by Malakoff (2000), the use of mice in biomedical research and drug testing is expected to increase dramatically in the next few years, especially as the sequencing of the mouse genome approaches completion, and the increased use of mice will create opportunities in developmental toxicology.

Gene-Deletion Transgenics for the Study of DMEs

Only within the last 5 years have transgenic methodologies been brought to bear on the reciprocal relationship between the animal's genetic constitution and its susceptibility to developmental toxicants. A particularly important and approachable set of genes are those encoding enzymes that metabolize exogenous chemicals. As discussed in Chapter 5, there are two major categories, the oxidizing enzymes (mostly P450 proteins) and the conjugation enzymes. There might be tens of these enzymes that metabolize most chemicals and hundreds more that metabolize a few chemicals each. In addition, several known transcription factors activate the expression of genes encoding these enzymes, and some of these factors themselves bind exogenous chemicals.

Much remains to be learned about the role of the various enzymes in generating and removing active toxicants, and the gene-deletion approach in the mouse has great advantages. At least in some cases the knockout mice are viable and fertile. One pioneering example in which gene-deletion transgenics was used concerns the dioxin-inducible mouse gene battery (Nebert and Duffy 1997), a group of genes believed to play an important role in developmental toxicity. Much more of this analysis can and should be done, both to understand the role of these enzymes in potentiation and detoxification and to define human and mouse differences for better-informed extrapolations of animal-test data.

Overexpression Transgenics for the Study of DMEs

Another approach to the study of developmental toxicity using transgenic animals is to overexpress, or ectopically express, a gene of interest in the embryo and fetus. For example, one could ask whether overexpression or ectopic expression of a gene encoding a particular oxidizing or conjugating enzyme either sensitizes or protects the embryo and fetus from the adverse effects of developmental toxicants. Expression of transgenes can be driven by ubiquitous, constitutive promoters, such as β-actin, to induce expression in most, if not all, cells of the embryo. Alternatively, transgene expression can be limited to specific tissues by using tissue-specific promoters. Finally, transgene expression can be limited to specific stages of development and specific tissues by using one of several inducible expression systems such as the Cre-recombinase described above.

Transgenes as Biomarkers for the Activation of Stress, Checkpoint, and Apoptosis Pathways

Transgenic mice might also play a role in testing drugs and chemicals for potential developmental toxicity (see further discussion in Chapters 8 and 9). As scientists learn more about normal development and about the mechanisms of developmental toxicity, in part from studies using gene deletion or overexpression approaches, sufficient information should become available to construct transgenic mouse lines designed to contain biomarkers of the animal's toxic response to a chemical. For example, a transgenic mouse could be constructed that contained a transgene consisting of the heat-shock promoter linked to an appropriate reporter gene, as has been done in *Drosophila*. This reporter, in turn, could be used to determine whether specific drugs and chemicals induce a stress response, which is often associated with developmental toxicity. Such a biomarker transgene could be used for in vivo developmental toxicity studies, or cells from appropriate tissues could be used for in vitro testing. Other stress and checkpoint pathways could be connected to reporter genes as well, as could components of apoptosis.

Signaling Pathways as Developmental Targets of Toxicants

As noted above, the mouse is at present the animal of choice for the selective knockout or replacement of genes. A large number of genes encoding components of many signaling pathways have now been eliminated, one at a time, as summarized in Chapter 6 (Tables 6-4 and 6-5), and the phenotypes of the single-gene homozygous null mutants have been examined. Surprisingly, many of these mutants achieve advanced development, living past birth, and some reach fertile adulthood. Many have discrete developmental defects. The current explanation of the viability of these mutants is that vertebrates have large, partially diversified families of genes for most signaling components, and the genes, which are expressed at different times and places in the embryo and fetus, encode proteins of partially redundant function. Defects tend to occur at times and places where there is no overlap of expression. A number of double and triple mutants have been made as well, and they have more severe effects.

Sensitized test mice can certainly be prepared for the testing of toxicant effects on development (i.e., mice with a particular pathway operating at or near the limiting rate in a designated developing tissue). Animals can also be prepared with *lacZ* or GFP reporter genes to give enhanced readout of effects on pathways.

Assays of Altered Behavioral Development in the Mouse

As noted in Chapter 2, developmental defects include functional as well as structural defects, but functional defects are less well diagnosed and less well understood. Behavioral defects comprise a large category of functional defects. The testing of environmental agents in animals for behavioral defects as an outcome of prenatal exposure is sometimes done, but not routinely. If an agent is suspected to have neurotoxic effects, a developmental neurotoxicity battery is required, and the suspected agents are mostly pesticides, fungicides, and rodenticides (see Makris et al. 1998; EPA 1998f).

In the developmental neurotoxicity tests, pregnant rodents are treated with the agent from day 6 after conception until 10 days after birth to include transfer of the chemical in the milk, and then the pups are tested in various ways from day 4 to day 60. Tests include a functional observational battery (FOB) and specialized behavioral tests.

The FOB includes cage-side observations of arousal; autonomic signs; convulsion; tremors; gait; mobility; posture; rearing; stereotypy; responses to touch, approach, tail pinch, and clicks; foot splaying; grip strength; righting reflex; body temperature; and body weight. The specialized tests include those of motor function, sensory function, and cognitive function. Motor-function tests involve tests of grip strength, swimming endurance, use of the suspension rod and rotorod, discriminative motor function, gait, righting reflex, and various ratings of spasms or tremors. Sensory-function evaluations include discrimination conditioning and reflex modification related to auditory, visual, somatosensory, and olfactory inputs and pain sensitivity. Cognitive-function tests include habituation of the startle reflex and classical conditioning related to the nictitating membrane, flavor aversion, passive avoidance, and olfactory conditioning.

Twelve chemicals, mostly pesticides, were tested by the above protocol, reviewed retrospectively as a group, and various developmental neurotoxic effects were distinguishable (Makris et al. 1998). Concordance was found between structural alterations of the nervous system and behavioral defects. The difficulties of interpreting the developmental causes of behavioral defects were discussed.

In the effort to assess functional defects, the committee favors greater inclusion of behavioral tests with other tests of toxicants. It would be useful to calibrate these tests with mice of known genetic defects affecting aspects of neurodevelopment. Null knockout mutants in homozygous and heterozygous condition could be used. The importance of modifier genes in the genetic background of each transgenic mouse line cannot be underestimated. It would also be useful to calibrate the tests with animals treated with doses of toxicants just below the level giving detectable structural defects in CNS development.

TABLE 7-3 Advantages and Disadvantages of Four Model Animals for Toxicant Testing

Organism	Advantages	Disadvantages
C. elegans	Relatively inexpensive, short life cycle, small size, large populations, extensive information on genetics, genomics known	Some signaling pathways are apparently not represented (e.g., Hedgehog); short life cycle, so test compounds might persist long enough to complicate the study of sensitive developmental periods and of primary versus secondary effects
Drosophila	Relatively inexpensive, short life cycle, small size, large populations, extensive information on genetics, genomics known	During pupal development organism is inaccessible to toxicants; short life cycle, so test compounds might persist long enough to complicate the study of sensitive developmental periods and of primary versus secondary effects
Zebrafish	Simple vertebrate, transparent, external embryos, best organogenesis model	Relatively expensive, long life cycle, genetics and genomics research in progress
Mouse	Mammalian model, most similar to humans, targeted gene replacement possible	Relatively expensive, long life cycle, embryonic development in utero and not as easily accessible for manipulation as other model animals

Comparative Utility of Model Organisms

In summary, each of the four genetically tractable model organisms—*C. elegans*, *Drosophila*, zebrafish, and mouse—offers promise for development of tests to identify potential developmental toxicants, and each has advantages and disadvantages, as summarized in Table 7-3.

A useful approach, which the committee will present more fully in Chapter 8, might be to develop the more-rapid and inexpensive systems, namely, *C. elegans* and *Drosophila*, for preliminary large-scale assays, which can be followed up by more definitive tests using the vertebrate models, zebrafish and mouse, which, although slower and more costly, are more likely to give results relevant to humans.

GENE EXPRESSION AS DETERMINED BY IN VITRO AND ENGINEERED CELL TECHNOLOGIES

Mammalian cells in culture, as an inexpensive and fast complement to whole-animal (e.g., mouse and rat) studies, can be used for a wide variety of specific purposes related to mechanisms of developmental toxicity and polymorphic variation in susceptibility of toxicants. The incisive use of cultured cells has been enhanced greatly by molecular methods in which genes of choice (from any organism including humans) can be introduced into such cells, for example, to express variant forms of DMEs and to investigate the relation of function to allelic variation.

Human allelic variants code for gene products different from those encoded by the wild-type allele, and it is important to characterize and, if possible, quantify any functional alterations. The fundamental techniques available for studying gene expression include (1) expression in vitro where DNA from the allelic variant or wild-type allele is transcribed to the mRNA and then translated into a functional, active protein in a cell-free extract; (2) high-level expression in cells from which proteins can be purified with relative ease; and (3) expression in eukaryotic cell culture in which the cell or the cell's DNA has been modified (i.e., genetically engineered). These techniques rely on the gene (or at least the cDNA-encoded "coding region" responsible for generating the amino acid sequence of the protein). The cDNA encoding the allele being studied must first be cloned into an appropriate plasmid vector that includes regulatory sequences that drive and terminate transcription and a selectable marker gene. As outlined below, each of the aforementioned expression strategies is used by the investigator to meet specific needs. In general, these in vitro and cell-culture systems are very useful, because they are relatively quick, efficient, inexpensive, and use simple technologies when compared with the development of mouse lines or other animal-model systems.

There is a major shortcoming, however, in the in vitro and cell-culture expression systems. Single-nucleotide polymorphisms (SNPs) outside the amino-acid-determining region of the gene (e.g., splice junctions; promoter sequence; 5′ and 3′ untranslated regions; and enhancers upstream, downstream, and inside the gene) can have striking effects on expression of the gene under study, and the effects would generally not be realized, characterized, or quantitated by the in vitro and cell-culture expression systems (Nebert 1999).

Expression In Vitro

There are both reticulocyte lysate and wheat-germ lysate combined transcription and translation kits now commercially available to assess the interaction of the protein under study with another purified protein. To use these systems, a plasmid containing the cDNA of interest and a radiolabeled amino acid are added to the lysate. Although the protein of interest is not expressed at high levels relative to the total amount of proteins in the lysate, it is the only radiolabeled protein in the reaction mixture.

High-Level Expression in Cells

Proteins or enzymes of interest can be expressed at very high levels, which can be advantageous for studying certain functional alterations. The most commonly used systems include plasmid-transformed bacteria (Oudenampsen et al. 1990), baculovirus-infected insect cells (Kost and Condreay 1999), and vaccinia-virus-infected mammalian cells (Chakrabarti et al. 1985; Eckert and Merchlinsky 1999). All three systems allow high-level expression of cloned genes. In general, bacterial expression is the easiest to use. However, both the baculovirus-infected and the vaccinia-virus-infected expression systems are eukaryotic, and it might be particularly important to use them if post-translational processing or intracellular accessory factors are needed in the production of a functional gene product. Moreover, proteins expressed in any of the above systems can be given "tag sequences" to allow for rapid, specific purification.

Transient and Stable Transfections of Mammalian Cells in Culture

Yeast (Oeda et al. 1985), African green monkey kidney fibroblast COS cells (Zuber et al.1986), and vaccinia virus (Battula et al. 1987) were among the earliest expression systems in vivo. They have been successfully used to study DMEs. Allelic DNA can be transfected (passed into the cell) by microinjection or by chemical-DNA aggregation methods including calcium phosphate precipitation and liposome-mediated transfection. By using such methods, followed by antibiotic treatment to isolate cells that house the plasmid containing the selectable marker gene, cells can be transfected either transiently or stably. In transient transfections, expression of the gene is generated from extrachromosomal copies of the transfected plasmid and persists until the expression plasmid is degraded or diluted by cell passage. In general, 5% to 50% of all cells in culture contain the incoming gene, the DNA is not stably "integrated" in the cell's genome, and the transfected cells contain many copies of the new genetic material. In contrast, for stable transfections, the incoming DNA is integrated (albeit randomly) into the cell's genome. Cells expressing the gene under study are initially selected on the basis of co-expression of a gene that provides antibiotic resistance. After antibiotic selection, continued cellular propagation in the presence of the antibiotic will ensure that the gene of interest is expressed in a more or less permanent (i.e., stable) fashion—remaining after the cells are passaged on through many additional generations. The copy number of integrated genes is highly variable and can range from one to several dozen, and the genes are generally arranged in tandem (head to tail) at an "integration site" that normally cannot be directed, or controlled for, on an experimental basis.

For better gene expression in culture, one or more introns (even when they are artificially inserted) have been found to enhance gene activity in cultured cells (Palmiter et al. 1991); alternatively, small genes with fewer than 10 and usually fewer than 5 exons and introns can often be transfected transiently or stably into cells in culture as the complete gene. Dozens of promoters have been studied over the past 15 years. They include the *Drosophila* heat-shock promoter (Hsp), the mouse mammary tumor virus long-terminal repeat (MMTV LTR) having a glucocorticoid response element (GRE), enhancer sequences from simian virus 40 (SV40) and herpes simplex virus type 1 (HSV-1), human cytomegalovirus (hCMV), thymidine kinase (Tk), and locus control regions of the metallothionein gene (LTR-MT) displaying variable potencies in driving transcription (Blackwood and Kadonaga 1998; Makrides 1999).

Both transient and stable expression of genes in mammalian cells have many advantages. First, genes are expressed in a native environment so post-translational modifications and subcellular targeting are authentic. Second, many transformed mammalian cell types are available for transfection. This allows for the selection of cell types—with the proper intracellular accessory factors or proteins—for enzyme activities that most closely resemble their in vivo counterparts. Finally, expression of gene products can be controlled by an increasing number of eukaryotic promoters. Thus, the levels of expression, including inducible expression, can be controlled.

Generations of cell lines that stably express DME genes represent the new generation of pharmacological and toxicological test systems (Langenbach et al. 1992). Expression of stably transfected DME genes in Epstein-Barr virus (EBV)-transformed human B-lymphoblastoid cell lines has provided cell lines that scientists can use to study and categorize numerous foreign compounds. Those drugs or chemicals can be classified according to the intermediates formed by way of the different metabolic pathways. More than a dozen EBV-transformed human B-lymphoblastoid cell lines are already commercially available (Crespi and Miller 1999). They contain anywhere from 1 to 12 human P450 and other DME genes (i.e., cDNAs), which retain their substrate specificity with regard to the metabolism of particular drugs or classes of pharmaceutical agents.

Some of these cell lines are already being used by pharmaceutical companies to determine whether a newly developed drug is metabolized by a particular DME. If, for example, the new drug is shown to be a CYP2D6 substrate, it is already known that humans differ by 10-fold to more than 30-fold in the activity of that enzyme (see Chapter 5). Poor metabolizers (the PM phenotype, which comprises 6% to 10% of Caucasian populations) would thus be expected to metabolize the new drug more slowly than extensive metabolizers (EM phenotype) and much more slowly than ultra-metabolizers (UM phenotype). If the parent drug is toxic, the PM individual would be at increased risk; if the CYP2D6-mediated metabolite is toxic, then the EM and especially the UM individual would be at increased risk. Dosage of the drug can therefore be adjusted to the patient's genotype before the physician prescribes the new drug. For drugs already on the market, molecular epidemiologists can search for possible associations between reported differences in birth defects or other developmental problems and genotype of the mother (or child, fetus, or embryo)—keeping in mind the range of drug doses that might have been given.

During the next decade, dozens more of the DME genes (alone and in combination) are likely to be expressed in the human B-lymphoblastoid cell lines or in other similar stably transfected cell backgrounds to determine which enzymes are responsible for either detoxification or metabolic potentiation of a particular drug under study. This information will be useful in the future of developmental pharmacology and toxicology, as well as molecular epidemiology.

SUMMARY

The committee has evaluated the state of the science for elucidating mechanisms of developmental toxicity and concludes that such elucidation, although not yet realized, can be achieved in the next decade for the following simple reasons:

1. Developmental biology has reached the molecular level of mechanistic explanations.
2. The accumulation of new and relevant information about vertebrate development is rapid (assisted greatly by research on model organisms, such as *Drosophila* and *C. elegans*).
3. The accumulation of genome sequence data for humans, mouse, rat, and *Drosophila* is rapid, adding to that already available for *C. elegans*, yeast, and many prokaryotes. Information on human polymorphisms and rare variants and their disease relatedness is increasing rapidly, as are data on the ever-increasing library of mouse mutants.
4. The methods are powerful and widely applicable, and the species barriers to comparative study have been greatly reduced in the past few years.

The committee begins in this chapter, and continues in the next, the third charge to evaluate how this information can be used to improve qualitative and quantitative risk assessment. In this chapter, the committee summarizes some of the techniques for modifying model organisms, including the mouse, for effective use in assays evaluating agents for potential developmental toxicity and for elucidating mechanisms of toxicity. The committee concludes that the methods and background knowledge are at hand to make incisive comparisons of humans and model animals so that the extrapolation of results from model animals to humans can be more accurate and useful for risk assessment.

8

A Multilevel Approach to Improving Risk Assessment for Developmental Toxicity

In this chapter, the committee addresses its third charge to evaluate how the new information and opportunities described in Chapters 4, 5, 6, and 7 can be used to improve qualitative and quantitative risk assessment for developmental effects. To make such improvements, the committee envisions exploiting the insights and opportunities from two kinds of research efforts:

- First, research advances can be made in the understanding of mechanisms of developmental toxicity. As discussed in Chapter 4, this entails an understanding of the toxicokinetics of delivery of the chemical to a target site with an understanding of all the subsequent toxicodynamic steps. These steps include the chemical's interaction with the target molecule(s), the consequence of altered target molecule activity for one or more developmental processes, and the subsequent emergence of a particular pathogenesis (developmental defect).
- Second, research advances can be made that would increase our ability to reliably extrapolate from test results from model test animals to humans and to all members of the heterogeneous human population. As discussed in Chapter 3, risk assessors often resort to applying large default corrections to the animal data to estimate allowable human exposure concentrations of chemicals because of uncertainties about the relevance of animal data to humans. These uncertainties underlie the quantitative limitations of current risk-assessment approaches.

In the committee's judgment, a research agenda should be prepared that addresses these gaps in knowledge about mechanisms of developmental toxicity and in the ability to extrapolate the results of animal tests to humans. The new information about development and genomics should be incorporated into the agenda. The agenda would address basic scientific questions about which there is inadequate information. Specific areas of opportunity include:

- *Intraspecies differences in sensitivity to toxicants.* Genetic differences are suspected to be a major factor in intraspecies differences within human populations and within test-animal populations. Do individual differences concern mostly genetic variation in toxicokinetics, particularly in DMEs? Do they also include genetic variation in developmental components, such as those of the 17 intercellular signaling pathways used throughout development? Do multiple genetic differences have additive effects for the individual's toxicant susceptibility? Is genetic variation in components of molecular-stress pathways also important in the individual's response to toxicants? What is the contribution to individual susceptibility of nongenetic differences, such as those of age, history of disease, nutrition, and exposure to other chemicals and pharmaceuticals?
- *Cross-species extrapolation.* What are the toxicokinetic differences, particularly in the activities of drug-metabolizing enzymes (DMEs), between test animals and humans? Are differences in susceptibility to developmental toxicants due to differences in developmental molecular components and processes? Can some of the differences between test animals and humans be reduced or eliminated?
- *Extrapolation of high-dose exposure of small populations of test animals to low-dose, long-term exposure of a large human population.* Do chemicals have different toxicokinetic and toxicodynamic effects at high doses versus low doses? Does the organism rely on different protective responses to chemicals at different doses, such as enzymatic detoxification at low doses, molecular-stress reactions at intermediate doses, and the apoptotic response at high doses?
- *Expanded test information for numerous chemicals and, especially, mixtures of chemicals.* Can structure-activity relationships be obtained for a larger variety of chemicals by using in vitro tests with purified components (e.g., DMEs and developmental components of signaling pathways and transcriptional regulation)? Can model animals be genetically modified so that more mechanistic information can be obtained from them than from standard animal tests? As mechanisms are better understood for certain chemicals, can the effects of related chemicals be better predicted?

THE MULTIDISCIPLINARY, MULTILEVEL, INTERACTIVE APPROACH

The committee will outline in the remainder of this chapter a multidisciplinary, multilevel, interactive approach in which recent and future advances in developmental biology and genomics can be integrated with developmental toxicology to improve risk assessment for human developmental defects. This approach is not simply an alternative to current practices, but represents a novel approach to assess risk for developmental defects. In Chapter 9, the interface of risk assessment and developmental toxicology is further explored within the fourth charge to the committee to develop recommendations for research in developmental toxicology and developmental biology to assist in risk assessment.

The approach is multidisciplinary and multilevel because it invokes a wide variety of sources of information for risk assessment, including not only the assessment of toxicity and mechanism of action of chemicals in a variety of model systems (in vitro assays, nonmammalian models, and mammalian models), but also the assessment of toxicity, susceptibility, and exposure in human populations. The approach is appropriate to the risk assessment task, because assessing developmental toxicity is a broad and difficult area. The understanding of toxicity mechanisms entails both toxicokinetics and toxicodynamics, ranging from molecules to pathogenesis. Furthermore, the analysis of the differences between model systems (e.g., test organisms) and humans, as needed to improve extrapolation of test results, will require extensive comparative work at a variety of levels. These advances will depend on knowledge from chemistry, biochemistry, molecular biology, cell biology, developmental biology, genetics, ecogenetics, anatomy and organ physiology, genomics, and even systematics and evolutionary biology (e.g., finding conserved and nonconserved processes of development). A multidisciplinary, multilevel approach is needed to "bridge the gap" between the emerging scientific information and the assessment of human risk. At the same time, an interactive approach is needed for the dynamic interplay between the sources of new information and the needs of risk assessors.

The committee's multilevel approach should not be mistaken for a multitiered approach, where a specific order of evaluation and types of testing are specified. In a strict tiered approach, screening data at a low tier (low-cost, high-throughput tests, low assurance of relevance to humans) are first used to estimate risks of a large number of agents, and an agent with a high potential risk estimate at this tier triggers more rigorous and relevant tests, with higher associated costs, at successively higher tiers. For example, the Endocrine Disruptor Screening Program (EDSP) will use a tiered approach (EPA 1998g). Although the committee's multilevel approach also involves tests ranging from inexpensive, high-throughput tests to slow, rigorous, and expensive tests, the approach differs in several respects. First, there is no uni-directional triggering of higher level testing by results obtained from a lower level test. Testing can be done independently at any level or at several levels, depending on the particular compound, the risk assessment questions, the anticipated human exposures, and the commercial uses considered for that compound. Results at one level could lead to tests at lower levels rather than higher levels. Second, the different testing levels yield different kinds of information, all with the potential to contribute to the knowledge of toxicity mechanisms, from molecular interactions to pathogenesis, and the understanding of the basis for extrapolation. All levels are designed to provide information useful for human developmental toxicity risk assessments. For example, if a chemical (parent compound or metabolite) shows toxicity in a mammalian test, with hints of interference with a particular cell signaling pathway, the next step might be to analyze the mechanism of toxicity in a genetically sensitized invertebrate model system specifically designed to evaluate that cell signaling pathway. Likewise, observations from receptor-binding assays might prompt a re-examination of the overall impacts of a compound on organ development in a mammal. Thus, assessments are anticipated to have implications for analyses in both directions in the information levels. As new scientific observations are made, this approach allows the incorporation of new data into the risk-assessment framework.

Ultimately, risk assessment has much to gain from the multidirectional flow of information across these information levels. In Figure 3-1, the committee introduced two-way arrows to indicate the importance of the responsiveness of the whole process to issues and ideas raised not only by science but also by risk-assessment needs. In light of the gaps in knowledge of toxicant effects, risk assessment is most likely to improve when research and risk assessment reinforce each other. Understanding mechanisms of toxicity can then be useful for predicting which other potential toxicants might act by the same mechanism, improving the ability to develop structure-activity relationships. Understanding the basis for extrapolations between test animals or in vitro assays and humans will give risk assessment greater validity. An iterative and interactive process for risk assessment was first defined in the National Research Council (NRC) report *Science and Judgment* (1994). However, such a process has yet to be fully implemented in risk assessment for developmental toxicology.

Table 8-1 summarizes the committee's multilevel approach. There are two components, each with four sources of information: (A) assessment in model systems of toxicity and mechanism of action of developmental toxicants (Table 8-1A), and (B) assessment in human populations of toxicity, susceptibility, and exposure to toxicants (Table 8-1B). The left column in both tables lists for each information level the experimental description of the tests, the application of the tests, the number of tests that can be done per year, and the value of the test information for risk assessment.

Toxicity Assessment in Model Systems

Information levels 1 and 2 of model systems in Table 8-1A generally involve relatively inexpensive and fast characterizations of chemicals and developmental effects. They should provide valuable information about which developmental pathways (signaling pathways and transcriptional regulatory circuits) are affected by which toxicants. Although extrapolations to human risk would be very limited without additional toxicokinetic and toxicodynamic information, the testing capacity should be available to characterize a large number of chemicals, and indeed, most of the several million chemicals in the environment, including chemi-

TABLE 8-1A Information Levels for Assessment of Toxicity and Mode of Action of Chemicals in Model Systems

Information Level	Information Level 1	Information Level 2	Information Level 3	Information Level 4
Experimental description	High-throughput biochemical and cellular assays. Identify potentially vulnerable pathways in development, e.g., the 17 signaling pathways, molecular stress pathways, and the main toxicokinetic pathways (DMEs). Identify molecular interactions of toxicants and target molecules, and activity changes	Developmental assays with genetically optimized and sensitized nonmammalian animals. Identify potentially vulnerable developmental pathways, target organs, and times of susceptibility in development.	In vivo mammalian developmental toxicity testing. Use genetically optimized animals for relevant sensitive tests. Identify potentially vulnerable pathways, target organs, and times of susceptibility. Link dose-response relationships to both developmental and functional (e.g., behavioral) changes.	Detailed mechanistic studies to understand mode of action for selected developmental toxicants. Toxicokinetic and dynamic processes of chemicals would be identified and quantitatively assessed. Animal tests of toxicant sensitivities of polymorphisms. Analysis of differences of test animals and humans, regarding DMEs and signaling components.
Application	Use to assess most chemicals (>100,000) used by humans or released into the environment. Assess combinations of chemicals. Collect dose-response data.	Use to assess many chemicals, doses, and combinations. Use to assess relative toxicity of related compounds.	Use for hazard assessment of chemicals of likely high exposure or concern to humans. Can generate useful dose-response data relevant for mammalian evaluation. Gain information on toxicant sensitivity of polymorphisms.	Use for a few chemicals where significant human exposure is likely or where knowledge about toxicant mechanism or about animal vs. human differences would be useful for human risk assessments.
Assays per year	10^5-10^6	10^3-10^4	10^2	10 studies

continues

TABLE 8-1A continued

Information Level	Information Level 1	Information Level 2	Information Level 3	Information Level 4
Database output	Linked databases of chemical information and effects on signaling pathways, molecular-stress pathways, and conversions by DMEs.	SAR (structure-activity relationship) database for chemical effects on developmental and signaling pathways, end points, and susceptibility factors.	Developmental toxicity database with both qualitative and quantitative information.	Mode-of-action database for developmental toxicants.
Risk assessment information available from each assay type	Identification of potentially vulnerable signaling pathways. Identification of chemical properties associated with potential to alter signaling and stress pathways.	Improvement of SAR data for developmentally relevant impacts. Identification of potentially sensitive target organs. Information on periods of sensitivity.	Dose-response characterization. Mammalian relevancy information. Special toxicant sensitivity of genetic variants.	Improved mechanistic understanding. Quantitative information for predicting human susceptibility. Better validation of animal-to-human extrapolation.

TABLE 8-1B Databases for Assessing Toxicity, Susceptibility, and Chemical Exposure in Human Populations

Database	Human Developmental Outcomes Database	Human Genome and Genetic Polymorphism Database	Biomarker Database for Exposure, Effect, and Susceptibility	Human Gene-Environment Interactions Database
Experimental description	Human epidemiology and surveillance databases relevant for birth-defects research will be linked for morphological and functional impacts.	Profiling of human populations for polymorphisms of developmentally relevant genes, such as those encoding DMEs, signaling components, and stress-response components.	Link human biomarkers of exposure, susceptibility, and effect for development.	Detailed investigation of genotype-environment interactions for toxicant effects on development.
Application	Identify, characterize, and link human databases: case reports, active and passive birth defects surveillance databases, post-market surveillance databases, link with chemical exposure and toxicity databases, link with known genetic birth defects syndrome databases, link with human genome project.	Prioritize human genomic profiling for genes encoding components of pathways relevant to developmental toxicity, i.e., DMEs, molecular-stress pathways, cell-cell signaling pathways and developmentally relevant signal pathways.	Improved understanding of critical signaling pathways should allow for improved linkage of chemically induced early cell biological effects with impacts on development (organogenesis and behavior). Knowledge about susceptibility biomarkers such as DMEs, signaling components, and developmentally critical systems will allow for linkage of susceptibility biomarkers with birth defects.	Identification of critical interactions between genes and environment, e.g., additivity of defects of several signaling components in determining toxicant sensitivity.
Risk assessment information	Linkage of relevant databases, improved hazard identification, improved surveillance.	Understanding of basic genetic variability across human populations. Identification of potential susceptibility genes relevant for development.	Improved biomarkers and improved understanding of biomarker data in risk assessment.	Characterization of susceptibility profiles for human developmental toxicants. Identification of variability in the susceptibility of human populations for birth defects from environmental factors. Improved quantitative information.

cals from natural products such as plants, have never been characterized for developmental toxicity. (It must be noted, however, that the EDSP plans to test 87,000 chemicals.) Of course, it will take years to develop, standardize, and validate these assays; however, the potential utility in screening with validated tests is high and justifies the effort. At information levels 3 and 4, the cost is higher and the test times are longer, but extrapolation to human risk potential can be more direct.

Information levels 1 and 2 make use of model systems of far less complexity than humans. The results from these specialized cell assays and model organisms would be useful to organize chemicals according to their effects (e.g., to reveal chemicals that bind to the same protein—e.g., a nuclear-hormone receptor—or interfere with the same conserved cell signaling pathway). Assays in information levels 3 and 4 are likely to improve in their relevance to humans in the near future as differences between rodents and humans are better understood and as genetically modified model animals become available. These models will more closely resemble humans with respect to toxicant uptake, metabolism, and developmental response.

The information level approach integrates risk assessment information from a variety of sources, both model systems and humans, and incorporates steadily improving methods into these sources of information. The recent advances in developmental biology that reveal the conservation of cell signaling pathways and genetic regulatory circuits across species, even phyla, gives a new demonstration to the toxicological principle that chemical impacts in humans can be predicted from animal systems. Further research will clarify the similarities and differences between model animals and humans and will improve the ability to extrapolate risk across species.

How the test results will inform risk assessment will depend in part on the questions asked. The bottom row of Table 8-1A describes the information available to risk assessment from each assay type. Until scientists gain a better understanding of embryonic development and the mechanisms of toxicity, especially the effects of chemicals on the highly conserved signaling pathways, the approach in risk assessment should be to use combined information about predicted chemical activity, bioassays on model animals, and identification of individuals with susceptible genotypes to predict potential risk for developmental defects in humans. For example, knowing that a chemical disrupts the activity of a component of the Hedgehog signaling pathway in a high-throughput cell assay (a level 1 result) has limited value for direct human risk assessment. However, from level 1, it might be useful to know that four structurally related compounds all cause inhibition of a specific kinase involved in the phosphorylation of an intermediate of the Hedgehog pathway but with widely varying potency. If the most potent compound is the one proposed for widespread use and release into the environment, the level 1 information would prioritize testing of that compound for effects in mammals in vivo. Thus, information on molecular and biochemical actions of chemicals at a cellular level, when considered in light of specific risk questions (human use and exposure paths), can be informative to human risk assessment.

The four information levels for toxicity assessment in model systems (Table 8-1A) are presented below in detail.

Model Systems Information Level 1

This level includes molecular, biochemical, and cell-based assays. Assays should be designed for a high throughput of chemicals, perhaps 10^5-10^6 assays per year, to provide basic information on the types of chemicals that disrupt signaling pathways and activate molecular-stress pathways and on the conversion of chemicals by DMEs. Such information will inform hazard identification and the evaluation of modes of action in risk assessment. For hazard identification, such assays will provide

- structure-activity information,
- relative potency information for chemicals evaluated in the same assays,
- information about the activity of chemical mixtures, and
- some quantitative information across assay end points for estimating relative potency across chemical classes.

When coupled with estimates of actual or impending human exposure, such assay information would be useful in prioritizing chemicals for in vivo assessments at information levels 2 and 3.

With the recent advances in molecular techniques, results from 10^5-10^6 tests of chemicals or chemical mixtures are feasible within 1 year. Assays already exist for estrogen receptors (EPA 1998g), and they could be readily modified to include other nuclear-hormone receptors, including the orphan members. Related assays could be devised for the transmembrane receptors, the various transduction intermediates, and the other 16 signaling pathways and their genetic regulatory proteins. Several pharmaceutical companies have active programs to evaluate chemicals in relationship to retinoic acid receptor binding and their pharmaceutical-versus-developmental-toxicity activities. Biochemical assays could involve purified human proteins expressed in bacteria or insect cells. Cell assays could involve mammalian cell lines or yeast into which human receptors (e.g., various receptor tyrosine kinases (RTK)) have been introduced. Molecular-stress pathways, cell-cycle checkpoint pathways, and the apoptosis pathway have already been shown to be especially relevant for environmental toxicants because of their roles in the cell's response to chemically induced deoxyribonucleic acid (DNA) damage, impaired DNA synthesis, spindle damage, and kinetochore malfunction. The genes encoding these pathways could be introduced into yeast or cell lines equipped with reporter genes for easy assessment.

At this high-throughput level, many chemicals should be assayed as substrates or inhibitors of human DMEs, both the oxidases and the converting enzymes, in biochemical assays or in cell lines carrying a variety of human DMEs. Such approaches have begun (for a review, see Crespi and Miller 1999). Thus far, this information has found limited use in risk assessment for several reasons. First, many assays have not utilized consistently stabilized transfected cell lines, and therefore the responses have been variable. (This variability has also been problematic for the EDSP.) Second, gene-induction profiles for specific compounds have shown tremendous variability when inducers are compared across established human cell lines. Thus, although comparisons can be made across compounds within some cell assays, comparisons across cell lines and with animals remain problematic.

Ongoing research efforts in these areas should help clarify and resolve these issues. However, risk assessors need to understand what types of information these assays can provide for assessment. Recent conferences have summarized key information available from the use of human cell- and tissue-based assays (Society of Toxicology (SOT) Workshop on In Vitro Human Tissue Models in Risk Assessment, September 1999). Additional work on such systems would prove especially useful, as such recent conferences attest.

As data are obtained, they would be entered in a widely accessible database (e.g., the recent Science magazine Web site for cell signaling pathways, www.stke.org). Compounds with high activity in such tests could be prioritized for higher levels of testing, especially if human exposure is current or pending. Compounds that do not show effects in these assays would still need testing at other levels of assessment if human exposure or environmental release is likely. At this information level, false positives are preferable to false negatives, and high sensitivity is preferable to low sensitivity (see discussions on how to use such information to strategize test applications by Lave and Omenn 1986).

It is expected that comprehensive gene expression assays will soon become routinely available with DNA microarrays. Some arrays now include 6,000-10,000 DNA sequences in order to detect changes in messenger ribonucleic acid (mRNA) levels (after conversion to complementary (c) DNAs) in cells or tissues exposed to a chemical. Libraries of yeast strains, each carrying one of the 6,000 genes on a plasmid with a reporter gene, will soon be available. These libraries will represent a full-spectrum profile of the effects of chemicals on gene expression. The committee expects that the assays at this level will be better and broader in the near future and emphasizes the need to expeditiously put the results from these assays in the context of temporal patterns, and dose and downstream responses.

Model Systems Information Level 2

At this level, nonmammalian animals should be used to assess the potential for chemicals to affect developmental processes. The animals are small, inexpensive, and fast developing. Information about their development is abundant, and genetic manipulations are easy. They can be genetically optimized to contain various sensitized signaling pathways and molecular-stress pathways and can often be coupled to reporter genes for enhanced observation of effects. Also, the animals could be genetically modified to reduce their differences from humans in various ways, such as their array of drug-metabolizing enzymes. (The committee acknowledges, however, that unrecognized differences between humans and these test animals may exist and may invalidate comparison. For example, humans and test animals may differ in unknown proteins of trans-epithelial transport of the toxicant or in unknown serum proteins that bind the toxicant. Therefore, validation studies would have to be done with a set of toxicants to establish cross-species concordance.) Assays would be designed for a medium throughput of chemicals, perhaps 10^3-10^4 assays per year. Some combinations of chemicals could be tested, and various doses could be examined in some cases to discern low-concentration and threshold effects for specific developmental pathways. The fruit fly and the nematode are currently the most favorable organisms for use. The zebrafish will probably be the most favorable vertebrate for use.

Genetic modifications can include the following:

- Sensitization of animals (e.g., individual signaling pathways are made rate-limiting for some aspect of development, such as eye or wing formation in *Drosophila*, so that a slight increase or decrease in the pathway's function due to a chemical would give an altered phenotype). The signaling pathways would be those used repeatedly in early development and conserved across many phyla, namely, the RTK, transforming growth factor (TGF) ß, Wnt, Notch, Hedgehog, and nuclear-hormone receptor pathways.
- Introduction of various reporter genes to enhance the readout of effects.
- Introduction of human signaling components into the animals to reduce the extrapolation.
- Introduction of human DME genes into the test organism so that, whenever possible, animals are presented with the same metabolized form of chemicals that humans would produce.

General toxicity caused by a chemical can be distinguished from specific effects on development in the animals by evaluating general lethality, growth, and developmental effects versus specific effects on the particular locally sensitized pathway of development.

An argument against the use of these model organisms is that the amount of information relevant to chemical effects on human organogenesis will be small, because the organs of model organisms, such as the fruit fly and nematode, differ substantially from those of humans. However, it is important to note that the choice of model organisms reflects the new insights about conserved processes of development, and emerging opportunities to directly assay for processes that are dependent upon each specific pathway. The signaling pathways and genetic regulatory circuits used in the development of the fruit fly and nematode organs are largely the same as those used in human organ formation but are used in different combinations and express different genes to form species-specific organs. At information level 2, the fruit fly organ can be used for assaying for chemical effects on conserved signaling pathways and gene regulation. In addition, some fruit fly organs are now thought to have deep evolutionary similarities to particular vertebrate organs, including human organs, in their combination of pathways and circuits. Examples of organs and structures include the development of the heart, gut, eyes, appendages, blood vessels and tracheae, somites and body segments, and several structures associated with neurogenesis. Thus, for some organs and structures, the human and fruit fly differences are not great.

Chemicals showing effects on development can be identified for additional testing if human exposure or use is anticipated. Presumably, some of the chemicals active at information level 1 would also be active at level 2. In fact, effects at level 1 should allow predictions of effects in assays at levels 2 and 3. For example, if particular signaling-pathway components are affected in biochemical and cell assays, it should be possible to predict which kinds of organogenesis will be affected in the fruit fly, nematode, and zebrafish. False positives and high sensitivity are still to be preferred over false negatives and low sensitivity for useful information at level 2. All the test information would be entered in a database linking chemical-structure information and toxicity.

As described for information level 1, assays in level 2 will provide

- structure-activity information,
- relative potency information for chemicals evaluated in the same assays,
- information about the activity of chemical mixtures, and
- some quantitative information across assay end points for estimating relative potency across chemical classes.

In addition, these assays will give information on affected organs and organ systems. Impacts on specific cell signaling pathways can be connected to defects in organogenesis and perhaps to functional deficits (e.g., see the behavioral assessments in model organisms discussed in Chapter 7).

It is expected that DNA microarray techniques will have an increasing role in level 1 and 2 assessments. At present, the methods are sensitive and reliable enough to detect manifold differences in mRNA profiles in cells and tissues exposed to different conditions (e.g., growth medium, ploidy, and temperature), however, interpreting the changes remains a problem. In the near future, there will undoubtedly be microarrays representing certain classes of genes and gene products, for example, those for signaling components, by which the changes in mRNA profiles can be related to particular developmental processes altered in the presence of toxicants. Also, as information accumulates, it will be possible to correlate the various patterns of altered gene expression caused by chemicals with unknown mechanisms with those changes known to be caused by toxicants acting via known mechanisms or by identified mutations in developmental processes. In this way, the effects of known and unknown compounds could be compared, as well as the relative strength of particular compounds in eliciting particular modes of action.

Model Systems Information Level 3

Due to expense and time, mammalian tests have a lower throughput of chemicals than described for information levels 1 and 2, perhaps 10^2 per year. (For reference, several hundred rat tests are now done per year in preparation for evaluations of human risk.) In the context of risk assessment (i.e., exposure pathways, human use, and environmental release), the tested chemicals would be those requiring a quantitative assessment of human risk. Chemicals that alter signaling pathways, molecular-stress pathways, or checkpoint pathways in assays at information levels 1 and 2 would be scrutinized at level 3 to ascertain in vivo mammalian effects.

Test animals would likely be the mouse and rat. At present, the mouse has the advantage of much greater available genetic information and ease of genetic manipulation, but the rat genome project will soon make that information available as well. Although it is widely recognized that mice and rats differ from humans in their metabolism of chemicals by DMEs and in their developmental responses to chemicals, these differences have not been well analyzed or characterized. Assays at level 3 will benefit greatly in the near future from research comparing mice and humans with respect to the uptake, conversion, and clearance of chemicals. The experimental means to analyze the metabolic differences are available, and the genetic means are available to reduce or eliminate the differences, thereby improving the accuracy of extrapolation across species. For example, mice could be provided with human transgenes to give a human-like profile of DMEs and other proteins of toxicokinetic importance. In cases in which differences cannot be eliminated, the differences should be well characterized so that bounds can be set on default corrections. Test animals can also be genetically modified to approximate the more sensitive rather than less sensitive members of the human population. Research is focusing on the analysis of human polymorphisms, and many of them could be approximated by genetically modified test animals. Although it can be argued that genetically modified animals might give false positives, the opposite can be argued—namely, that when a chemical has no negative effect in the sensitized test animal, the result would not require so many orders of magnitude of default correction when the extrapolation to humans is made at the time of risk assessment.

The molecular-stress and checkpoint pathways are expected to be activated by a wide range of toxicants and to show toxicant effects. Recently, these pathways for oxidative stress, protein unfolding, and checkpoints (including those involving p21 or p53) have been used for developmental toxicity assessment (Wilson et al. 1999; Wubah et al 1997). Test animals might express a reporter gene in response to a chemical without suffering overt developmental damage. This result could be taken as an indicator of the need to retest the chemical at other doses and times, or in combinations with certain other chemicals, for synergistic detrimental effects. Mouse strains are available with sensitized signaling pathways (see Tables 6-4 and 6-5 on targeted disruption of genes encoding signaling components) in which null alleles are used in heterozygous or homozygous states and in various combinations. Animals could also be made to possess a variety of reporter genes by which the responses through specific signaling pathways could be assessed easily. Similar programs are already under way with sensitized mice for carcinogen assays (Eastin et al. 1998). Until recently, the lack of pathways with clearly identified developmental relevance has limited similar programs for developmental toxicity assessment.

Level 3 assessments provide the highest level of information routinely available to a regulatory agency for evaluation of risk—that is, information requiring the least default correction for estimating human risk, although some default corrections are still likely to be needed. The new information on development and genomics implies that different genetically modified mammals would be optimal for testing different chemicals. The choices could be guided by the results of levels 1 and 2 tests. The following kinds of information available from mammalian systems would include

- structure-activity information, with activity carried to the level of effects on mammalian organogenesis,
- relative in vivo potency information,
- some information about activity of chemical mixtures,
- mechanistic information from sensitized animals, and
- quantitative information on shape of dose-response relationships in in vivo organ systems.

Model Systems Information Level 4

Perhaps only 10 chemicals per year can be studied at this level. Those studied should be those for which further research would give important information about (1) chemical effects on development, (2) basic mechanisms of developmental toxicity, and (3) significant clues for human risk assessment (e.g., analyzing differences between test animals and humans). Such chemicals might be prototype members of families of chemicals for which other derivatives would deserve testing at lower levels for relative toxicity, where mechanisms of action can be elucidated or where the effects are difficult to score, as in behavioral and other functional assessments. These chemicals might also represent compounds for which widespread environmental exposure is anticipated or is occurring. Presumably, this information would feed back into improvements of the level 1, 2, and 3 assays and would provide valuable input for the chemical databases. The animals used in these studies would probably be mammals, especially genetically optimized rodents.

The committee recognizes that the action of some toxicants might fall outside the realm of current understanding of development (i.e., of areas emphasized in this report). Studies of such mechanisms would necessarily fall into level 4 assays (i.e., basic research). Two approaches of note are (1) DNA microarray surveys of changes in gene expression of cells or test organisms treated with potential toxicants, which then require the researcher to interpret the changes and substantiate the interpretation; and (2) phage display methods to screen and isolate cellular proteins that bind particular toxicants, which then require the researcher to identify the role of the protein in the cell and its relevance to a toxicity mechanism.

Assessment of Toxicity, Susceptibility, and Chemical Exposure in Human Populations

The committee believes that the quality and the accessibility of human epidemiological information need re-examination, in light of its present and increasing relevance for developmental toxicity risk assessment. The committee considered ways to link data from human surveillance studies with data from in vitro studies and in vivo animal studies and discussed how new biomarkers of exposure and susceptibility in humans could be linked more effectively with new biomarkers of effect, in order to improve the assessment of human risk for developmental toxicity. The committee defined four informational databases as domains of information about humans. These databases were not referred to as "levels," because they provide different kinds of information and cannot be ranked, as the model systems can, by remoteness or immediacy of human relevance. All the databases contain information of use to risk assessors.

Database of Human Developmental Outcomes

This database is the domain of information from epidemiology and surveillance. The quality of various case reports of birth defects and possible toxicant exposure varies widely. Many are incomplete and of unknown accuracy. However, most known human developmental toxicants were first identified by case reports, including thalidomide, diphenylhydantoin, diethylstilbestrol (DES), and valproic acid. At the outset, case report information is of unknown value to risk assessment and cannot be used without extensive follow-up, a situation similar to information from the level 1 assays of the model systems. When correlations are strong enough or the developmental effects are incontrovertible, the prospective toxicant should be brought forward for further investigation, such as more rigorous epidemiological characterization of the correlation of exposure and birth defect and more incisive animal testing to ascertain dose-response relationships.

In addition to case reports, there are both active and passive birth defects surveillance systems in place, both nationally and internationally (NBDPN 2000). They vary in quality and completeness. A number of established non-governmental databases are used to monitor drug exposures and developmental defects, and these are frequently used in pharmacoepidemiology studies (Strom 1994).

A recent study entitled "Healthy from the Start: Why America Needs a Better System to Track and Understand Birth Defects and the Environment" and conducted by the Environmental Health Commission (Goldman et al. 1999) evaluated the quality of state tracking systems for birth defects. The study analyzed existing data from those systems and looked at the connection between environmental agents and birth defects. The authors concluded that the majority of states either do not have a tracking system or have one that is inadequate. They concluded that the data are inadequate to draw conclusions about the role of environmental exposures in causing birth defects and recommended that a national effort for tracking birth defects and a national approach for monitoring environmental exposures be established.

Those state tracking systems currently in place are usually either active or passive systems. In an active surveillance system, such as that administered by the Centers for Disease Control and Prevention, trained personnel actively seek data from sources such as vital records and hospital reports. Passive case identification involves relying on patients and health care providers to voluntarily report exposures (usually drug exposures) and outcomes (Strom 1994); follow-up is minimal or nonexistent. Passive, voluntary reporting of systems have several limitations, including under-reporting of adverse events, incomplete information on cases, and the retrospective nature of most adverse event reports (i.e., the reports are made after an adverse pregnancy outcome has occurred), and the true denominator for exposed pregnancies is not known.

There are a number of pregnancy registries currently being conducted (Weiss et al. 1999). For example, the North American Registry for Epilepsy and Pregnancy is a surveillance program to monitor pregnancy outcomes in women taking antiepileptic drugs (NAREP 1998). Some prospective post-surveillance follow-up studies have been directed by pharmaceutical companies evaluating post-marketing impacts of drugs. The committee suggests that the data from pharmaceutical company studies be made available and that the existing efforts to track human developmental outcomes be better characterized and recognized. Improvements could include making various information from epidemiology databases accessible via the Internet.

There are databases for which known genetically based malformations are linked with databases for specific clinically described syndromes. One example of such a database is the On Line Mendelian Inheritance in Man (OMIM) database (see Appendix B for Internet address). As of January 2000, the results from OMIM for genetically based malformations and pediatric diseases (out of 11,080 entries) are 10,362 (93%) with autosomal transmission; 620 (5%) with X-linked transmission; 38 (0.3%) with Y-linked transmission; and 60 (0.5%) with mitochondrial transmission. Information from levels 1 and 2 of the model systems (e.g., the numerous targeted gene disruptions in the mouse) has already pointed to molecular and cellular processes that might be affected in numerous human clinical cases with similar organ dysmorphogenesis. Linking human databases on clinical syndromes with information from model system levels 1 and 2 would be especially valuable, and earlier examples in this report reveal the importance of establishing such linked databases.

Database of the Human Genome and Genomic Polymorphisms

The primary focus of this database is to provide information about the frequency and distribution of genetic polymorphisms in humans. Such characterization would allow developmental toxicologists and risk assessors to include human variation in their definition of a human response to a developmental toxicant. One approach to address this would be to study offspring with developmental defects for alterations in the genes of interest (e.g., DMEs and morphogenetic pathways). Such populations (infants with malformations and surgical candidates for correction of malformations) and tissues from collections around the world are available, and have been well defined for the anomalies of interest. Determining the frequency, distribution, and correlation for various human polymorphisms of consequence for susceptibility to toxicants (e.g., polymorphisms of the DMEs and of developmental targets, such as components of signaling pathways and genetic regulatory circuits) with observed developmental defects would be especially useful in defining human responses. The new information on human polymorphisms is seen as providing entrance to the realm of gene-environment causes of developmental defects. The technologies for approaching this are now available for the first time. Such approaches can be useful for determining the correlation between genetic variants in human populations with those observed in model systems for developmental defects. There are probably large, undefined differences and understanding these differences will be invaluable.

Information from the Human Genome Project and the National Cancer Institute (NCI) will be relevant for understanding human polymorphisms of disease-susceptibility traits and cancer-susceptibility traits. Most of these traits will probably be complex and have several genes and modifiers. Human populations are expected to contain a large number of sequence polymorphisms (e.g., single nucleotide polymorphisms occur at least at 1 in 500 bases, recent estimates being as high as 1 in 25 bases); therefore, population samples cannot be selected only on the basis of a shared sequence difference. Indeed, there are too many, and some polymorphisms probably have no effect on phenotype (so-called synonymous and neutral mutations). Instead, the population sample must be selected on the basis of a shared sequence difference known to result in the altered function of a gene product of suspected relevance to toxicant exposure and susceptibility. Information from model systems (e.g., mice) will be useful to establish altered function and suspected susceptibility.

Toxicant susceptibility genes are not well identified, but the current favorites for attention are those encoding products involved in the toxicokinetic aspects of exogenous chemicals (uptake, distribution, metabolic conversion, and clearance), particularly the DMEs. There are at least one thousand of these gene products, although several dozen probably metabolize 90% of chemicals. Individuals and ethnic groups are already known to have substantial differences in these genes, and a few such polymorphisms are associated with altered developmental toxicity (e.g., a polymorphism in the epoxide hydrolase gene that might result in fetal hydantoin syndrome in susceptible individuals; see Chapter 5 for details). However, most DME polymorphisms have not yet been associated with increased (or decreased) susceptibility to a chemical.

Polymorphisms of genes involved in toxicodynamics need to be investigated. As evident from the extensive discussions of conserved cell signaling pathways and genetic regulatory circuits, the committee suggests that polymorphisms in components of these pathways and circuits be tracked for the following reasons:

- The pathways and circuits are used widely in embryonic development.
- Polymorphisms of the genes encoding some components correlate with particular kinds of cancers (e.g., Patched (a component of the Hedgehog pathway) heterozygosity and basal-cell carcinoma; adenomatous polyposis coli (a component of the Wnt pathway) loss and colon carcinoma).
- A few correlates already exist, such as higher frequency of cleft palate in humans who smoke cigarettes and have TGF variants. Identification of *Hox A1* polymorphisms in autistic populations is also progressing, as described in Chapter 4.

Information from level 2 model-system studies and from basic developmental biology on model organisms will become ever more important for human evaluations because many aspects of embryonic development are conserved across phyla. It would be useful for epidemiologists interested in studying developmental defects to interact with their counterparts involved in NCI projects for profiling cancer-susceptibility polymorphisms (several of which concern signaling components). Some information from the epidemiological analysis of polymorphisms might be directly relevant to understanding birth defects that have mainly a genetic rather than a genotype-environment basis.

Information from the database on the human genome and human polymorphisms would benefit risk assessment by providing information on human diversity in relation to sensitivity to potential toxicants. This information would be used in level 3 of the model systems, because test animals can be made more similar to humans in their DMEs and signaling-pathway components, in order to improve the extrapolation from test animals to humans. The rat and mouse genome projects are expected to reach completion not long after the time of completion of the Human Genome Project, and useful information about gene identity, location, and mutants should flow both ways between researchers focusing on humans and those focusing on rodents.

Database of Human Biomarkers

This domain would provide the best information on human exposure to chemicals and on human susceptibility to developmental effects from chemical exposure. Biomarkers of exposure indicate the actual level of a chemical in the individual (e.g., lead concentrations in blood or dentine in children; organophosphate metabolites in urine). Biomarkers of effect allow researchers to examine dose-response relationships at environmentally relevant exposures in humans, and biomarkers of susceptibility allow researchers to identify sensitive subpopulations of humans. Used in combination, such biomarkers are essential for improving human risk assessment. There remains a need to improve biomarkers based on incisive new information about toxicokinetics and toxicodynamics. Thus, a linked database on the Internet for biomarkers of exposure, effect, and susceptibility in humans should be developed.

Advances in DNA microarray technology discussed in information level 1 of the model systems will help immensely in the development of human biomarkers. Eventually, all the gene expression profiles of all organs and embryonic parts at different developmental stages will be catalogued; this will provide the control condition for detecting chemical-induced departures in gene expression. The potential to monitor thousands of gene expression changes simultaneously in regard to dose-response relationships and temporal patterns will provide a critical link of exposure biomarkers with early effect biomarkers. Applications in human birth defects research could be immense. However, the relevant developing databases must be linked so that the numerous changes are connected to functional effects in a developmental framework—that is, a framework organized around the temporal and spatial changes of the embryo and fetus.

Given the sensitivity of reverse transcription polymerase chain reaction (RT-PCR) techniques, small maternal and embryonic or fetal samples could be simultaneously monitored for gene-expression changes. The gene expression changes could serve as biomarkers of effect, and responses in utero could be compared with maternal responses to evaluate differential sensitivity. If successful, the amount of information in such databases will challenge the organizational abilities of scientists. When temporal changes in developmental patterns of gene expression are combined with changes at different doses for thousands of genes, the data set grows to immense size. Obviously, the storage of data in a retrievable form, as well as analyzing such data, pose serious issues in bioinformatics. Current efforts are ongoing in cancer research to store and retrieve vast amounts of data; however, there is no equivalent example for birth defects and developmental toxicology.

Biomarkers of susceptibility would include polymorphism sequence data for which susceptibility has been correlated with protein function, as in the case of decreased DME activity. The use of data on DME genetic polymorphisms has lagged in carcinogen risk assessment, and in developmental toxicity risk assessment, their use is even more delayed. The committee believes that DME polymorphisms can serve as excellent biomarkers of susceptibility and encourages development of programs to ensure these applications. Biomarkers of toxicodynamic (developmental) susceptibility are less advanced but might include polymorphisms of genes encoding components of signaling pathways, genetic regulatory circuits, or molecular-stress pathways. Biomarkers of effect might include indicators of early activation of molecular-stress pathways or signaling pathway inhibition (e.g., due to exposure of a person to environmental chemicals and pharmaceuticals).

Biomarkers would, in principle, provide risk assessors with better information than that available from the human developmental outcome and human genome database to link human exposure, developmental effect, and heritable susceptibility. To conclude, the benefits of biomarkers for risk assessment include (1) identification of susceptible populations (addressing intraspecies variability); (2) improved dose-response information where subtle changes of biomarkers of effect can be linked with biomarkers of exposure (addressing issues of extrapolation from high to low doses); and (3) improved linkage between biological effects in humans and mechanisms of toxic action as developed from human and animal studies.

Database of Human Gene-Environment Interactions

Similar to level 4 of the model systems, this database would be a domain of research inquiry in a few cases where detailed epidemiological information might yield widespread value, and where assessment of gene-environment interactions in birth defects is feasible given the available resources. Investigative epidemiological inquiries, such as those on endocrine disruptors, methylmercury, lead, and organophosphates, might provide adequately robust data for linkage with adverse birth outcomes. Several of these databases have provided information supporting interesting proposals about genetic polymorphisms (e.g., on lead and organophosphates) or exposure conditions (e.g., on lead and mercury). In contrast to pharmaceuticals, many environmental agents interact with a wide range of targets and with different targets at different doses. Yet, the identification of critical early biological effects that dominate or initiate subsequent disease states is essential for risk assessment. Thus, some of these environmental agents are candidates for an exhaustive research analysis. The multifactorial view of developmental defects accepts the possibility of complex and subtle combinations of circumstances leading to developmental defects, and except for a few special cases, some agents undoubtedly need to be illuminated by research, and databases will be needed to store and cull the large amount of information.

The Importance of Linking Databases

The committee's multidisciplinary, multilevel, interactive approach to improving risk assessment assumes that the recent research advances in development and genomics have the potential, not yet realized, to improve cross-species extrapolations and cross-assay extrapolations and to ascertain the developmental targets of toxicants.

Although more relevant information will become available for human risk assessment, a significant challenge facing risk assessors who want to use this information is the informatics problem. Most of the information relevant for human risk assessment will exist in the separate databases that the committee has described and will be organized according to discipline-focused applications. For example, the field of medical genetics has databases containing information on birth defect syndromes, but lacks databases containing epidemiological information on chemical impacts on development. It was the consensus opinion of this committee that efforts are now needed to link these diverse databases.

In this section, the committee describes the need for integrated databases that link information from model systems and human populations. Relevant databases for such purposes would include the following:

- Chemical databases with metabolic pathways, structure-activity correlates, and bioassay results.
- Genome databases of humans, mouse, rat, zebrafish, *Drosophila*, *C. elegans*, and yeast.
- Developmental databases containing information about components of developmental processes and their functions and interactions in model organisms (e.g., in *Drosophila*, *C. elegans*, zebrafish, frog, and chick).
- Functional genomics databases on expressed genes, including their time and place of expression in the embryo, and on the function of the encoded proteins (specific function or categorization of function by motif).
- Databases on human polymorphisms and disease associations and on human and mouse mutants, including all the targeted disruption mutants of mice and their phenotypes.
- Databases recording DNA microarray results of the simultaneous changes of expression of thousands of genes (currently as many as 10,000 simultaneously) following the exposure of cells, tissues, or organisms to various conditions.

For toxicologists, databases must be searchable by chemical names, chemical classes, structural characteristics, and reactivities. Approaches already in place for the structure-activity relationship databases could be used to begin identifying relevant information. Current initiatives are under way to reanalyze archived and frozen tissues from chemical toxicity bioassays for gene expression changes. Any changes are then evaluated specifically for the originally tested chemical and for structurally and functionally related compounds. Such initiatives should provide the types of linked databases the committee believes should be developed. This information should be linked with developmental toxicity bioassay databases as well as with general toxicity information.

Several databases are available that contain developmental toxicity and general toxicity information. For example, general toxicity information as well as some developmental toxicity information on selected chemicals are contained in the Agency for Toxic Substances and Disease Registry (ATSDR) Toxicological Profiles, the International Agency for Research (IARC) Monographs on the Evaluation of Carcinogenic Risks to Humans, the Integrated Risk Information System (IRIS), and the Registry of Toxic Effects of Chemical Substances (RTECS). There are additional specific sources for evaluating chemicals for developmental toxicity. These sources include the California EPA Hazard Identification Documents, the Evaluative Process Documents on Lithium and Boric Acid (Moore et al. 1995, 1997), REPROTOX, REPROTEXT, the Teratogen Information Service, the Developmental and Reproductive Toxicological Database, and the National Toxicology Program Teratology Studies. The newly established National Institute of Environmental Health Sciences (NIEHS) Center for the Evaluation of Risks to Human Reproduction will also provide detailed evaluations on developmental effects of chemicals (no completed reports are yet available). Some of the above-mentioned databases and reports contain detailed toxicity evaluations of chemicals; others provide less-detailed summaries or contain only bibliographic information. Some databases also include information on human exposure. It is important to connect these types of data in a way that is ultimately useful for human risk assessments.

Genome databases are available on humans, mice, rats, zebrafish, *Drosophilia*, *C. elegans*, and yeast with sequence information on open reading frames (ORFs), introns and exons, *cis*-regulatory sequences, and relatedness to other genes. Of particular use for developmental research would be the ability to search and identify all genes with known developmental relevance, and to link that search with gene expression and temporal and spatial developmental information by organism, as well as organ and tissue development across species. Efforts by the National Institutes of Health through the NCI Cancer Genome Anatomy Project and NIEHS through the Environmental Genome Project are directed toward identifying genes of interest for cancer and toxicology. A similar effort for genes of developmental and toxicological relevance could be initiated and linked. In both cases, profiles of gene expression changes can be determined for tissues with specific developmental defects and for affected tissues after toxicant exposure.

These gene expression profiles can be linked with specific tissue changes using microdissection techniques. Again, emphasis on developmental characteristics is important. Such characteristics include temporal aspects of gene expression and multiple organisms, organs, and tissues of interest. Functional genomics databases on expressed genes are particularly relevant for developmental toxicologists. Expressed sequence tags (ESTs), ORFs, and the time and place of expression of each gene in the embryo and on the function of encoded proteins (specific function or categorization by motifs) would be essential information. For example, signal transduction pathways and the interactions of pathways are in the process of being summarized on the Web site www.stke.org. Journal articles, as well as methods protocols, will be posted (announced in *Science*, April 1999). Likewise, databases on human polymorphisms and disease associations, human and mouse mutants, including all the targeted disruption mutants and phenotypes in mice, would be important. Online access to databases listing all known human genetic syndromes relevant for development would also be essential.

Within a decade, most genes encoding components of signaling pathways and genetic regulatory circuits important in development will probably be identified in humans and mice (the extensive synteny among vertebrates will be valuable here), and their times and places of expression will be known. Many human polymorphisms will be identified and correlated with heritable diseases. At levels 1 and 2 of the model system toxicity assays, this database information will be useful for choosing proteins to use in simple biochemical or cell assays or to modify in test animals, such as *Drosophila*. It is safe to say that almost any human gene and, hence, its encoded protein, can be put into such an animal assay. The question will be which proteins are most relevant to the identification of developmental toxicants in humans. If 100,000 assays can be done per year at level 1 and 10,000 at level 2, the results would have to be preserved in an immense database. Over the same period, all genes for proteins of the major toxicokinetic pathways of chemical uptake, distribution, metabolic conversion, and elimination in humans, mice, and rats will probably be identified, as will polymorphisms of these genes. This information will be preserved in databases, and should be widely available. Catalogs of phenotypes of mouse null mutants for individual genes in the heterozygous and homozygous states, and in combination with other gene disruption, will be useful for comparison with phenotypes of human birth defects in order to gain inferences about what is affected in human development. The level 2 and 3 model system assays for developmental defects will draw from these genomic databases and contribute to them. Several of the databases of epidemiological characterizations will benefit as well.

Chemical databases might be more difficult to organize than genome databases. Whereas there are about 140,000 genes in humans, the universe of possible chemicals is unlimited (the number of possible human allele combinations is also unlimited). New compounds can always be synthesized, and their effects on developmental mechanisms are rarely predictable, at least so far. Also, a new generation of specific and complex chemicals will be introduced in the near future, because of new strategies of rational drug design based on knowledge of protein three-dimensional structure and chemical mechanisms, and because of high-throughput function-based screens of huge combinatorial chemical libraries. Still, mechanistic evaluations of the biological effects of chemicals should serve as an organizing principle for grouping compounds in a database and predicting their risk.

In the committee's proposed hierarchy of four information levels for model animal systems, each higher level provides increasingly complex information concerning higher biological complexity and responses to environmental chemicals. Systems of lower complexity might be used to organize chemicals, for example, to reveal those chemicals that bind to the same protein (e.g., a nuclear-hormone receptor) or act in the same signaling pathway. This information would be useful for predicting effects at the next level, for example, on a particular kind of organogenesis in which a particular signaling pathway is used.

SUMMARY

The committee has developed a multilevel, multidisciplinary, interactive approach for improving risk assessment for developmental toxicity. Model animal systems and human epidemiological studies are shown to be valuable sources of information for risk assessment, and it is emphasized that the multilevel approach is not a tiered approach. To meet the goals of this approach, the committee has described what information is available from model systems of differing complexity, from in vitro assays to whole animals, for the assessment of toxicity and mechanism of action of chemicals. The committee has also described databases and database needs for assessing chemical exposure, susceptibility, and developmental effects in human populations and the continuing value of human epidemiological data for risk assessment. For each database, the committee has identified the type of information provided and how that information answers risk assessment questions. Examples are given of the anticipated interactive aspect of the approach, whereby new data and methods can be incorporated into the risk assessment process. Finally, the committee has described integrated approaches essential for the linkage of numerous relevant databases across chemicals, times of development, descriptions of toxic effects, and applications.

9

Conclusions and Recommendations

The general conclusions regarding the four charges to the committee are given in the first part of this chapter. In response to the fourth charge, specific recommendations are given in the second part. Several of these recommendations concern a multilevel, multidisciplinary approach for obtaining data relevant to risk assessment for developmental toxicity, as described in Chapter 8.

CONCLUSIONS IN RELATION TO THE CHARGE

Charge 1: Evaluate the evidence supporting hypothesized mechanisms of developmental toxicity.

Issues in developmental toxicology have been clarified incisively in the past 30 years, and many experimental advances have been made. Still, there are only a few compounds for which developmental toxicity is partially explained and no compound for which it is fully explained in an inclusive hypothesis supported by strong evidence. Several reasons for this limited understanding should be cited:

- Development is complicated. Only recently have developmental processes and molecular components been elucidated. Many steps are likely to be involved between the toxicant's initial interaction and the ultimate developmental defect.
- The etiology of developmental toxicity is complex. Many developmental defects of individuals might be the outcome of a multifactorial impact with overlap of genetic variation, exposure variation, and variation in other factors, such as nutrition and disease. Genotype-environment interactions, especially those involving several genes and environmental factors, are difficult to study, and have usually been avoided in basic laboratory research. The complexity of develop-

ment is manifest by both temporal- and tissue-specific sensitivities; thus, assessing a toxicant's potential effects for development requires a dynamic and multilevel assessment strategy.

- Environmental toxicants represent a broad spectrum of agents, probably working by a variety of mechanisms. Some toxicants probably have one or a few targets in the conceptus (the embryo or fetus, plus the embryo-derived extraembryonic tissues). Others probably have numerous targets ("broad specificity") in the mother and conceptus, and others probably affect the mother, whose altered health secondarily affects the conceptus.
- Without a thorough understanding of basic mechanisms of development and knowledge about variability in responses across species to toxicants, insights from animal studies have largely been only of assumed validity for human mechanisms.

The analysis of mechanisms of toxicity requires advanced interdisciplinary information and approaches of the kind that have only recently become available.

In considering hypothesized mechanisms, the committee discussed the different scopes and levels of understanding implied by the term "mechanism," as used by different researchers in biochemistry, molecular biology, genetics, developmental biology, toxicology, and epidemiology. If the emphasis on toxicant action is exclusively molecular, some members felt that the mechanism misses the scope of potential linked impacts of a toxicant on overall development and morphogenesis. Additionally, most felt that a mechanism lacking molecular detail is inadequate for explaining the action of toxicants. Realizing this complexity, the U.S. Environmental Protection Agency (EPA 1996b) and International Programme on Chemical Safety (IPCS Workshop on Developing a Conceptual Framework for Cancer Risk Assessment, 16-18 February 1999, Lyon, France) have defined chemical "modes of action" in addition to "mechanisms of action." In these definitions, "mechanism of toxicant action" is taken to refer to a detailed understanding of the overall toxic response. In contrast, "mode of action" usually refers to a more limited description of the overall process of toxicity that focuses on defining possible cascades of biological events that can occur following exposure to a toxic agent. To preserve the full range of causes and effects relevant to risk assessment of human developmental toxicity, the committee sought to designate "levels of information" obtainable from various model systems (including in vitro assays and mammalian and nonmammalian assays) to illuminate mechanisms of action (see Chapter 8). Hypotheses about toxicant action in humans, based on the information from animal models, can then be strengthened or dismissed by using information obtained from various types of human data. Chapter 8 provides suggestions on how these different types of data can specifically improve our ability to predict potential developmental toxicity in humans.

The committee believes that it is impossible to provide the most scientifically defensible risk assessments without understanding mechanisms of action.

The committee generally agreed that a complete description of the mechanism of action of a developmental toxicant should include the following types of mechanistic information:

- The toxicant's kinetics and means of absorption, distribution, metabolism, and excretion throughout the mother and conceptus.
- The toxicant's interactions (or those of a metabolite(s) derived from it) with specific molecular components of cellular or developmental processes in the conceptus or with maternal or extraembryonic components of processes supporting development.
- The consequences of those interactions for the function of components in a cellular or developmental process.
- The consequences of an altered process for the developmental outcome, namely, the generation of a defect, functional changes, or altered growth and development.

The committee acknowledges that a complete explanation of mechanism of action is not currently available for any chemical and that having even partial mechanistic information of the kind described above can improve the ability to predict adverse human developmental outcomes.

Toxicokinetics describes the steps of toxicant entry and absorption, distribution, metabolism, and excretion throughout an organism or, in this case, throughout mother and conceptus. Toxicodynamics, in the context of this report, describes the steps of the toxicant's effects and interaction with the developmental processes. Both are important. Toxicokinetics explains whether, when, and how much of a potential toxicant reaches the embryo or fetus. The understanding of the toxicokinetic steps of detoxification or metabolic potentiation of a chemical holds great promise for safe drug design and preclusion of toxicant effects in humans from environmental agents. Furthermore, human individual differences in susceptibility to toxicants might in large measure result from differences in toxicant uptake and metabolism, and some of the problems of extrapolating toxicity test results from animals to humans can certainly be attributed to differences of laboratory animals and humans in the metabolism of chemicals.

In toxicokinetics, researchers have identified routes and rates of exposure of the conceptus to certain toxicants, and the recent information on drug-metabolizing enzymes (DMEs; the numerous P450 heme oxidases and conjugating enzymes) is very significant. Researchers have been able to verify the presence of parent compounds and metabolites in the mother and the conceptus during development. They have successfully explained some species differences in toxicity responses based on metabolism differences and have explained some human variations in drug responsiveness. Nevertheless, such knowledge about critical metabolites and their reactivity with specific target tissues is lacking for most agents, and much remains to be done in this promising area.

In toxicodynamics, the mechanistic picture of toxicity is less complete. This limitation has been inherent to the field, because so little has been known until recently about the identity and activity of specific molecular components of the developmental processes or about the roles of the processes in the development of the embryo. Hence, little could be said about the developmental consequences of a toxicant's reduction or exaggeration of a component's activity. In the absence of such information, hypothesized mechanisms of toxicant action (and evidence for these mechanisms) have had limited ability to ascribe developmental defects to failures of specific components and processes. Furthermore, species differences in developmental components have been poorly discerned, as has human variation in these components.

In a few cases, toxicodynamic hypotheses of mechanism emphasize molecular components and activities. Multiple retinoic acid (RA) receptors have been identified for retinoids in animal models. The molecular function of these receptors in regulating gene expression has been ascribed, and the altered time and place of gene expression has been detected in the presence of the toxicant. The availability of multiple structurally related analogs of RA has helped these investigations. Diethylstilbestrol is also known to bind to other nuclear hormone receptors and alter gene expression, and cyclopamine (a plant alkaloid) is known to bind to signaling components of the Hedgehog signaling pathway and alter inductive responses. Such information, coming from recent molecular studies, has greatly furthered the understanding of the toxicity of those agents. Still lacking, however, is full understanding of the developmental processes affected by this altered gene expression or altered signaling and, hence, the generation of the developmental defect.

Of relevance to the committee's later proposals, these three examples of advanced toxicodynamic hypotheses concern signaling proteins and transcriptional regulators, the kinds of molecular components the committee recommends for greater attention in future analysis and testing of toxicants. In the near future, developmental toxicology will likely provide more comprehensive explanations of toxicity, but at present, mechanistic information is available for only a small number of toxicants and these have had limited application for risk assessment.

Charge 2: Evaluate the state of the science on testing for mechanisms of developmental effects.

The state of the science has improved greatly in the past decade, indeed even since this committee was first formed. Relevant advances have occurred in developmental biology and genomics (gene sequencing and gene identification), built upon advances in genetics, cell biology, molecular biology, and biochemistry. Developmental processes have been illuminated for the first time in a number of animals at the level of identification of molecular components and their activities, especially of the signaling pathways and genetic regulatory circuits of these processes. The same molecular components are used repeatedly at different times and places in an animal's development, and these same components are used across widely different animal phyla. Species differences of development seem to be largely differences in the combinations and sequences of use of conserved molecular components and differences in the final target genes of the conserved genetic regulatory circuits. The committee believes that the newly accessed level of molecular components and processes of development is the level that will provide incisive understanding of mechanisms of toxicity and improved predictability of toxicant effects. Developmental processes are increasingly understood in terms of the activities and ordered interactions of molecular components. Because of the recent advances in developmental biology and genomics, the committee is optimistic about the ability to improve testing procedures and the interpretation of data in the future.

Recent and ongoing studies of mammalian development have benefitted greatly from the study of nonmammalian model organisms that are genetically tractable and suited for rapid systematic analysis, such as *Drosophila* and *Caenorhabditis elegans*. Although not anticipated at the outset, the information from these model organisms proved to be extensively transferable to mammals, because molecular components and many developmental processes are deeply conserved across animal phyla. In particular, development in diverse animals, including mammals, depends on cell-cell signaling at all stages before cytodifferentiation, and this signaling involves repeated use of the same 10 intercellular signaling pathways. (Seven additional conserved pathways are used in the functions of cytodifferentiated cells.) Genetic regulatory circuits, involving certain transcription factors, are also conserved from *Drosophila* to mammals, as are components of most basic cellular processes, such as the cell cycle, secretion, and motility. The success in using nonmammalian organisms to illuminate mammalian development suggests that the same organisms could be useful in illuminating toxicodynamic mechanisms in mammals and useful in testing certain kinds of toxicants. Based on the past decade of progress in cellular and developmental biology research, it seems likely that this information will be relevant to mammals, including humans.

Charge 3: Evaluate how that information can be used to improve qualitative and quantitative risk assessment for developmental effects.

Throughout its deliberations, the committee kept in mind that the decisions of risk assessors about a chemical's potential for developmental toxicity are ideally derived from mechanism-based, quantitative data on test animals for which the validity of extrapolation to humans is known. Unfortunately, such data are rarely available. As currently designed, rodent tests for developmental toxicity are limited in their capacity to provide both qualitative and quantitative mechanistic information for human health risk assessment. They are costly in time and resources, and therefore only a small fraction of the more than 80,000 chemicals in commercial use (and the much larger number, about 6 million, of natural products) have been tested. Chemicals are usually assessed for effects on growth, morphology, and viability of the newborn rodent but not for functional (e.g., behavioral), molecular (e.g., toxicant metabolism or transcriptional changes), or cellular (e.g., mitosis defects and apoptosis) effects. Because the validity of extrapolation from animal results to humans is often assumed, as is the relevance of some of the animal exposure conditions, risk assessment often includes large default corrections in the extrapolation to "safe" human exposure concentrations.

To improve qualitative risk assessment, a better understanding of the mechanisms of toxicity of a reference set of compounds would provide predictions for related, but less well-tested, compounds. Toxicity information on a wider range of compounds would also help qualitatively. Results from current developmental toxicity assessments reveal both qualitative and quantitative differences in animal responses to toxicants. A complete understanding of these species differences is not known, particularly, the proportion of these differences that is due to toxicokinetic versus toxicodynamic differences. Thus, to improve quantitative risk assessment, a better understanding of the differences and similarities between test animals and humans with regard to toxicokinetics and toxicodynamics would allow better extrapolations to be made. Use of default corrections could be limited or be made on the basis of knowledge rather than assumptions. An understanding of genetic differences (polymorphisms) among test animals, and a deliberate effort to control the differences, would reduce and rationalize the wide range of responses of test animals to a toxicant. An understanding of human polymorphisms of molecular components of toxicokinetic and toxicodynamic importance would allow better estimation of safe exposures of individuals over the whole range of human susceptibility to a toxicant. Finally, an understanding of toxicity mechanisms, including toxicant effects on cell division and cell function, as reflected in molecular-stress and cell-cycle checkpoint pathways, and an understanding of the polymorphisms of toxicokinetic and toxicodynamic components, would allow better estimation of low-dose exposure effects, as extrapolated from high-dose results. The committee believes that the possibilities for improvements in data for risk assessment are great.

To date, little of the new information has been brought to bear on the tests and interpretations needed for qualitative and quantitative risk assessment, simply because the problems are difficult and it takes time and resources to do so. The committee believes the new information can be incorporated extensively in the next decade.

Because molecular components and processes of development are best understood in genetically modifiable model organisms, such as *Drosophila*, *C. elegans*, the zebrafish, and the mouse, and because the conservation of components is so pervasive that it extends to humans, the committee recommends that these organisms be used more effectively for analyzing mechanisms of developmental toxicity at the molecular level and for assaying the developmental toxicity of the numerous never-tested chemicals and chemical combinations. These organisms can be used inexpensively and rapidly. Through straightforward genetic manipulation, they can be sensitized to toxicant effects and provided with reporter genes so that the impacts of chemicals on development can be easily scored. The organisms can be analyzed to define their toxicokinetic and toxicodynamic differences from humans, hence improving extrapolation, and they can be genetically modified to reduce some of the differences. Finally, the suspected genotype-environment source of many human developmental defects can initially be explored in model organisms for which the genotype can be controlled.

At the same time, human individual differences in drug-metabolism components and developmental components are being identified and quantified. As polymorphisms are related to individual susceptibility to toxicants, the difficult domain of human genotype-environment interactions will be entered. The knowledge of human variation will improve quantitative risk assessment.

Charge 4: Develop recommendations for future research in developmental biology and developmental toxicology; focus on those areas most likely to assist in risk assessment for developmental effects.

The committee concludes that recent advances in the fields of developmental biology and genomics provide unprecedented opportunities to understand the molecular mechanisms of action of toxicants, the differences in the developmental responses of test animals and humans to toxicants, the extrapolation of high-dose effects to low-dose effects, and the differences in the individual human susceptibility effects of toxicants on development. These advances can lead to improved animal tests for toxicants, improved extrapolation of animal results to humans, and through the means of better data, improved risk assessment.

A multilevel, multidisciplinary approach to toxicity assessment involves the use of genetically tractable model organisms for which development is well understood and the genome is completely sequenced (or soon will be), namely, *C. elegans*, *Drosophila*, zebrafish, and mouse. Because some of these animals are inexpensive to use, testing could be expanded to cover a larger number of chemicals, chemical combinations, and testing conditions, including various genotypic backgrounds of the test organism and scoring of toxic effects at developmental stages before organogenesis. The validity of extrapolation of test results can be tested as well. Risk assessors would benefit, the committee believes, from the use of a variety of toxicity data, combined with human exposure and susceptibility information. The recommendations in the next section are summarized in a text box.

RECOMMENDATIONS

Recommendation 1. To improve the understanding of the mechanisms of action of toxicants, the committee recommends that critical molecular targets of toxicants be identified among the components of developmental processes.

How can progress be made in the analysis of mechanisms of toxicity? What variety of mechanisms is there? Are there certain molecular components and processes of development that seem to be frequent targets of toxicants? The recent advances in developmental biology, cell biology, and genomics provide information about components, functions, and processes for the first time and for a wide range of animals, including mammals. Fortunately, many of the components and processes are conserved across phyla, making it possible to study them initially in experimentally convenient organisms, such as *C. elegans* and *Drosophila*, or even yeast and bacteria in some cases, and then carry the information across species to the studies of mammalian development, with great success. Many aspects of mammalian development resemble arthropod and nematode development.

The committee's recommendation for future studies on mechanisms of developmental toxicity is that greater use be made of model organisms, taking advantage of the shared features of development with mammals. Fruit flies, nematodes, and mice have different organs and body organization, and yet they use many of the same molecular components and basic molecular interactions, although in different combinations and sequences. The analysis of toxicity mechanisms would have to bear the similarities and dissimilarities in mind.

Mechanisms of toxicity can now be understood more fully at the level of affected molecular components and the consequences of the altered activity of this component for developmental processes. Previously, toxicity analysis often ended with defining the time window of sensitivity of the conceptus to the toxicant and the profile of affected organs. This information is still important to have. Because toxicants affect the activity of molecular components, the recommended emphasis would be to identify components with which toxicants interact, determine the toxicant's effect on the component's function (the functions of many components are now understood), and fit the altered function into the effects on a developmental process, the abnormal operation of which leads to a structural or functional defect. The committee recommends that model organisms be used extensively to access this level of analysis so that researchers can benefit from genetic tractability, experimental convenience, and a background of information about molecular components and developmental processes.

1.1. In searching for mechanisms of developmental toxicity, the committee recommends research on how toxicants perturb evolutionarily conserved molecular targets and pathways of development.

To learn about mechanisms of developmental toxicity, special attention should be given to research in the areas of (1) signaling pathways and their associated genetic regulatory circuits (a group of key developmental targets); (2) molecular-stress and checkpoint pathways (a group of key cellular targets); and (3) the DMEs (a group of key toxicokinetic components).

Recommendations

Recommendation 1. To improve the understanding of the mechanisms of action of toxicants, the committee recommends that critical molecular targets of toxicants be identified among the components of developmental processes.

- In searching for mechanisms of toxicity, it will be important to explore how toxicants perturb evolutionarily conserved molecular targets and pathways of development.
  - Conserved signaling pathways and genetic regulatory circuits as potential targets of toxicants.
  - Molecular-stress and checkpoint pathways as potential targets of toxicants, and
  - Toxicokinetic components, especially drug-metabolizing enzymes.
- In pursuing mechanisms of toxicity, it is important to explore how molecular perturbations lead to dysmorphogenesis and other adverse outcomes of development in different species.
- Define the genetic and epigenetic basis of variability in human response to developmental toxicants.
  - Individual toxicokinetic differences, especially in the metabolism and transport of chemicals.
  - Toxicodynamics: individual differences in developmental components.
  - Individual differences in molecular-stress and checkpoint pathways, which normally operate to counteract failure of cell function.
- In seeking to understand molecular mechanisms of toxicity, it is important to clarify how these approaches and this information can be applied to a comprehensive assessment of human developmental risk.
  - The metabolism of developmental toxicants.
  - Toxicodynamics: toxicant effects on developmental components-information about mechanism and susceptibility.
  - Molecular-stress and checkpoint pathways.

1.1.1. Conserved signaling pathways and genetic regulatory circuits as potential targets of toxicants.

Approximately 10 kinds of signaling pathways are repeatedly used at all times and places of development before cytodifferentiation (i.e., during cleavage, gastrulation, neurulation, and organogenesis). Development is built on coupled signaling and genetic regulation. As differentiated cell function begins during cytodifferentiation, seven other pathways are added. Thus, in the mammalian fetus, all 17 pathways are in use. Almost all activities of cells of multicellular organisms, including all embryonic stages, are contingent on extracellular signals. These activities include secretion, motility, adhesion, proliferation, differ-

Recommendation 2. The committee recommends that it is important to study how the new information about development and developmental toxicity can address the uncertainties in quantitative and qualitative risk assessment.

- Qualitative: Testing a larger variety of chemicals and chemical mixtures.
- Qualitative: Assessing toxicant effects across all stages of development.
- Quantitative: Toxicokinetic differences of test animals and humans should be characterized in order to improve extrapolations.
- Quantitative: Toxicodynamic differences of test animals and humans should be characterized to improve extrapolations.
- Quantitative: Low-dose effects of toxicants and chemical mixtures should be better detected and characterized.
  - Low-dose cellular responses as reflected in molecular-stress and checkpoint pathways.
  - Genetically sensitized animals should be tested for low-dose toxicant effects.
- Quantitative: Modeling the extrapolation from test animals to humans

Recommendation 3. To improve the interdisciplinary advances in developmental toxicology, the committee recommends that the databases of developmental toxicology, developmental biology, and genomics be better linked on the Internet, and that multidisciplinary outreach programs be established for the effective exchange of information and techniques related to the analysis of developmental defects and to the assessment of toxicity for risk assessment.

- Development of cross-disciplinary, linked databases of relevance for developmental toxicity.
- Enhancement of multidisciplinary research interactions.

entiation, and transcription. The components of signaling pathways and genetic regulatory circuits are key components to investigate.

Further arguments for emphasizing these components as potential targets are the following:

- When signaling is interrupted in mutant animals, development is affected, with specific effects depending on the component that is inactivated.
- A few toxicants currently are known to affect signaling in mammals, such as cyclopamine (the Hedgehog signaling pathway; Chapter 4), retinoic acid and its many derivatives (the nuclear hormone signaling pathway and transcriptional

regulator; Chapter 4), smoking and transforming growth factor (TGF) variants in humans (the receptor tyrosine kinase pathway; Chapter 5), and smoking and MSX/BMP variants in humans (the TGF pathway; Chapter 5).

The committee acknowledges that signaling and gene regulation have not been proved to be particularly vulnerable points of development. An argument can be made that, because these pathways are so essential for all aspects of development, evolutionary selection has made them particularly resistant to perturbation, including that by toxicants. However, data from experimentally constructed mutant animals contradict this argument. In invertebrates, such as *Drosophila* and *C. elegans*, null mutants are often lethal, and hypomorphic mutants (i.e., ones having partial activity) show specific developmental defects, probably reflecting the times and places where the inadequate component is required at highest activity. In vertebrates, including mammals, the picture is somewhat different. The genes encoding most components have undergone duplication and slight diversification, so there are functionally equivalent members for each kind of component. This is termed "redundancy". The members are expressed at different times and places in development. In vertebrates, null mutants frequently show, not lethality, but specific local losses, reflecting the few times and places where a lost component's function is not overlapped by a related component. Some of these null phenotypes in mice closely resemble human developmental defects (e.g., see Tables 6-4 and 6-5). Thus, the committee's hypothesis, namely, that signaling pathways and the associated genetic regulatory circuits are critical molecular points of susceptibility of development to toxicants, has not been proved.

A final point should be made about the appropriate use of model organisms for studies of molecular mechanisms of toxicity. As mentioned above, although these organisms have different organs from mammals, the developmental processes involved in their organogenesis are similar to those of mammals. Organogenesis in various species represents different combinations and orderings of conserved processes, such as signaling pathways and the responses they engender (e.g., proliferation, locomotion, and secretion). Thus, for example, in the course of a study of toxicity mechanisms, *Drosophila* might be scored for the effects of a chemical on its wing development, but what is really being scored is the effect of the chemical on the kinds of signaling pathways and genetic regulatory circuits also used in human development of different organs. Wing development serves as a well understood set of conserved molecular components and interactions. Effects on the wing can be readily recognized, and because of the advanced understanding of its development, the targeted developmental processes can be surmised. From the fly results, predictions can be made of the effects of those chemicals on mammalian organogenesis, for organs in which these same components operate. Zebrafish, as a vertebrate test animal, are expected to share many developmental processes with mammals (e.g., more details of specific organogenesis) and can be used as an intermediate model. All these model systems share with mammalian embryos the need to coordinate and integrate the many temporally and spatially distinct cell regulatory networks operating during development. A fully developed, functional fly wing, fish fin, or human arm are all products of comparable hierarchically organized processes.

1.1.2. Molecular-stress and checkpoint pathways as potential targets of toxicants.

There is evidence that some toxicants primarily damage basic cell functions, such as those of cellular reproduction, and that the damaged cells subsequently fail to participate in development. Such toxicants can act indirectly on development, although the failing of basic cell functions can soon enough impede development. Molecular-stress and checkpoint pathways are the cell's defenses for counteracting such damage. At least 10 of these pathways are known. They respond to different kinds of damage and decisions to make different counteractions. Apoptosis is the ultimate pathway, by which the cell is self-destroyed. This is thought to happen when the other counteractions are not sufficient to restore the cell to a minimal state of basic function.

Especially with regard to environmental toxicants, these pathways might deserve special attention, because some toxicants react widely with cellular components. As mentioned in Chapter 6, many environmental toxicants set off an enlarged domain of apoptosis in the embryo. The particular spectrum of activated pathways would give an indication of the impacted cellular processes and components. Also, low- and high-dose effects could be discriminated. At low doses, cells might recover or a fraction might die and be replaced by the proliferation of others, in which case, development might continue normally. At high doses, recovery might be exceeded by impaired proliferation or apoptosis and defects occur.

The investigation of toxicant targets in these pathways can exploit *C. elegans*, *Drosophila*, zebrafish, and mouse. The molecular-stress and checkpoint pathways have been well analyzed in these organisms (in fact, discovered in them in some cases). The pathways are nearly identical in all those organisms, although mammals have a greater variety of closely related components. In general, there has been less study of these pathways in embryos than in adults, so there might be some unexpected differences.

1.1.3. Toxicokinetic components, especially drug-metabolizing enzymes.

As noted above, there are many steps in absorption, distribution, metabolism, and excretion that determine whether the conceptus will be exposed to a toxicant. The molecular components of absorption, distribution, and excretion are not well known, but those of metabolism, the DMEs, have been extensively elucidated in recent years. There might be a few hundred kinds of DMEs, with different substrate specificities and time and place of expression in the mother and conceptus. Studies are well under way to analyze the roles of DMEs to metabolic potentiation or detoxification of chemicals, as part of the research program of ecogenetics, pharmacogenetics, and the Environmental Genome Project. Much study of the roles of DMEs is done in the mouse, for example, with null mutants of individual enzymes. The committee fully favors this direction of study and the support for it. As described later, it will also be valuable for animal assays of toxicants to know the DME similarities and differences from humans, as a part of the validation of the extrapolation of data from animals to humans. Some studies with other model organisms are probably of use. Some DMEs are widely conserved among animals (e.g., CYP1A1 in fish) and would be easier to study in other model organisms.

1.2. In pursuing mechanisms of toxicity, the committee recommends research to explore how molecular perturbations lead to dysmorphogenesis and other adverse outcomes of development in different species.

A decade ago it would have seemed impossible to analyze how a toxicant's initial interaction with a molecule is connected with the ultimate developmental defect. The complexity of development seemed daunting, and there was little knowledge of the activities and interactions of components and their roles in developmental processes. In the past decade, however, the situation for analysis has improved greatly. The activities of numerous components are known, and there are insights into the organization and coupling of processes. The understanding of the early developmental steps of axis formation, gastrulation, and neurulation is increasing, and various examples of organogenesis are available for study. The new information on development provides the framework for analysis of toxicant effects on developmental processes and the connection of dysfunctional processes to structural and functional defects in the newborn. The analysis of development in model organisms has been crucial to the progress in mammals, specifically the mouse. Some efforts in connecting molecular effects to dysmorphologies have been successful. For example, the analysis of cyclopamine (a plant alkaloid) in causing cyclopia (a diminished head and single median eye) in cattle was possible once it was realized that mouse mutants of some components of the Hedgehog signaling pathway are also cyclopic; once the basic developmental studies showed the importance of sonic Hedgehog signal for inhibiting eye development in the ventral midline of the prospective diencephalon, leaving bilateral eye development, the mechanism of the cyclopamine-induced birth-defect became better understood.

Scrutiny of the developmental defects of mutants of developmental components in genetically favorable model organisms, and the scrutiny of toxicant-caused developmental defects, might provide informative parallels. In some mutants, a component is inactive due to a mutated gene; in others, a mutated component can be overactive or underactive, but not absent. If a toxicant interacts with one component and modifies its activity, single-gene mutant studies serve as a guide for the interpretation of how molecular perturbations result in dysmorphogenesis and other adverse outcomes. The phenotypes for the large variety of mouse null mutants, prepared by targeted gene disruption, are a resource for such analysis. Many of these show well-defined developmental defects at birth, or prenatal death at various stages.

If a toxicant affects several developmental components, the mutant comparisons are not as good as those of one component, although some multiple mutations have been prepared. If a toxicant affects cellular activities, setting off molecular-stress and checkpoint pathways, and causes cell death, comparison mutants could be generated and analyzed to understand the consequences for development.

The committee recommends basic research on toxicant-affected developmental processes in well-understood model organisms. *Drosophila* mutants can be prepared with a chosen sensitized signaling pathway in a chosen kind of organogenesis, such as the wing or compound eye. In such animals, the effects of toxicants can be most favorably associated with specific processes of organogenesis.

1.3. The committee recommends research to define the genetic and epigenetic basis of variability in human response to developmental toxicants.

Variability is a large problem, covering a variety of issues. It is apparent even in the heterogeneous developmental response of individuals in a group of inbred test animals (rodents) exposed to a toxicant under controlled conditions. Two approaches to address this large problem might be useful: (1) a human epidemiological approach making use of genome information, and (2) a model animal research approach making use of new molecular biological techniques and insights to learn about the sources of variability. The following aspects of individual variability in response to developmental toxicants deserve study:

1.3.1. Individual toxicokinetic differences, especially in the metabolism and transport of chemicals.

Human individuals are known to differ substantially in their levels of DMEs of both the P450 oxidative group and the group of conjugating enzymes. For example, differences are found in ethnic groups from different parts of the world and in individuals with varied lifestyles (smoking and alcohol intake) and nutrition. These differences are being explored rapidly, and data indicate that, in cases of genetic variability, the genes encoding these enzymes might be unusually polymorphic compared with other kinds of genes. In current research, which this committee favors, the differences are being defined in terms of base sequence, protein function (e.g., loss of function, reduction of function, increase of function, and change of function), and the basis for the change (e.g., altered time and place of expression and altered catalytic activity). Differences in metabolism of chemicals by individuals with different gene combinations are also being analyzed, because some allele combinations are strongly synergistic. Current genetic, epidemiological, and genome data can greatly assist in identifying human polymorphisms, and the Environmental Genome Project has such identification as a goal, as do the National Cancer Institute and various pharmaceutical companies. Genotype-environment interactions are suspected to underlie a variety of developmental defects, and identification of human polymorphism will shed light on those interactions.

As these polymorphic genes are identified, experiments are being done to eliminate them by targeted inactivation in the mouse and to assess the animal's sensitivity or resistance to various chemicals. Studies are also being done to observe the protective or sensitizing effects of overexpression and ectopic expression of enzymes encoded by these genes. The committee supports this work enthusiastically.

The information about individual differences in the metabolism of exogenous chemicals is being extended to research on model organisms (the mouse, *Drosophila*, and *C. elegans*), because the validity of animal test results for predicting human toxicity depends not only upon an understanding of the toxicodynamics of the animals' developmental process but also on an understanding of the drug-metabolizing capacity of the animal and how it is similar or dissimilar to that of humans.

1.3.2. Individual toxicodynamic differences in developmental components.

Another large domain of possible human variability concerns genetic differences in components of developmental processes. Toxicants might interact with these components. Although signaling pathways and genetic regulatory circuits are particularly attractive for study, there is little information on their genetic differences at this time compared with that on DMEs. However, many mouse mutants in genes encoding such components show developmental defects similar to those in humans, and a few human polymorphisms of components are known to correlate with disease (e.g., Patched mutations and a predisposition to basal-cell carcinoma) and predisposition to developmental defects from certain toxicants (e.g., TGF and tobacco smoke). Some signaling component polymorphisms behave as complex traits with sensitivity to genetic background (e.g., APC mutants in mice represent a Wnt pathway intermediate). Gene locations for many signaling components will soon be known through genetic studies of mouse and zebrafish development, cancer databases, and genome studies. This information will be deposited in widely accessible databases on the Internet.

The current experimental situation is favorable for determining the relationship between mutations in these components and susceptibility to toxicants. Mice offer an opportunity for a survey of the importance of genetic background for toxicant sensitivity and for toxicant specificity of effects. Genetic variants, usually produced by targeted gene knockout (but also by gene replacement to make hypomorphs and gain-of-function types and ectopic expression types), can be made with relative ease in mice, which can be used in developmental toxicity studies. When a null allele of any of a variety of kinds of signaling components is present in the heterozygous condition, the animal is usually viable and apparently normal, although sometimes with minor developmental defects. These animals could be tested for toxicant sensitivity, to see if the genetic background biases certain outcomes. In general, the connection made between mutational alterations in developmental components and chemical induced developmental defects should be strong.

A further determination to be made is how much of the apparent specificity of the outcome of a toxicant-induced developmental defect is due to the specificity of the toxicant and how much is due to the particular genetic background of the exposed animal. This determination should be applied especially to broad-acting toxicants (e.g., ones causing widespread cell death). In model animals, such as *Drosophila*, *C. elegans*, or the mouse, various genetic constructs that are sensitized in various ways (e.g., a slightly reduced Hedgehog pathway or a slightly reduced TGF pathway) could be exposed to the same toxicants to see how the outcomes differ. Already there are human toxicant-induced differences in profiles of birth outcomes in Tp53 (-/-) knockout mice that vary with genetic background. The full implication of these observations has not yet been exploited for assessing developmental toxicants.

Ultimately, it is not known how similar a group of animals can ever be in individual responses to toxicants. In the standard rodent tests for toxicants the response of a group of animals exposed to a single intermediate dose is heterogeneous. The developmental outcome is affected in some animals and not affected at all in others. How much of the heterogeneity is genetic and how much is "epigenetic" (i.e., associated with variable histories of nutrition, disease, stress, or other chemical exposures) is an important issue. Conventional inbred strains, which are more than 98% genetically identical after 20 generations of inbreeding, are a valuable and easily available resource for studying epigenetic contributions.

1.3.3. Individual differences in molecular-stress and checkpoint pathways, which normally operate to counteract failure of cell function.

These components are part of the organism's line of defense and might be activated by broad-specificity toxicants. The extent to which individual humans differ in their molecular-stress and checkpoint pathway components and, hence, in their responses to environmental chemicals, is unknown. These pathways are known to have important roles in other organisms and to be conserved across phyla. Individual differences in these pathways among humans should be explored. Model animals should be prepared with mutated components of these pathways in order to determine whether their sensitivity to toxicants is increased or decreased. Mouse and *Drosophila* mutants are already available.

1.4. In seeking to understand molecular mechanisms of toxicity, it is important to clarify how the approaches and information can be applied to a comprehensive assessment of human developmental risk.

The committee has stated that the developments in developmental biology and genomics present an unprecedented opportunity to understand the mechanisms of action of toxicants at a molecular level and that, at some time in the future, perhaps a decade, risk assessors will have primarily "mechanism-based data" from test animals to use in arriving at human toxicity estimates. However, standard toxicant bioassays on mammals do not yet yield comprehensive data on mechanisms and consequences valid for extrapolation to humans.

What can be done in the interim to build on recent advances? In Chapter 8, the committee outlined a multilevel approach to risk assessment that incorporated various assays intended to provide information ranging from molecular interactions to developmental consequences. Advantages and disadvantages of the assays of each level were outlined as well. This approach is briefly described here. There are two domains of information. One contains results from model systems and model-animal tests of toxicant effects on development and of genetic alterations affecting toxicant susceptibility. The results need to be extrapolated to humans. The second domain contains results from human studies of toxicant exposure, toxicant susceptibility (including polymorphisms), and toxicant effect.

There are four levels of model systems for providing information for assessing the effects of toxicants (or the absence of effects): (level 1) in vitro tests and cell tests; (level 2) nonmammalian animal tests of development and the role of genotype; (level 3) mammalian tests of development and the role of genotype; and (level 4) mammalian tests of mechanism and susceptibility. In general, expense increases with each level, and the number of chemicals that can be tested decreases. The questions of extrapolation to humans are greatest at the low levels. As described in Chapter 8, the committee did not develop a tiered approach, but rather showed how information from each level could be used to improve developmental toxicity risk assessments.

Various examples of this information follow.

1.4.1. The metabolism of developmental toxicants.

Knowledge of DMEs is sufficient to devise level 1 tests of the conversions of a large variety of chemicals by a large variety of enzymes. For example, human metabolism genes have been introduced into test cells, such as yeast or human lymphoblast lines, to generate assay systems. Once the specificity range of human enzymes is well characterized, it should be possible to make predictions about the capacity of a battery of enzymes to modify yet-untested chemicals. It is also clear that various animals (*Drosophila*, *C. elegans*, and mice) can be constructed with deficiencies or excesses of various metabolizing enzymes to determine whether the developmental toxicity of a chemical is increased or decreased. Much of this work is under way. There is substantial research inquiry about DMEs, their roles, and their synergisms, especially among oxidases and conjugating enzymes. Although the study of the DMEs is relatively advanced, research is still needed on other aspects of absorption, distribution (including the multidrug transporter proteins), and excretion.

In general, this research is an exemplary area for the exchange of information across levels with the ever closer approximation of the fast, inexpensive level 1 tests to human metabolizing conditions.

1.4.2. Toxicodynamics: toxicant effects on developmental components—information about mechanism and susceptibility.

Other assays of toxicants should be used to focus on their effects on developmental processes, particularly on the intercellular signaling pathways and genetic regulatory circuits, which operate repeatedly and pervasively in the development of animals of all phyla. Some assays can be done at level 1, since many signaling components have been identified and their genes isolated. Relevant proteins can be produced for in vitro tests or single-cell tests. For example, to test for agents interfering with Hedgehog signaling, the signal transduction intermediates could be introduced into cultured cells. To some extent, cell functions—such as secretion, entry into mitosis, motility, or specific gene expression—that depend on signaling could be scored for interruption by chemicals. The availability of components for nonanimal assay systems is already considerable. Many signaling components have been mapped on the mouse genome, and their corresponding location in the human genome is predictable and will soon be sequenced.

The committee recognized at the outset that the information about chemical impacts on cell signaling components is of little use for risk assessment if there is no organ and mammalian relevancy. Therefore, a comprehensive approach was envisioned to allow this initial information to be placed directly into an overall assessment framework. For testing the effects of toxicants on the activity of these components in development, level 2 assays on *Drosophila*, *C. elegans*, and zebrafish development are incisive. Multiple pathways are used in the development of complex organs. One pathway at a time can be sensitized in a specific aspect of development by genetic means. An animal, when exposed to a toxicant affecting that pathway, will have altered development of that organ, whereas the rest of the animal will probably be unaffected or less affected. Altered development is, then, the scored end point of the toxicant's effect on the specific pathway. Although the model organisms have different organs from those of humans, the signaling pathways and genetic regulatory circuits that operate in the development of that organ also operate in the development of mammalian organs of other kinds. Thus, the effects of chemicals on fundamental processes, such as signaling and transcription, can be detected. The signaling pathways operating in the various kinds of organogenesis in mammals are known; therefore, a prediction can be made and tested in mammals, using level 3 testing approaches. Because level 2 assays are inexpensive and fast, many compounds can be tested, and patterns of toxicity effects can be recognized in advance of the rodent tests. In the multilevel approach, level 3 tests are ones with specially modified mice (containing sensitized pathways and reporter-gene constructs), and more information is obtained with them than with standard animals. In this way, the levels are connected to each other and kept relevant to risk assessment.

1.4.3. Molecular-stress and checkpoint pathways.

Other assays should be directed toward the detection and characterization of cellular responses to chemicals by way of these defense pathways. Approximately 10 of these conserved pathways are now known. Their activation might be relevant to detecting the effects of broadly acting toxicants on maternal and embryonic cells—that is, toxicants such as antimitotic agents or inhibitors of replication, transcription, or translation—that interact with many targets in many cells.

Research remains to be done to connect the damaging effects of toxicants on cells to the disruption of particular steps of development. The connection could be established with level 2 organisms genetically sensitized in one or more of their molecular-stress and checkpoint pathways in a particular organ. The established connection might provide leads for level 3 tests with mice.

Recommendation 2. The committee recommends investigating how the new information about development and developmental toxicity can address the uncertainties in quantitative and qualitative risk assessment.

The committee believes that the new information and approaches of developmental biology and genomics will be useful in improving the quantitative as well as qualitative components of risk assessment. As they are currently designed, the rodent tests for developmental toxicity are limited in their capacity to provide mechanistic information. They are costly in time and resources, and, therefore, only a small percentage of the more than 80,000 chemicals in commercial use (or the even larger number—about 6 million—of natural products) can be tested. Rodent-test end points are frequently limited to effects on growth, organogenesis, and viability of the conceptus and do not include functional, molecular, or cellular effects, nor do they include early developmental losses. The relevancy of animal toxicity outcomes for humans is often questioned, as is the significance of high dose animal exposure conditions for human exposures. Hence, risk assessors must often resort to large default corrections when extrapolating animal results to define safe exposure concentrations for humans. The validity of the extrapolation of particular test results from animals to humans is itself usually not assessed.

The committee envisions that research directions included in the informational framework established in Chapter 8 will address various existing limits on the data available to risk assessors and have the potential to provide the scientific basis to reduce the magnitude of or replace defaults using mechanism-based extrapolation approaches.

2.1. Qualitative risk assessment: testing a larger variety of chemicals and chemical mixtures.

Expanding the number of tested chemicals is an enterprise in qualitative risk assessment. The new rapid and inexpensive model assay systems have an important potential use in such an enterprise. At level 1, which involves in vitro and single-cell assays, tens of thousands of assays could be run per year to test chemicals as substrates for DMEs, as agonists and antagonists of signaling components and genetic regulators of the kind used pervasively in development, and as triggers of molecular-stress and checkpoint pathways. At level 2, which involves tests of chemical effects on the development of nonmammalian animals, thousands of assays could be run per year. Genetically sensitized model organisms (*Drosophila*, *C. elegans*, and zebrafish) equipped with various reporter genes would facilitate analysis. Mammalian relevancy and human applications would be further defined in level 3 tests involving mammals—the mouse being the most favorable because of its ease of genetic modification, its vast libraries of mutants, and the advanced knowledge (among mammals) of its development.

2.2. Qualitative risk assessment: assessing toxicant effects across all stages of development.

As noted in Chapter 2, early fetal loss in human development is frequent (20-30% of initial pregnancies). Although many of these losses may be due to chromosomal aberrations for which there are good chemical assay methods, other mechanisms of early loss are less well understood. Recent observations have demonstrated that, contrary to what was previously believed, toxicant exposures during early times in development can not only result in fetal loss but also specific birth defects and adverse functional impacts. In addition, functional impacts occurring as a result of post-organogenesis toxicant exposure have also not been clearly delineated. Further tests of toxicant impacts during these early and late developmental time points are needed and can build on the rapidly expanding body of knowledge about early events such as axis formation, primitive streak formation, and node regression as well as an expanded understanding of functional deficits.

2.3. Quantitative risk assessment: the toxicokinetic differences of test animals and humans should be characterized to improve extrapolations.

The committee recommends that test animals be better characterized with regard to their differences from humans in DMEs and other toxicokinetic variables. With better characterizations, it can be known whether the test conceptus and the hypothetical human conceptus are indeed exposed to the same chemicals at corresponding concentrations and intervals of development. Many DMEs have been identified, and others are known to exist. The profile of activity of mice (level 3 assays) and humans should be determined so that their similarities and differences are known. Some of this effort is already under way, and the comparison will be accelerated by the availability of mouse and human genome data. Proteins involved in chemical uptake across the gut, distribution in the fluid space, multidrug transport of chemicals in and out of cells, and excretion from the body are less well known and should also be characterized and compared.

Moreover, once the differences are recognized, test animals such as mice can perhaps be modified genetically to reduce their differences from humans. Thus, cross-species extrapolations could be improved in this respect. It may not be feasible to eliminate all toxicokinetic differences between the mouse and human, however, the differences will be better known with such approaches when extrapolations are invoked.

In level 2 assays involving nonmammalian organisms, such as *Drosophila*, *C. elegans*, and zebrafish, it is also important to know the differences between humans and rodents in terms of drug metabolism to increase the accuracy of extrapolations to mammals. Transgenesis and mutagenesis can be done at high frequency in these animals to make them less different from mammals in their drug metabolism.

2.4. Quantitative risk assessment: toxicodynamic differences of test animals and humans should be characterized to improve extrapolations.

The differences in development of various organisms mostly reflect differences in the time, place, order, and combinations of use of conserved developmental components, such as those of the signaling pathways and genetic regulatory circuits. The committee's recommendation to make more use of nonmammalian model animals in developmental toxicology is based on the recognition of the conservation, although with the caveat that the scoring of toxicant effects is done in these animals at the molecular level of conserved components, and not at the diversified tissue and organ levels, which are obviously not conserved across phyla.

The extent to which developmental components of different animals differ in their interactions with toxicants is not known. Some components can be exchanged between flies and mice without loss of function, but most have not been tested for interchangeability. Vertebrates also have large genomes containing two or more duplicated and slightly diversified genes for many components for which nonvertebrates have a single gene. These diversified components might differ in their toxicant interactions. The recognition of differences will be aided by the genome databases and by further genetic substitutions in test animals. Once toxicokinetic differences are minimized between mice and humans, modified mice should be tested with a battery of toxicants known to affect humans in order to make sure that equivalent developmental outcomes are obtained. If equivalent outcomes are not obtained, the difference is grounds for further analysis of toxicodynamic comparisons.

Current research in developmental biology, which includes mouse development as the exemplar of mammalian development, will increasingly use comparisons between mice and primates. The tests of the toxicant susceptibility of developmental components of mouse mutants will greatly guide research in humans.

2.5. Quantitative risk assessment: low-dose effects of toxicants and chemical mixtures should be better detected and characterized.

Risk assessors need data on toxicant effects covering a wide range of doses. A chemical might affect a variety of development processes at high doses but only one critically sensitive pathway at low doses, making that single pathway the most relevant for overall risk assessment. Studies of model systems—such as *Drosophila*, *C. elegans*, zebrafish, and the mouse—could provide quantitative information to improve understanding of such dose distinctions and their basis.

In risk assessment, animal test results obtained at high doses of a toxicant and with a small population of animals are frequently used to estimate the consequences of low doses in large populations. Furthermore, when a group of animals is exposed to the lowest dose of toxicant for which there is an effect, the individuals of the group usually respond in a heterogeneous way. Thus, there are many uncertainties about extrapolation to low doses. Basic questions to be explored in this area include the following: (1) What is the shape of the dose-response curve for developmental toxicants at low, environmentally relevant doses? (2) Can the increased attention on key cell-signaling pathways and genetic-regulatory circuits identify biomarkers useful for defining low-dose responses caused by developmental toxicants? (3) Do the low-dose responders represent variants with genetic susceptibility? and (4) Is there an inescapable nongenetic variability to development?

2.5.1. Low-dose cellular responses revealed through the molecular-stress and checkpoint pathways.

Some developmental toxicants might act primarily by interfering with basic cellular reproduction (e.g., DNA synthesis or mitosis). The conceptus might be more sensitive than the adult because it has a higher frequency of cell division (and fewer cells are in a nondividing differentiated state). Dose effects might be nonlinear. High doses of a toxicant might cause so much cell death that local development is impaired, but low doses might cause so little cell death that cell proliferation by the unaffected cells can restore the population and development is not detectably abnormal. At even lower doses, the various molecular-stress and checkpoint pathways might protect individual cells so that none dies, and development is completely normal. Nonetheless, the activation of the recovery pathways might be detected as an indicator of effect in this low-dose range. The committee recommends that these toxicants be explored over a range of doses, to detemine whether responses can be found in doses too low to cause developmental defects and to reveal the capacity of the conceptus for recovery.

2.5.2. Genetically sensitized animals should be tested for low-dose toxicant effects.

Does the heterogeneous response of a population of test animals to toxicants give a detectable effect that reflects genetic heterogeneity? To determine the role of genetic differences, various mouse strains can be produced with limiting levels of activity of particular developmental components, and these strains can be tested for their sensitivity to a standard set of toxicants representing a variety of suspected mechanisms of action. Such tests would be a measure of whether animals that are genetically close to abnormal development due to their genetic constitution are more sensitive than normal animals to toxicants, and if so, whether a general sensitivity or a specific one is related to the particular limiting component.

Because genetically sensitized models can approximate more closely potentially susceptible members of the human population, risk assessors can use the toxicity data more comfortably. Developmental toxicologists might want to explore the relationship between genetic variation of toxicodynamic components and toxicant sensitivity in model animals before human variants of developmental components are identified in the future. (The association of cigarette smoking, TGF variants, and cleft palate is already an example.)

Improved information on human exposures gained from other areas of study (e.g., improved biomarkers) should provide additional information to predict more accurately the human exposure range for any given chemical or mixture. Such information could then be used to set exposure concentrations for model-animal assessments in levels 2 and 3.

2.6. Quantitative risk assessment: modeling extrapolation from test animals to humans.

The informational framework in Chapter 8 should provide a guide for obtaining the kinds of test-animal data that are needed for a comprehensive cross-species toxicokinetic and toxicodynamic model of exposure and development of test animals, such as mice, to humans. As outlined in Table 8-1B, information from improved human biomarkers of exposure, susceptibility (both genetic and nongenetic), and effect would be used in the model as well. Such models, difficult as they are to devise and fill with satisfactory data, are needed if a chemical's potential for developmental effects are to be extrapolated to humans in a meaningful way. For example, complex computational models and abundant data are needed to estimate in utero and postnatal exposures in mammals and to link this information with toxicological impacts.

Interest in such models is demonstrated by the efforts of the National Institute of Environmental Health Sciences to link exposure information with mechanistic toxicity data. However, only a few such specialized models exist, and none adequately combines toxicokinetic and toxicodynamic information, especially the information on molecular and cellular impacts. The lack of an adequate framework into which mechanistic data can be incorporated has limited the usefulness of the recent advances in developmental biology for quantitative risk assessment. Instead, default corrections continue to be used, despite recommendations such as those published in *Science and Judgment* (NRC 1994) calling for the incorporation of new scientific information into the risk assessment process. The committee believes that the framework laid out in this report has the potential to bring the information gained from recent advances in developmental biology into developmental toxicity risk assessment.

Recommendation 3. To improve the interdisciplinary advances in developmental toxicology, the committee recommends that the databases of developmental toxicology, developmental biology, and genomics be better linked on the Internet and that multidisciplinary outreach programs be established for the effective exchange of information and techniques related to the analysis of developmental defects and to the assessment of toxicity for risk assessment.

The committee concludes that increased multidisciplinary efforts and exchanges of information in chemistry and biology are essential to improve risk assessment for developmental toxicity. As mentioned at the beginning of this chapter, developmental toxicology is a broad and complex field.

In recognition of the interdisciplinary nature of future work in developmental toxicology, this interdisciplinary NRC committee was formed. It has been a struggle for the scientists from different relevant disciplines to communicate the research and public-health challenges in developmental toxicology within the committee. Members of this committed group were unfamiliar with each other's discipline and with the differing connotations of such terms as "mechanism." The committee soon realized the need for future activities fostering communication and joint research efforts within the scientific community.

3.1. Development of cross-disciplinary, linked databases of relevance for developmental toxicity.

To support the growth of knowledge in developmental toxicology and organize information in a way useful for risk assessment, this committee proposes that cross-disciplinary, linked databases of relevance for developmental toxicology be established with entries from industry, academia, and government. To capture and collate information about chemical toxicants that are important as developmental toxicants, internal organization of the data should reflect knowledge about chemical structure and should include known molecular targets, organotypic effects, and defined associations with developmental anomalies primarily from animal tests but also, when available, from humans. The database should link to genomics databases, for example, with epidemiological information on human variation, such as the large number of human DME polymorphisms. Another important link would be the historical control database for developmental and reproductive toxicity.

Ideally, a separate but linked relational database would be established, grouped by signaling pathways and genetic regulatory circuits, and referred to when chemicals are identified as interacting with an element of the pathway. The database could be helpful in identifying potential biological interactions of a chemical with other chemicals that affect components of the same pathway. A signal-transduction database was recently activated at www.stke.org. This or a similar database should keep track of the involvement of signaling pathways and genetic regulatory circuits in all aspects of development for a wide range of organisms. This information should also be connected to the large and growing database of phenotypes of mouse mutants, many of which are being generated by targeted gene disruption and transgenesis of signaling components or combinations of components. The mouse mutant collection represents the most systematic library of mammalian birth defects associated with known genetic defects. In addition to homozygous null mutants, the library should include phenotypes of hypomorphs, heterozygous null mutants, and suppressor loci, as they become available.

3.2. Enhancement of multidisciplinary research interactions.

The challenges that investigators face when trying to work across fields, such as developmental biology, developmental toxicology, and risk assessment, are a key issue that the committee identified early in its deliberations. This issue previously impeded the successful application of the new scientific information to improve developmental toxicity risk assessment. For the successful application of this report's findings, the committee believes that multidisciplinary educational and research programs must be conducted. Programs, such as workshops and professional meetings, should be organized so that researchers of developmental toxicology, developmental biology, genomics, medical genetics, epidemiology, and biostatistics can come together to exchange new insights, approaches, and techniques related to the analysis of developmental defects and to risk assessment. By accelerating the necessary research, cooperative research projects would move forward the recommendations of this report.

References

Abbott, B.D., and L.S. Birnbaum. 1990. TCDD-induced altered expression of growth factors may have a role in producing cleft palate and enhancing the incidence of clefts after coadministration of retinoic acid and TCDD. Toxicol. Appl. Pharmacol. 106(3):418-432.

Abbott, B.D., J.J. Diliberto, and L.S. Birnbaum. 1989. 2,3,7,8-Tetrachlorodibenzo-p-dioxin alters embryonic palatal medial epithelial cell differentiation in vitro. Toxicol. Appl. Pharmacol. 100(1):119-131.

Abbott, B.D., M.W. Harris, and L.S. Birnbaum. 1992. Comparisons of the effects of TCDD and hydrocortisone on growth factor expression provide insight into their interaction in the embryonic mouse palate. Teratology 45(1):35-53.

Abbott, B.D., G.H. Perdew, and L.S. Birnbaum. 1994. Ah receptor in embryonic mouse palate and effects of TCDD on receptor expression. Toxicol. Appl. Pharmacol. 126(1):16-25.

Abbott, B.D., M.R. Probst, G.H. Perdew, and A.R. Buckalew. 1998. AH receptor, ARNT, glucocorticoid receptor, EGF receptor, EGF, TGF alpha, TGF beta 1, TGF beta 2, and TGF beta 3 expression in human embryonic palate, and effects of 2,3,7,8-tetrachlorodibenzo-p-dioxin (TCDD). Teratology 58(2):30-43.

Abbott, B.D., G.A. Held, C.R. Wood, A.R. Buckalew, J.G. Brown, and J. Schmid. 1999a. AhR, ARNT, and CYP1A1 mRNA quantitation in cultured human embryonic palates exposed to TCDD and comparison with mouse palate in vivo and in culture. Toxicol. Sci. 47(1):62-75.

Abbott, B.D., J.E. Schmid, J.G. Brown, C.R. Wood, R.D. White, A.R. Buckalew, and G.A Held. 1999b. RT-PCR quantification of AHR, ARNT, GR, and CYP1A1 mRNA in craniofacial tissues of embryonic mice exposed to 2,3,7,8-tetrachlorodibenzo-p-dioxin and hydrocortisone. Toxicol. Sci. 47(1):76-85.

Abel, E.A. 1995. An update on incidence of FAS: FAS is not an equal opportunity birth defect. Neurotoxicol. Teratol. 17(4):445-462.

Adams, M.D., et al. 2000. The genome sequence of Drosophila melanogaster. Science 287(5461):2185-2196.

Allen, B.C., R.J. Kavlock, C.A. Kimmel, and E.M. Faustman. 1994a. Dose-response assessment for developmental toxicity. III. Statistical models. Fundam. Appl. Toxicol. 23(4):496-509.

Allen, B.C., R.J. Kavlock, C.A. Kimmel, and E.M. Faustman. 1994b. Dose-response assessment for developmental toxicity. II. Comparison of generic benchmark dose estimates with no observed adverse effect levels. Fundam. Appl. Toxicol. 23(4):487-495.

Allen, B.C., P.L. Strong, C.J. Price, S.A. Hubbard, and G.P. Daston. 1996. Benchmark dose analysis of developmental toxicity in rats exposed to boric acid. Fundam. Appl. Toxicol. 32(2):194-204.

Alles, A.J., and K.K. Sulik. 1989. Retinoic-acid-induced limb-reduction defects: Perturbation of zones of programmed cell death as a pathogenetic mechanism. Teratology 40(2):163-171.

Alon, U., N. Barkai, D.A. Notterman, K. Gish, S. Ybarra, D. Mack, and A.J. Levine. 1999. Broad patterns of gene expression revealed by clustering analysis of tumor and normal colon tissues probed by oligonucleotide arrays. Proc. Natl. Acad. Sci. U.S.A. 96(12):6745-6750.

Alonso-Aperte, E., N. Ubeda, M. Achon, J. Perez-Miguelsanz, and G. Varela-Moreiras. 1999. Impaired methionine synthesis and hypomethylation in rats exposed to valproate during gestation. Neurology 52(4):750-756.

Amsterdam, A., S. Burgess, G. Golling, W. Chen, Z. Sun, K. Townsend, S. Farrington, M. Haldi, and N. Hopkins. 1999. A large-scale insertional mutagenesis screen in zebrafish. Genes. Dev. 13(20):2713-2724.

Anderson, N.G., and L. Anderson. 1982. The Human Protein Index. Clin. Chem. 28(4 Pt. 2):739-748.

Anderson, N.L., and N.G. Anderson. 1998. Proteome and proteomics: New technologies, new concepts, and new words. Electrophoresis 19(11):1853-1861.

Andrews, J.E., M. Ebron-McCoy, U. Bojic, H. Nau, and R.J. Kavlock. 1995. Validation of an in vitro teratology system using chiral substances: Stereoselective teratogenicity of 4-yn-valproic acid in cultured mouse embryos. Toxicol. Appl. Pharmacol. 132(2):310-316.

Andrews, J.E., M.T. Ebron-McCoy, U. Bojic, H. Nau, and R.J. Kavlock. 1997. Stereoselective dysmorphogenicity of the enantiomers of the valproic acid analogue 2-N-propyl-4-pentynoic acid (4-yn-VPA): cross-species evaluation in whole embryo culture. Teratology 55(5):314-318.

Arman, E., R. Haffner-Krausz, Y. Chen, J.K. Heath, and P. Lonai. 1998. Targeted disruption of fibroblast growth factor (FGF) receptor 2 suggests a role for FGF signaling in pregastrulation mammalian development. Proc. Natl. Acad. Sci. U.S.A 95(9):5082-5087.

Artavanis-Tsakonas, S., M.D. Rand, and R.J. Lake. 1999. Notch signaling: Cell fate control and signal integration in development. Science 284(5415):770-776.

Baggs, R.B., R.K. Miller, and C.L. Odoroff. 1991. Carcinogenicity of diethylstilbestrol in the Wistar rat: effect of postnatal oral contraceptive steroids. Cancer Res. 51(12):3311-3315.

Bailey, A., A. Le Couteur, I. Gottesman, P. Bolton, E. Simonoff, E. Yuzda, and M. Rutter. 1995. Autism as a strongly genetic disorder: Evidence from a British twin study. Psychol. Med. 25(1):63-77.

Baker, N.E., and S.Y. Yu. 1997. Proneural function of neurogenic genes in the developing Drosophila eye. Curr. Biol. 7(2):122-132.

Baker, K., K.S. Warren, G. Yellen, and M.C. Fishman. 1997. Defective "pacemaker" current (Ih) in a zebrafish mutant with a slow heart rate. Proc. Natl. Acad. Sci. U.S.A. 94(9):4554-4559.

Baldwin, H.S., and M. Artman. 1998. Recent advances in cardiovascular development: Promise for the future. Cardiovasc. Res. 40(3):456-468.

Bantle, J.A., R.A. Finch, D.T. Burton, D.J. Fort, D.A. Dawson, G. Linder, J.R. Rayburn, M. Hull, M. Kumsher-King, A.M. Gaudet-Hull, and S.D. Turley. 1996. FETAX interlaboratory validation study: Phase III—Part 1 testing. J. Appl. Toxicol. 16(6):517-528.

Barber, C.V., and A.G. Fantel. 1993. The role of oxygenation in embryotoxic mechanisms of three bioreducible agents. Teratology 47(3):209-223.

Bargmann, C.I. 1998. Neurobiology of the Caenorhabditis elegans genome. Science 282(5396):2028-2033.

Bargmann, C.I., and I. Mori. 1997. Chemotaxis and termotaxis. Pp. 717-737 in C. elegans II, D.L. Riddle, T. Blumenthal, B.J. Meyer, and J.R. Priess, eds. Plainview, NY: Cold Spring Harbor Laboratory Press.

Barker, D.J. 1999. Fetal origins of cardiovascular disease. Ann. Med. 31(Suppl. 1):3-6.

Barr, M., Jr. 1997. Growth as a manifestation of teratogenesis: Lessons from human fetal pathology. Reprod. Toxicol. 11(4):583-587.

Barth, L.G. 1939. The chemical nature of the amphibian organizer: III. Stimulation of the presumptive epidermis of ambystoma by means of cell extracts and chemical substances. Physiol. Zool. 12:22-29.

Battula, N., J. Sagara, and H.V. Gelboin. 1987. Expression of P1-450 and P3-450 DNA coding sequences as enzymatically active cytochromes P-450 in mammalian cells. Proc. Natl. Acad. Sci. U.S.A. 84(12):4073-4077.

Bayer, S.A., J. Altman, R.J. Russo, and X. Zhang. 1993. Timetables of neurogenesis in the human brain based on experimentally determined patterns in the rat. Neurotoxicology 14(1):83-144.

Beachy, P.A., M.K. Cooper, K.E. Young, D.P. von Kessler, W.J. Park, T.M. Hall, D.J. Leahy, and J.A. Porter. 1997. Multiple roles of cholesterol in hedgehog protein biogenesis and signaling. Cold Spring Harb. Symp. Quant. Biol. 62:191-204.

Becerra, J.E., M.J. Khoury, J.F. Cordero, and J.D. Erickson. 1990. Diabetes mellitus during pregnancy and risks for specific birth defects: A population-based case-control study. Pediatrics 85(1):1-9.

Bechter, R., G.D. Terlouw, M. Tsuchiya, T. Tsuchiya, and A. Kistler. 1992. Teratogenicity of arotinoids (retinoids) in the rat whole embryo culture. Arch. Toxicol. 66(3):193-197.

Behringer, R.R., D.A. Crotty, V.M. Tennyson, R.L. Brinster, R.D. Palmiter, and D.J. Wolgemuth. 1993. Sequences 5' of the homeobox of the Hox-1.4 gene direct tissue-specific expression of lacZ during mouse development. Development 117(3):823-833.

Bellen, H.J. 1998. The fruit fly: A model organism to study the genetics of alcohol abuse and addiction? Cell 93(6):909-912.

Bellinger, D. 1994. Teratogen update: Lead. Teratology 50:367-373.

Best, J.B., and M. Morita. 1991. Toxicology of planarians. Hydrobiologia 227:375-383.

Binns, W., L. James, J.L. Shupe, and W.T. Huffman. 1963. A congenital cyclopian-type malformation in lambs induced by maternal ingestion of a range plant, Veratrum californicum. Am. J. Vet. Res. 24(103):1164-1175.

Blackwood, E.M., and J.T. Kadonaga. 1998. Going the distance: A current view of enhancer action. Science 281(5373):61-63.

Bladt, F., D. Riethmacher, S. Isenmann, A. Aguzzi, and C. Birchmeier. 1995. Essential role for the c-met receptor in the migration of myogenic precursor cells into the limb bud. Nature 376(6543):768-771.

Bodey, B., B. Bodey, Jr., S.E. Siegel, and H.E. Kaiser. 1999. Molecular biological ontogenesis of the thymic reticulo-epithelial cell network during the organization of the cellular microenvironment. In Vivo 13(3):267-294.

Bojic, U., M.M. Elmazar, R.S. Hauck, and H. Nau. 1996. Further branching of valproate-related carboxylic acids reduces the teratogenic activity, but not the anticonvulsant effect. Chem. Res. Toxicol. 9(5):866-870.

Bojic, U., K. Ehlers, U. Ellerbeck, C.L. Bacon, E. O'Driscoll, C. O'Connell, V. Berezin, A. Kawa, E. Lepekhin, E. Bock, C.M. Regan, and H. Nau. 1998. Studies on the teratogen pharmacophore of valproic acid analogues: Evidence of interactions at a hydrophobic centre. Eur. J. Pharmacol. 354(2-3):289-299.

Bostrom, H., K. Willetts, M. Pekny, P. Leveen, P. Lindahl, H. Hedstrand, M. Pekna, M. Hellstrom, S. Gebre-Medhin, M. Schalling, M. Nilsson, S. Kurland, J. Tornell, J.K. Heath, and C. Betsholtz. 1996. PDGF-A signaling is a critical event in lung alveolar myofibroblast development and alveogenesis. Cell 85(6):863-873.

Bournias-Vardiabasis, N. 1990. Drosophila-melanogaster embryo cultures: An in vitro teratogen assay. ATLA, Altern. Lab. Anim. 18:291-300.

Brambilla, R., N. Gnesutta, L. Minichiello, G. White, A.J. Roylance, C.E. Herron, M. Ramsey, D.P. Wolfer, V. Cestari, C. Rossi-Arnaud, S.G. Grant, P.F. Chapman, H.P. Lipp, E. Sturani, and R. Klein. 1997. A role for the Ras signalling pathway in synaptic transmission and long- term memory. Nature 390(6657):281-286.

Brand-Saberi, B., and B. Christ. 1999. Genetic and epigenetic control of muscle development in vertebrates. Cell Tissue Res. 296(1):199-212.

Brandsma, A.E., D. Tibboel, I.M. Vulto, J.J. de Vijlder, A.A. Ten Have-Opbroek, and W.M. Wiersinga. 1994. Inhibition of T3-receptor binding by Nitrofen. Biochim. Biophys. Acta 1201(2):266-270.

Braun, A.G., C.A. Buckner, D.J. Emerson, and B.B. Nichinson. 1982. Quantitative correspondence between the in vivo and in vitro activity of teratogenic agents. Proc. Natl. Acad. Sci. U.S.A. 79(6):2056-2060.

Brenner, S. 1974. The genetics of Caenorhabditis elegans. Genetics 77:71-94.

Brewster, R., and N. Dahmane. 1999. Getting ahead of the organizer: Anterior- posterior patterning of the forebrain. Bioessays 21(8):631-636.

Bridgeman, A.M., D. Buck, J. Burgess, W.D. Burrill, K.P. O'Brien, et al. 1999. The DNA sequence of human chromosome 22. Nature 402(6761):489-495.

Briggs, R., and T.J. King. 1952. Transplantation of living nuclei from blastula cells into enucleated frog eggs. Proc. Natl. Acad. Sci. U.S.A. 38:455-462.

Brockerhoff, S.E., J.B. Hurley, U. Janssen-Bienhold, S.C. Neuhauss, W. Driever, and J.E. Dowling. 1995. A behavioral screen for isolating zebrafish mutants with visual system defects. Proc. Natl. Acad. Sci. U.S.A. 92(23):10545-10549.

Brody, L.C., and B.B. Biesecker. 1998. Breast cancer susceptibility genes. BRCA1 and BRCA2. Medicine (Baltimore) 77(3):208-226.

Brown, D.L., K.R. Reuhl, S. Bormann, and J.E. Little. 1988. Effects of methyl mercury on the microtubule system of mouse lymphocytes. Toxicol. Appl. Pharmacol. 94(1):66-75.

Brown, J.R., H. Ye, R.T. Bronson, P. Dikkes, and M.E. Greenberg. 1996. A defect in nurturing in mice lacking the immediate early gene fosB. Cell 86(2):297-309.

Brown, N.A. 1987. Teratogenicity testing in vitro: Status of validation studies. [Review]. Arch. Toxicol. (Suppl.)11:105-114.

Brown, N.A., and S.J. Freeman. 1984. Alternative tests for teratogenicity. ATLA, Altern. Lab. Anim. 12(1):7-23.

Brown, N.A., M.E. Coakley, and D.O. Clarke. 1987. Pp. 17-31 in Approaches to Elucidate Mechanisms in Teratogenesis, F. Welsch, ed. Washington, DC: Hemisphere Pub.

Brown, N.A., J. Kao, and S. Fabro. 1980. Teratogenic potential of valproic acid. [letter]. Lancet 1(8169):660-661.

Brown, N.A., H. Spielmann, R. Bechter, O.P. Flint, S.J. Freeman, R.J. Jelinek, E. Koch, H. Nau, D.R. Newall, A.K. Palmer, J.Y. Renault, M.F. Repetto, R. Vogel, and R. Wiger. 1995. Screening chemicals for reproductive toxicity: The current alternatives: The report and recommendations of an ECVAM/ETS workshop (ECVAM Workshop 12). ATLA, Altern. Lab. Anim. 23(6):868-882.

Brown, P.O., and D. Botstein. 1999. Exploring the new world of the genome with DNA microarrays. Nat. Genet. 21(1 Suppl.):33-37.

Brunet, L.J., J.A. McMahon, A.P. McMahon, and R.M. Harland. 1998. Noggin, cartilage morphogenesis, and joint formation in the mammalian skeleton. Science 280(5368):1455-1457.

Bryson, S.E., and I.M. Smith. 1998. Epidemiology of autism: Prevalence, associated characteristics and implications for research and service delivery. MRDD Res. Rev. 4:97-103.

Bryson, S.E., B.S. Clark, and I.M. Smith. 1988. First report of a Canadian epidemiological study of autistic syndromes. J. Child Psychol. Psychiatry 29(4):433-445.

Buehler, B.A. 1984. Epoxide hydrolase activity and the fetal hydantoin syndrome. Proc. Greenwood Genet. Ctr. 3:109-110.

Buehler, B.A., D. Delimont, M. van Waes, and R.H. Finnell. 1990. Prenatal prediction of risk of the fetal hydantoin syndrome. N. Engl. J. Med. 322(22):1567-1572.

Bui, L.M., M.W. Taubeneck, J.F. Commisso, J.Y. Uriu-Hare, W.D. Faber, and C.L. Keen. 1998. Altered zinc metabolism contributes to the developmental toxicity of 2-ethylhexanoic acid, 2-ethylhexanol and valproic acid. Toxicology 126(1):9-21.

Burbacher, T.M., P.M. Rodier, and B. Weiss. 1990. Methylmercury developmental neurotoxicity: A comparison of effects in humans and animals. Neurotoxicol. Teratol. 12(3):191-202.

Burke, A.C. 2000. Hox genes and the global patterning of the somitic mesoderm. Curr. Top. Dev. Biol. 47:155-181.

Cahill, G.M., M.W. Hurd, and M.M. Batchelor. 1998. Circadian rhythmicity in the locomotor activity of larval zebrafish. Neuroreport 9(15):3445-3449.

California Environmental Protection Agency. 1991. Draft Guidelines for Hazard Identification and Dose-Response Assessment of Agents Causing Developmental and/or Reproductive Toxicity. California Dept. of Health Services, Health Hazard Assessment Division. Reproductive and Cancer Hazard Assessment Section. April, 3.

Carney, E.W., A.B. Liberacki, M.J. Bartels, and W.J. Breslin. 1996. Identification of proximate toxicant for ethylene glycol developmental toxicity using rat whole embryo culture. Teratology 53(1):38-46.

Carpenter, E.M, J.M. Goddard, O. Chisaka, N.R. Manley, and M.R. Capecchi. 1993. Loss of Hox-A1 (Hox-1.6) function results in the reorganization of the murine hindbrain. Development 118(4):1063-1075.

Cash, D.E., C.B. Bock, K. Schughart, E. Linney, and T.M. Underhill. 1997. Retinoic acid receptor alpha function in vertebrate limb skeletogenesis: A modulator of chondrogenesis. J. Cell Biol. 136(2):445-457.

CDC (Centers for Disease Control). 1987. Table V. Estimated years of potential life lost before age 65 and cause-specific mortality, by cause of death-United States, 1985. MMWR 36(20):313.

CDC (Centers for Disease Control). 1995. Economic costs of birth defects and cerebral palsy-United States, 1992. MMWR 44:694-699.

C. elegans Sequencing Consortium. 1998. Genome sequence of the nematode C. elegans: A platform for investigating biology. Science 282(5396):2012-2018.

Chakrabarti, S., K. Brechling, and B. Moss. 1985. Vaccinia virus expression vector: Coexpression of beta-galactosidase provides visual screening of recombinant virus plaques. Mol. Cell Biol. 5(12):3403-3409.

Chambon, P. 1996. A decade of molecular biology of retinoic acid receptors. FASEB J. 10(9):940-954.

Chan, A.W., T. Dominko, C.M. Luetjens, E. Neuber, C. Martinovich, L. Hewitson, C.R. Simerly, and G.P. Schatten. 2000. Clonal propagation of primate offspring by embryo splitting. Science 287(5451):317-319.

Chang, C., D.R. Smith, V.S. Prasad, C.L. Sidman, D.W. Nebert, and A. Puga. 1993.
Ten nucleotide differences, five of which cause amino acid changes, are associated with the Ah receptor locus polymorphism of C57BL/6 and DBA/2 mice. Pharmacogenetics 3(6):312-321.

Chanut, F., and U. Heberlein. 1997. Role of decapentaplegic in initiation and progression of the morphogenetic furrow in the developing Drosophila retina. Development 124(2):559-567.

Chappell, P.E, J.P. Lydon, O.M. Conneely, B.W. O'Malley, and J.E. Levine. 1997. Endocrine defects in mice carrying a null mutation for the progesterone receptor gene. Endocrinology 138(10):4147-4152.

Chazaud, C., P. Chambon, and P. Dolle. 1999. Retinoic acid is required in the mouse embryo for left-right asymmetry determination and heart morphogenesis. Development 126(12):2589-2596.

Chen, J., A. Nachabah, C. Scherer, P. Ganju, A. Reith, R. Bronson, and H.E. Ruley. 1996. Germ-line inactivation of the murine Eck receptor tyrosine kinase by gene trap retroviral insertion. Oncogene 12(5):979-988.

Chervitz, S.A., L. Aravind, G. Sherlock, C.A. Ball, E.V. Koonin, S.S. Dwight, M.A. Harris, K. Dolinski, S. Mohr, T. Smith, S. Weng, J.M. Cherry, and D. Botstein. 1998. Comparison of the complete protein sets of worm and yeast: Orthology and divergence. Science 282(5396):2022-2028.

Chiang, C., Y. Litingtung, E. Lee, K.E. Young, J.L.Corden, H. Westphal, and P.A. Beachy. 1996. Cyclopia and defective axial patterning in mice lacking sonic hedgehog gene function. Nature 383:407-413.

Chisaka, O., T.S. Musci, and M.R. Capecchi. 1992. Developmental defects of the ear, cranial nerves and hindbrain resulting from targeted disruption of the mouse homeobox gene Hox-1.6. Nature 355(6360):516-520.

Chou, T.B., and N. Perrimon. 1996. The autosomal FLP-DFS technique for generating germline mosaics in Drosophila melanogaster. Genetics 144:1673-1679.

Christianson, A.L., N. Chesler, and J.G. Kromberg. 1994. Fetal valproate syndrome: clinical and neuro-developmental features in two sibling pairs. Dev. Med. Child Neurol. 36(4):361-369.

Chu, S., J. DeRisi, M. Eisen, J. Mulholland, D. Botstein, P.O. Brown, and I. Herskowitz. 1998. The transcriptional program of sporulation in budding yeast. Science 282(5389):699-705.

Clark, R.L., R.T. Robertson, D.H. Minsker, S.M. Cohen, D.J. Tocco, H.L. Allen, M.L. James, and D.L. Bokelman. 1984. Diflunisal-induced maternal anemia as a cause of teratogenicity in rabbits. Teratology 30(3):319-332.

Clarke, D.O., J.M. Duignan, and F. Welsch. 1992. 2-Methoxyacetic acid dosimetry-teratogenicity relationships in CD-1 mice exposed to 2-methoxyethanol. Toxicol. Appl. Pharmacol. 114(1):77-87.

Clarke, D.O., B.A. Elswick, F. Welsch, and R.B. Conolly. 1993. Pharmacokinetics of 2-methoxyethanol and 2-methoxyacetic acid in the pregnant mouse: A physiological based mathematical model. Toxicol. Appl. Pharmacol. 121:239-252.

Clarkson, T.W. 1987. The role of biomarkers in reproductive and developmental toxicology. Environ. Health Perspect. 74:103-107.

Clarkson, T.W. 1993. Mercury: Major issues in environmental health. Environ. Health Perspect. 100:31-38.

Coberly, S., E. Lammer, and M. Alashari. 1996. Retinoic acid embryopathy: Case report and review of literature. Pediatr. Pathol. Lab. Med. 16(5):823-836.

Collins, F.S., A. Patrinos, E. Jordan, A. Chakravarti, R. Gesteland, and L. Walters. 1998. New goals for the U.S. Human Genome Project: 1998-2003. Science 282(5389):682-689.

Collins, M.D., and G.E. Mao. 1999. Teratology of retinoids. Annu. Rev. Pharmacol. Toxicol. 39:399-430.

Collins, T.F., R.L. Sprando, D.L. Hansen, M.E. Shackelford, and J.J. Welsh. 1998. Testing guidelines for evaluation of reproductive and developmental toxicity of food additives in females. Int. J. Toxicol. 17(3):299-325.

Colussi, P.A., and S. Kumar. 1999. Targeted disruption of caspase genes in mice: What they tell us about the functions of individual caspases in apoptosis. Immunol. Cell Biol. 77(1):58-63.

Colvin, J.S., B.A. Bohne, G.W. Harding, D.G. McEwen, and D.M. Ornitz. 1996. Skeletal overgrowth and deafness in mice lacking fibroblast growth factor receptor 3. Nat. Genet. 12(4):390-397.

Conlon, R.A., and J. Rossant. 1992. Exogenous retinoic acid rapidly induces anterior ectopic expression of murine Hox-2 genes in vivo. Development 116(2):357-368.

Conlon, R.A., A.G. Reaume, and J. Rossant. 1995. Notch1 is required for the coordinate segmentation of somites. Development 121(5):1533-1545.

Cooper, M.K., J.A. Porter, K.E. Young, and P.A. Beachy. 1998. Teratogen-mediated inhibition of target tissue response to Shh signaling. Science 280(5369):1603-1607.

Couture, L.A., B.D. Abbott, and L.S. Birnbaum. 1990. A critical review of the developmental toxicity and teratogenicity of 2,3,7,8-tetrachlorodibenzo-p-dioxin: Recent advances toward understanding the mechanism. Teratology 42(6):619-627.

Crabb, D.W. 1990. Biological markers for increased risk of alcoholism and for quantitation of alcohol consumption. J. Clin. Invest. 85(2):311-315.

Crespi, C.L., and V.P. Miller. 1999. The use of heterologously expressed drug metabolizing enzymes—state of the art and prospects for the future. Pharmacol. Ther. 84(2):121-131.

Cresteil, T. 1998. Onset of xenobiotic metabolism in children: Toxicological implications. Food Addit. Contam. 15(Suppl.):45-51.

Crowley, C., S.D. Spencer, M.C. Nishimura, K.S. Chen, S. Pitts-Meek, M.P. Armanini, L.H. Ling, S.B. MacMahon, D.L. Shelton, A.D. Levinson, et al. 1994. Mice lacking nerve growth factor display perinatal loss of sensory and sympathetic neurons yet develop basal forebrain cholinergic neurons. Cell 76(6):1001-1011.

Crump, K.S. 1984. An improved procedure for low-dose carcinogenic risk assessment from animal data. J. Environ. Pathol. Toxicol. Oncol. 5(4-5):339-348.

Cunha, G.R., J.G. Forsberg, R. Golden, A. Haney, T. Iguchi, R. Newbold, S. Swan, and W . Welshons. 1999. New approaches for estimating risk from exposure to diethylstilbestrol. Environ. Health Perspect. 107(Suppl. 4):625-630.

Cushnir, J.R., S. Naylor, J.H. Lamb, P.B. Farmer, N.A. Brown, and P.E. Mirkes. 1990. Identification of phosphoramide mustard/DNA adducts using tandem mass spectrometry. Rapid. Commun. Mass Spectrom. 4(10):410-414.

Dareste, M.C. 1877. La production artificielle des monstruosites. C. Reinwald et Cie, Paris.

Dasen, J.S., and M.G. Rosenfeld. 1999. Signaling mechanisms in pituitary morphogenesis and cell fate determination. Curr. Opin. Cell Biol. 11(6):669-677.

Daston, G.P. 1996. The theoretical and empirical case for in vitro developmental toxicity screens, and potential applications. Teratology 53(6):339-344.

Daston, G.P., and L.D. Lehman-McKeeman. 1996. Constitutive and induced metalothionein expression in development. Pp. 1139-1151 in Toxicology of Metals, L.W. Chang, ed. Boca Raton, FL: CRC Press.

Daston, G.P., G.J. Overmann, M.W. Taubeneck, L.D. Lehman-McKeeman, J.M. Rogers, and C.L. Keen. 1991a. The role of metallothionein induction and altered zinc status in maternally mediated developmental toxicity: Comparison of the effects of urethane and styrene in rats. Toxicol. Appl. Pharmacol. 110(3):450-463.

Daston, G.P., J.M. Rogers, D.J. Versteeg, T.D. Sabourin, D. Baines, and S.S. Marsh. 1991b. Interspecies comparisons of A/D ratios: A/D ratios are not constant across species. Fundam. Appl. Toxicol. 17(4):696-722.

Daston, G.P., G.J. Overmann, D. Baines, M.W. Taubeneck, L.D. Lehman-McKeeman, J.M. Rogers, and C.L. Keen. 1994. Altered Zn status by alpha-hederin in the pregnant rat and its relationship to adverse developmental outcome. Reprod. Toxicol. 8(1):15-24.

Daston, G.P., D. Baines, E. Elmore, M.P. Fitzgerald, and S. Sharma. 1995. Evaluation of chick embryo neural retina cell culture as a screen for developmental toxicants. Fundam. Appl. Toxicol. 26(2):203-210.

DeChiara, T.M., A. Efstratiadis, and E.J. Robertson. 1990. A growth-deficiency phenotype in heterozygous mice carrying an insulin-like growth factor II gene disrupted by targeting. Nature 345(6270):78-80.

Deng, C., M. Bedford, C. Li, X. Xu, X. Yang, J. Dunmore, and P. Leder. 1997. Fibroblast growth factor receptor-1 (FGFR-1) is essential for normal neural tube and limb development. Dev. Biol. 185(1):42-54.

DeSesso, J.M. 1979. Cell death and free radicals: A mechanism for hydroxyurea teratogenesis. Med. Hypotheses 5(9):937-951.

DeSesso, J.M., and G.C. Goeringer. 1990. The nature of the embryo-protective interaction of propyl gallate with hydroxyurea. Reprod. Toxicol. 4(2):145-152.

DeSesso, J.M., and G.C. Goeringer. 1991. Amelioration by leucovorin of methotrexate developmental toxicity in rabbits. Teratology 43(3):201-215.

DeSesso, J.M., and G.C. Goeringer. 1992. Methotrexate-induced developmental toxicity in rabbits is ameliorated by 1-(p-tosyl)-3,4,4-trimethylimidazolidine, a functional analog for tetrahydrofolate-mediated one-carbon transfer. Teratology 45(3):271-283.

DeSesso, J.M., A.R. Scialli, and G.C. Goeringer. 1994. D-mannitol, a specific hydroxyl free radical scavenger, reduces the developmental toxicity of hydroxyurea in rabbits. Teratology 49(4):248-259.

Dickson, D. 1999. Gene estimate rises as US and UK discuss freedom of access. Nature 401(6751):311.

Downward, J., Y. Yarden, E. Mayes, G. Scrace, N. Totty, P. Stockwell, A. Ullrich, J. Schlessinger, and M.D. Waterfield. 1984. Close similarity of epidermal growth factor receptor and v-erb-B oncogene protein sequences. Nature 307(5951):521-527.

Dubnau, J., and T. Tully. 1998. Gene discovery in Drosophila: New insights for learning and memory. Annu. Rev. Neurosci. 21:407-444.

Duboule, D. 1998. Vertebrate hox gene regulation: Clustering and/or colinearity? Curr. Opin. Genet. Dev. 8(5):514-518.

Dudley, A.T., and E.J. Robertson. 1997. Overlapping expression domains of bone morphogenetic protein family members potentially account for limited tissue defects in BMP7 deficient embryos. Dev. Dyn. 208(3):349-362.

Duester, G. 1991. A hypothetical mechanism for fetal alcohol syndrome involving ethanol inhibition of retinoic acid synthesis at the alcohol dehydrogenase step. Alcohol Clin. Exp. Res. 15(3):568-572.

Dunham, I., N. Shimizu, B.A. Roe, S. Chissoe, A.R. Hunt, J.E. Collins, R. Bruskiewich, D.M. Beare, M. Clamp, L.J. Smink, R. Ainscough, J.P. Almeida, A. Babbage, C. Bagguley, J. Bailey, K. Barlow, K.N. Bates, O. Beasley, C.P. Bird, S. Blakey,

Dupe, V., N.B. Ghyselinck, V. Thomazy, L. Nagy, P.J.A. Davies, P. Chambon, and M. Mark. 1999. Essential roles of retinoic acid signaling in interdigital apoptosis and control of BMP-7 expression in mouse autopods. Dev. Biol. 208(1):30-43.

Durbin, J.E., R. Hackenmiller, M.C. Simon, and D.E. Levy. 1996. Targeted disruption of the mouse Stat1 gene results in compromised innate immunity to viral disease. Cell 84(3):443-450.

Dwivedi, R.S., and P.M. Iannaccone. 1998. Effects of environmental chemicals on early development. Pp. 11-46 in Reproductive and Developmental Toxicology, K.S. Korach, ed. New York: Marcel Dekker.

Eastin, W.C., J.K. Haseman, J.F. Mahler, and J.R. Bucher. 1998. The National Toxicology Program evaluation of genetically altered mice as predictive models for identifying carcinogens. Toxicol. Pathol. 26(4):461-473.

Ebert, J.D., and M. Marois. 1976. Tests of Teratogenicity in Vitro: Proceedings of the International Conference on Tests of Teratogenicity In Vitro, Woods Hole, Mass., 1-4 April, 1975. New York: North-Holland Publishing Co.

Eckert, D., and M. Merchlinsky. 1999. Vaccinia virus-bacteriophage T7 expression vector for complementation analysis of late gene processes. J. Gen. Virol. 80(Pt. 6):1463-1469.

Edlund, T., and T.M. Jessell. 1999. Progression from extrinsic to intrinsic signaling in cell fate specification: A view from the nervous system. Cell 96(2):211-224.

Eisen, M.B., P.T. Spellman, P.O. Brown, and D. Botstein. 1998. Cluster analysis and display of genome-wide expression patterns. Proc. Natl. Acad. Sci. U.S.A. 95(25):14863-14868.

Eisenhardt, E.U., and M.H. Bickel. 1994. Kinetics of tissue distribution and elimination of retinoid drugs in the rat. II. Etretinate. Drug Metab. Dispos. 22(1):31-35.

Ellsworth, D.L., D.M. Hallman, and E. Boerwinkle. 1997. Impact of the human genome project on epidemiologic research. Epidemiol. Rev. 19(1):3-13.

Elmazar, M.M., U. Reichert, B. Shroot, and H. Nau. 1996. Pattern of retinoid-induced teratogenic effects: Possible relationship with relative selectivity for nuclear retinoid receptors RAR alpha, RAR beta, and RAR gamma. Teratology 53(3):158-167.

Elmazar, M.M., R. Ruhl, U. Reichert, B. Shroot, and H. Nau. 1997. RARalpha-mediated teratogenicity in mice is potentiated by an RXR agonist and reduced by an RAR antagonist: Dissection of retinoid receptor-induced pathways. Toxicol. Appl. Pharmacol. 146(1):21-28.

Emmert-Buck, M.R., R.F. Bonner, P.D. Smith, R.F. Chuaqui, Z. Zhuang, S.R. Goldstein, R.A. Weiss, and L.A. Liotta. 1996. Laser capture microdissection. Science 274(5289):998-1001.

Ensenbach, U., and R. Nagel. 1995. Toxicity of complex chemical mixtures: Acute and long-term effects on different life stages of zebrafish (Brachydanio rerio). Ecotoxicol. Environ. Saf. 30(2):151-157.

Ensenbach, U., and R. Nagel. 1997. Toxicity of binary chemical mixtures: Effects on reproduction of zebrafish. Arch. Environ. Contam. Toxicol. 32(2):204-210.

EPA (U.S. Environmental Protection Agency). 1990. Fungicide and Rodenticide Act (FIFRA): Good Laboratory Practice Standards. Final Rule. Fed. Regist. 54(158):34052-34074.

EPA (U.S. Environmental Protection Agency). 1991. Guidelines for Developmental Toxicity Risk Assessment; Notice. Part V. Fed. Regist. 56(234):63798-63826.

EPA (U.S. Environmental Protection Agency). 1995. Guidance for Risk Characterization. Science Policy Council. ORD. February. [Online]. Available: http://search.epa.gov/ordntrnt/ORD/spc/rcguide.htm

EPA (U.S. Environmental Protection Agency). 1996a. Guidelines for Reproductive Toxicity Risk Assessment. Fed. Regist. 61(212):56274-56322.

EPA (U.S. Environmental Protection Agency). 1996b. Proposed Guidelines for Carcinogen Risk Assessment. EPA/600/P-92/003C. Office of Research and Development. Washington, DC.

EPA (U.S. Environmental Protection Agency). 1997. TSCA Test Guidelines, 40 CFR Part 799. Fed. Regist. 62(158):43820-438649.

EPA (U.S. Environmental Protection Agency). 1998a. Chemical Hazard Data Availability Study: What Do We Really Know About the Safety of High Production Volume Chemicals? Office of Pollution Prevention and Toxics. [Online]. Available: http://www.epa.gov/opptintr/chemtest/hazchem.htm

EPA (U.S. Environmental Protection Agency). 1998b. Endocrine Disruptor Screening and Testing Advisory Committee (EDSTAC), Final Report. Office of Prevention, Pesticides, and Toxic Substances. [Online]. Available: http://www.epa.gov/opptintr/opptendo/finalrpt

EPA (U.S. Environmental Protection Agency). 1998c. Health Effects Test Guidelines, OPPTS 870.3800, Reproduction and Fertility Effects. EPA 712-C-98-208, August 1998. [Online]. Available: http://www.epa.gov/opptsfrs/OPPTS_Harmonized/

EPA (U.S. Environmental Protection Agency). 1998d. Health Effects Test Guidelines, OPPTS 870.3700, Prenatal developmental Toxicity Study. EPA 712-C-98-207, August 1998. [Online]. Available: http://www.epa.gov/opptsfrs/OPPTS_Harmonized/

EPA (U.S. Environmental Protection Agency). 1998e. Health Effects Test Guidelines, OPPTS 870.6300, Developmental Neurotoxicity Study. EPA 712-C-98-239, August 1998. [Online]. Available: http://www.epa.gov/opptsfrs/OPPTS_Harmonized/

EPA (U.S. Environmental Protection Agency). 1998f. EPA Guidelines for Neurotoxicity Risk Assessment FRL-6011-3. NTIS PB98-117831. Effective April 30, 1998. [Online]. Available: http://www.epa.gov/ncea

EPA (U.S. Environmental Protection Agency). 1998g. Endocrine Disruptor Screening Program. Fed. Regist. 63(248) (December 28).

Everson, R.B. 1987. A review of approaches to the detection of genetic damage in the human fetus. Environ. Health Perspect. 74:109-117.

Everson, R.B., E. Randerath, T.A. Avitts, H.A. Schut, and K. Randerath. 1987. Preliminary investigations of tissue specificity, species specificity, and strategies for identifying chemicals causing DNA adducts in human placenta. Prog. Exp. Tumor Res. 31:86-103.

Everson, R.B., E. Randerath, R.M. Santella, T.A. Avitts, I.B. Weistein, and K. Randerath. 1988. Quantitative associations between DNA damage in human placenta, maternal smoking and birth weight. J. Natl. Cancer Inst. 80(8):567-576.

Eyre-Walker, A., and P.D. Keightley. 1999. High genomic deleterious mutation rates in hominids. Nature 397(6717):344-347.

Fambrough, D., K. McClure, A. Kazlauskas, and E.S. Lander. 1999. Diverse signaling pathways activated by growth factor receptors induce broadly overlapping, rather than independent, sets of genes. Cell 97(6):727-741.

Fantel, A.G. 1996. Reactive oxygen species in developmental toxicity: Review and hypothesis. [Review]. Teratology 53(3):196-217.

Faustman, E.M. 1988. Short-term tests for teratogens. [Review]. Mutat. Res. 205(1-4):355-384.

Faustman, E.M., T.A. Lewandowski, R.A. Ponce, and S.M. Bartell. 1999. Biologically cased dose-responce models for developmental toxicants: Lessons from methylmercury. Inhalaton Toxicol. 11:559-572.

Faustman, E.M., R.A. Ponce, M.R. Seeley, and S.G. Whittaker. 1997. Experimental approaches to evaluate mechanisms of developmental toxicity. Pp.13-41 in Handbook of Developmental Toxicology, R. Hood, ed. Boca Raton, FL: CRC Press.

Faustman, E.M., S.M. Silbernagel, R.A. Fenske, T.M. Burbacher, and R.A. Ponce. 2000. Mechanisms underlying children's susceptibility to environmental toxicants. Environ. Health Perspect. 108(Suppl.1):13-21.

Fawcett, L.B., S.J. Buck, D.A. Beckman, and R.L. Brent. 1996. Is there a no-effect dose for corticosteroid-induced cleft palate? The contribution of endogenous corticosterone to the incidence of cleft palate in mice. Pediatr. Res. 39(5):856-861.

FDA (U.S. Food and Drug Administration). 1987. Good Laboratory Practice Regulation for Nonclinical Laboratory Studies, Regulatory Program. 21 CFR 58.

FDA (U.S. Food and Drug Administration). 1994. International Conference on Harmonisation: Guideline on Detection of Toxicity to Reproduction for Medicinal Products; Availability. Fed. Regist. 59(183):48746-48752. September 22, 1994.

Fekete, D.M. 1999. Development of the vertebrate ear: Insights from knockouts and mutants. Trends Neurosci. 22(6):263-269.

Feldman, B., W. Poueymirou, V.E. Papaioannou, T.M. DeChiara, and M. Goldfarb. 1995. Requirement of FGF-4 for postimplantation mouse development. Science 267(5195):246-249.

Feng, D-F., G. Cho, and R.F. Doolittle. 1997. Determining divergence times with a protein clock: Update and reevaluation. Proc. Natl. Acad. Sci. U.S.A. 94:13028-13033.

Fernandez-Salguero, P.M., D.M. Hilbert, S. Rudikoff, J.M. Ward, and F.J. Gonzalez. 1996. Arylhydrocarbon receptor-deficient mice are resistant to 2,3,7,8-tetrachlorodibenzo-p-dioxin-induced toxicity. Toxicol. Appl. Pharmacol. 140(1):173-179.

Ferretti, P., and C. Tickle. 1997. The limbs. Pp.101-132 in Embryos, Genes and Birth Defects, P. Thorogood, ed. Chichester: John Wiley & Sons.

Fex, G., K. Larsson, A. Andersson, and M. Berggren-Soderlund. 1995. Low serum concentration of all-trans and 13-cis retinoic acids in patients treated with phenytoin, carbamazepine and valproate. Possible relation to teratogenicity. Arch. Toxicol. 69(8):572-574.

Finnell, R.H., B. Bielec, and H. Nau. 1997a. Anticonvulsant drugs: Mechanisms and pathogenesis of teratogenicity Pp. 121-159 in Drug Toxicity in Embryonic Development II., Handbook of Experimental Pharmacology, Vol. 124/II, R.J. Kavlock, and G.P. Daston, eds. Heidelberg: Springer.

Finnell, R.H., B.C. Wlodarczyk, J.C. Craig, J.A. Piedrahita, and G.D. Bennett. 1997b. Strain-dependent alterations in the expression of folate pathway genes following teratogenic exposure to valproic acid in a mouse model. Am. J. Med. Genet. 70(3):303-311.

Fire, A., S. Xu, M.K. Montgomery, S.A. Kostas, S.E. Driver, and C.C. Mello. 1998. Potent and specific genetic interference by double-stranded RNA in Caenorhabditis elegans. Nature 391:806-811.

Flint, O.P. 1993. In vitro tests for teratogens: Desirable endpoints, test batteries and current status of the micromass teratogen test. [Review]. Reprod. Toxicol. 7(Suppl. 1):103-111.

Flint, O.P., and F.T. Boyle. 1985. An in vitro test for teratogens: Its application in the selection of non-teratogenic triazole antifungals. Concepts Toxicol. 3:29-35.

Floss, T., H.H. Arnold, and T. Braun. 1997. A role for FGF-6 in skeletal muscle regeneration. Genes Dev. 11(16):2040-2051.

Fong, G.H., J. Rossant, M. Gertsenstein, and M.L. Breitman. 1995. Role of the Flt-1 receptor tyrosine kinase in regulating the assembly of vascular endothelium. Nature 376(6535):66-70.

Fort, D.J., E.L. Stover, J.A. Bantle, J.R. Rayburn, M.A. Hull, R.A. Finch, D.T. Burton, S.D. Turley, D.A. Dawson, G. Linder, D. Buchwalter, M. Kumsher-King, and A.M. Gaudet-Hull. 1998. Phase III interlaboratory study of FETAX, Part 2: Interlaboratory validation of an exogenous metabolic activation system for frog embryo teratogenesis assay-Xenopus(FETAX). Drug Chem. Toxicol. 21(1):1-14.

Francis, N.J., and S.C. Landis. 1999. Cellular and molecular determinants of sympathetic neuron development. Annu. Rev. Neurosci. 22:541-566.

Francis, E.Z., C.A. Kimmel, and D.C. Rees. 1990. Workshop on the qualitative and quantitative comparability of human and animal developmental neurotoxicity: Summary and implications. Neurotoxicol. Teratol. 12(3):285-292.

Francis-West, P., R. Ladher, A. Barlow, and A. Graveson. 1998. Signalling interactions during facial development. Mech. Dev. 75(1-2):3-28.

Freese, E. 1982. Use of cultured cells in the identification of potential teratogens. Teratog. Carcinog. Mutagen. 2(3-4):355-360.

Fretts, R.C., J. Schmittdiel, F.H. McLean, R.H. Usher, and M.B. Goldman. 1995. Increased maternal age and the risk of fetal death. N. Engl. J. Med. 333(15):953-957.

Friedman, J.M., and J. E. Polifka. 1994. Teratogenic Effects of Drugs: A Resource for Clinicians: TERIS. Baltimore: Johns Hopkins University Press. 703pp.

Galitski, T., A.J. Saldanha, C.A. Styles, E.S. Lander, and G.R. Fink. 1999. Ploidy regulation of gene expression. Science 285(5425):251-254.

Gallo, M.A. 1996. History and scope of toxicology. Pp. 3-11 in Casarett and Doull's Toxicology, The Basic Science of Poisons, 5th Ed., C.D. Klaassen, M.O. Amdur, and J. Doull, eds. New York: McGraw-Hill.

Geisler, R., G.J. Rauch, H. Baier, F. van Bebber, L. Brobeta, M.P. Dekens, K. Finger, C. Fricke, M.A. Gates, H. Geiger, S. Geiger-Rudolph, D. Gilmour, S. Glaser, L. Gnugge, H. Habeck, K. Hingst, S. Holley, J. Keenan, A. Kirn, H. Knaut, D. Lashkari, F. Maderspacher, U. Martyn, S. Neuhauss, P. Haffter, et al. 1999. A radiation hybrid map of the zebrafish genome. Nat. Genet. 23(1):86-89.

Genbacev, O., T.E. White, C.E. Gavin, and R.K. Miller. 1993. Human trophoblast cultures: Models for implantation and peri-implantation toxicology. Reprod. Toxicol. 7(Suppl. 1):75-94.

Generoso, W.M., J.C. Rutledge, K.T. Cain, L.A. Hughes, and P.W. Braden. 1987. Exposure of female mice to ethylene oxide within hours after mating leads to fetal malformation and death. Mutat. Res. 176(2):269-274.

Gershon, M.D. 1999a. Endothelin and the development of the enteric nervous system. Clin. Exp. Pharmacol. Physiol. 26(12):985-988.

Gershon, M.D. 1999b. Lessons from genetically engineered animal models. II. Disorders of enteric neuronal development: Insights from transgenic mice. Am. J. Physiol. 277(2 Pt. 1):G262-267.

Ghyselinck, N.B., V. Dupe, A. Dierich, N. Messaddeq, J.M. Garnier, C. Rochette-Egly, P. Chambon, and M. Mark. 1997. Role of the retinoic acid receptor beta (RARbeta) during mouse development. Int. J. Dev. Biol. 41(3):425-447.

Goffeau, A., B.G. Barrell, H. Bussey, R.W. Davis, B. Dujon, H. Feldmann, F. Galibert, J.D. Hoheisel, C. Jacq, M. Johnston, E.J. Louis, H.W. Mewes, Y. Murakami, P. Philippsen, H. Tettelin, and S.G. Oliver. 1996. Life with 6000 genes. Science 274:546-567.

Goldberg, J.M., and T. Falcone. 1999. Effect of diethylstilbestrol on reproductive function. Fertil. Steril. 72(1):1-7.

Goldman, L.R., B.J. Apelberg, S. Koduru, C.E. Ward, and R. Sorian. 1999. Healthy from the Start: Why America Needs a Better System to Track and Understand Birth Defects and the Environment. Pew Environmental Health Commission. [Online]. Available: http://pewenvirohealth.jhsph.edu/html/home/home.html.

Golic, K.G., and S. Lindquist. 1989. The FLP recombinase of yeast catalyzes site-specific recombination in the Drosophila genome. Cell 59:499-509.

Goodrich, L.V., L. Milenkovic, K.M. Higgins, and M.P. Scott. 1997. Altered neural cell fates and medulloblastoma in mouse patched mutants. Science 277(5329):1109-1113.

Graillet, C., G. Pagano, and J.P. Girard. 1993. Stage-specific effects of teratogens on sea-urchin embryogenesis. Teratog. Carcinog. Mutagen. 13(1):1-14.

Granato, M., F.J. van Eeden, U. Schach, T. Trowe, M. Brand, M. Furutani-Seiki, P. Haffter, M. Hammerschmidt, C.P. Heisenberg, Y.J. Jiang, D.A. Kane, R.N. Kelsh, M.C. Mullins, J. Odenthal, and C. Nüsslein-Volhard. 1996. Genes controlling and mediating locomotion behavior of the zebrafish embryo and larva. Development 123:399-413.

Gregg, N.M. 1941. Congenital cataract following German measles in the mother. Trans. Ophthalmol. Soc. Aust. 3:35-45.

Gretchen, J.D. 1999. Molecular mechanisms of liver development and differentiation. Curr. Opin. Cell Biol. 11(6):678-682.

Gruda, M.C., J. van Amsterdam, C.A. Rizzo, S.K. Durham, S. Lira, and R. Bravo. 1996. Expression of FosB during mouse development: Normal development of FosB knockout mice. Oncogene 12(10):2177-2185.

Gurdon, J.B. 1960. Factors responsible for the abnormal development of embryos obtained by nuclear transplantation in Xenopus laevis. J. Embryol. Exp. Morphol. 8:327-341.

Guo, L., L. Degenstein, and E. Fuchs. 1996. Keratinocyte growth factor is required for hair development but not for wound healing. Genes. Dev. 10(2):165-175.

Gurtoo, H.L., C.J. Williams, K. Gottlieb, A.I. Mulhern, L. Caballes, J.B. Vaught, A.J. Marinello, and S.K. Bansal. 1983. Population distribution of placental benzo(a)pyrene metabolism in smokers. Int. J. Cancer 31:29-37.

Haffter, P., and C. Nüsslein-Volhard. 1996. Large scale genetics in a small vertebrate, the zebrafish. Int. J. Dev. Biol. 40(1):221-227.

Haffter, P., M. Granato, M. Brand, M.C. Mullins, M. Hammerschmidt, D.A. Kane, J. Odenthal, F.J. van Eeden, Y.J. Jiang, C.P. Heisenberg, R.N. Kelsh, M. Furutani-Seiki, E. Vogelsang, D. Beuchle, U. Schach, C. Fabian, and C. Nüsslein-Volhard. 1996. The identification of genes with unique and essential functions in the development of the zebrafish, Danio rerio. Development 123:1-36.

Hahn, H., L. Wojnowski, A.M. Zimmer, J. Hall, G. Miller, and A. Zimmer. 1998. Rhabdomyosarcomas and radiation hypersensitivity in a mouse model of Gorlin syndrome. Nat. Med. 4(5):619-622.

Hale, F. 1933. Pigs born without eye balls. J. Hered. 24(3):105-106.

Hall, J.C. 1998. Molecular neurogenetics of biological rhythms. J. Neurogenet. 12(3):115-181.

Hanselaar, A., M. van Loosbroek, O. Schuurbiers, T. Helmerhorst, J. Bulten, and J. Bernhelm. 1997. Clear cell adenocarcinoma of the vagina and cervix. An update of the central Netherlands registry showing twin age incidence peaks. Cancer 79(11):2229-2236.

Hanson, J.W., N.C. Myrianthopoulos, M.A. Harvey, and D.W. Smith. 1976. Risks to the offspring of women treated with hydantoin anticonvulsants, with emphasis on the fetal hydantoin syndrome. J. Pediatr. 89(4):662-668.

Harada, Y., and K. Noda. 1988. How the discovery of methyl mercury poisoning in Minamata district came about. Teratology 38(5):544.

Harborne, J.B., and H. Baxter. 1996. Dictionary of Plant Toxins. New York: Wiley. 540pp.

Harland, R., and J. Gerhart. 1997. Formation and function of Spemann's organizer. Ann. Rev. Cell Dev. Biol. 13:611-667.

Harmon, M.A., M.F. Boehm, R.A. Heyman, and D.J. Mangelsdorf. 1995. Activation of mammalian retinoid X receptors by the insect growth regulator methoprene. Proc. Natl. Acad. Sci. U.S.A. 92(13):6157-6160.

Hartl, D.L., and C.H. Taubes. 1998. Towards a theory of evolutionary adaptation. Genetica 102-103(1-6):525-533.

Hassett, C., L. Aicher, J.S. Sidhu, and C.J. Omiecinski. 1994. Human microsomal epoxide hydrolase: Genetic polymorphism and functional expression in vitro of amino acid variants. Hum. Mol. Genet. 3(3):421-428.

Hatasaka, H.H. 1994. Recurrent miscarriage: Epidemiologic factors, definitions, and incidence. Clin. Obstet. Gynecol. 37(3):625-634.

Hauck, R.S., and H. Nau. 1992. The enantiomers of the valproic acid analogue 2-n-propyl-4-pentynoic acid (4-yn-VPA): Asymmetric synthesis and highly stereoselective teratogenicity in mice. Pharm. Res. 9(7):850-855.

Heberlein, U., C.M. Singh, A.Y. Luk, and T.J. Donohoe. 1995. Growth and differentiation in the Drosophila eye coordinated by hedgehog. Nature 373(6516):709-711.

Hebert, J.M., T. Rosenquist, J. Gotz, and G.R. Martin. 1994. FGF5 as a regulator of the hair growth cycle: Evidence from targeted and spontaneous mutations. Cell 78(6):1017-1025.

Hemminki, K., and P. Vineis. 1985. Extrapolation of the evidence on teratogenicity of chemicals between humans and experimental animals: Chemicals other than drugs. Teratog. Carcinog. Mutagen. 5(4):251-318.

Henkemeyer, M., D.J. Rossi, D.P. Holmyard, M.C. Puri, G. Mbamalu, K. Harpal, T.S. Shih, T. Jacks, and T. Pawson. 1995. Vascular system defects and neuronal apoptosis in mice lacking ras GTPase-activating protein. Nature 377(6551):695-701.

Henry, E.C., and R.K. Miller. 1986. Comparison of the disposition of diethylstilbestrol and estradiol in the fetal rat. Correlation with teratogenic potency. Biochem. Pharmacol. 35(12):1993-2001.

Henry, E.C., R.K. Miller, and R.B. Baggs. 1984. Direct fetal injections of diethylstilbestrol and 17 beta-estradiol: A method for investigating their teratogenicity. Teratology 29(2):297-304.

Henry, T.R., J.M. Spitsbergen, M.W. Hornung, C.C. Abnet, and R.E. Peterson. 1997.
Early life stage toxicity of 2,3,7,8-tetrachlorodibenzo-p-dioxin in zebrafish (Danio rerio). Toxicol. Appl. Pharmacol. 142(1):56-68.

Herbst, A.L. 1981. Clear cell adenocarcinoma and the current status of DES-exposed females. Cancer 48(2 Suppl.):484-488.

Herbst, A.L., H. Ulfelder, and D.C. Poskanzer. 1971. Adenocarcinoma of the vagina. Association of maternal stilbestrol therapy with tumor appearance in young women. N. Engl. J. Med. 284(15):878-881.

Herrmann, K. 1993. Effects of the anticonvulsant drug valproic acid and related substances on the early development of the zebrafish (Brachydanio-rerio). Toxicol. in Vitro 7(1):41-54.

Hilberg, F., A. Aguzzi, N. Howells, and E.F. Wagner. 1993. c-jun is essential for normal mouse development and hepatogenesis. Nature 365(6442):179-181.

Hirayama, K. 1980. Effect of amino acids on brain uptake of methyl mercury. Toxicol. Appl. Pharmacol. 55(2):318-323.

Hirayama, K. 1985. Effects of combined administration of thiol compounds and methylmercury chloride on mercury distribution in rats. Biochem. Pharmacol. 34(11):2030-2032.

Hishida, R., and H. Nau. 1998. VPA-induced neural tube defects in mice. I. Altered metabolism of sulfur amino acids and glutathione. Teratog. Carcinog. Mutagen. 18(2):49-61.

Hiss, W., Sr. 1880-1885. Anatomie Menschlicher Embryonen. Leipzig: F.C.W. Vogel.

Hogan, B.L., C. Thaller, and G. Eichele. 1992. Evidence that Hensen's node is a site of retinoic acid synthesis. Nature 359(6392):237-241.

Holland, P.W., J. Garcia-Fernandez, N.A. Williams, and A. Sidow. 1994. Gene duplications and the origins of vertebrate development. Dev. (Suppl.):125-33.

Holme, R.H., and K.P. Steel. 1999. Genes involved in deafness. Curr. Opin. Genet. Dev. 9(3):309-314.

Holmes, L.B. 1997. Impact of the detection and prevention of developmental abnormalities in human studies. Reprod. Toxicol. 11(2/3):267-269.

Holtfreter, J. 1947. Neural induction in explants which have passed through a sublethal cytolysis. J. Exp. Zool. 106:197-222.

Honda, K., T. Hatayama, K. Takahashi, and M. Yukioka. 1991. Heat-shock proteins in human and mouse embryonic cells after exposure to heat shock or teratogenic agents. Teratog. Carcinog. Mutagen. 11(5):235-244.

Honeycutt, D., L. Dunlap, H. Chen, and G. al Homsi. 1999. The Cost of Developmental Disabilities. Final Report. Task Order No. 021-09. RTI Project No. 6900-009. Center for Economics Research, Research Triangle Institute, NC.

Hooghe, R.J., and D. Ooms. 1995. Use of the fluorescence-activated cell sorter (FACS) for in vitro assays of developmental toxicity. Toxicol. in Vitro 9(3):349-354.

Horster, M.F., G.S. Braun, and S.M. Huber. 1999. Embryonic renal epithelia: Induction, nephrogenesis, and cell differentiation. Physiol. Rev. 79(4):1157-1191.

Hrabe de Angelis, M., J. McIntyre, II, and A. Gossler. 1997. Maintenance of somite borders in mice requires the Delta homologue DII1. Nature 386(6626):717-721.

Hwang, S.J., T.H. Beaty, I. McIntosh, T. Hefferon, and S.R. Pany. 1998. Association between homeobox-containing gene MSX1 and the occurrence of limb deficiency. Am. J. Med. Genet. 75:419-423.

Hwang, S.J., T.H. Beaty, S.R. Panny, N.A. Street, J.M. Joseph, S. Gordon, I. McIntosh, and C.A. Francomano. 1995. Association study of transforming growth factor alpha (TGF alpha) Taq1 polymorphism and oral clefts: Indication of a gene-environment interaction in a population-based sample of infants with birth defects. Am. J. Epidemiol. 141(7):629-636.

ICBD (International Clearinghouse for Birth Defects Monitoring Systems). 1991. Congenital Malformations Worldwide. Amsterdam: Elsevier. 220pp.

Imura, N., K. Miura, M. Inokawa, and S. Nakada. 1980. Mechanism of methylmercury cytotoxicity: By biochemical and morphological experiments using cultured cells. Toxicology 17(2):241-254.

Incardona, J.P., W. Gaffield, R.P. Kapur, and H. Roelink. 1998. The teratogenic Veratrum alkaloid cyclopamine inhibits sonic hedgehog signal transduction. Development 125(18):3553-3562.

Ingham, P.W. 1989. Drosophila development. Curr. Opin. Cell Biol. 1(6):1127-1131.

Ingram, J.L., and P.M. Rodier. 1998. Valproic acid alters expression of HOX A1 in rat embryos: A mechanism of teratogenesis? Teratology 57(4-5):191.

Iulianella, A., and D. Lohnes. 1997. Contribution of retinoic acid receptor gamma to retinoid-induced craniofacial and axial defects. Dev. Dyn. 209(1):92-104.

Iyer, N.V., S.W. Leung, and G.L. Semenza. 1998. The human hypoxia-inducible factor 1alpha gene: HIF1A structure and evolutionary conservation. Genomics 52(2):159-165.

Iyer, V.R., M.B. Eisen, D.T. Ross, G. Schuler, T. Moore, J.C.F. Lee, J.M. Trent, L.M. Staudt, J. Hudson, Jr., M.S. Boguski, D. Lashkari, D. Shalon, D. Botstein, and P.O. Brown. 1999. The transcriptional program in the response of human fibroblasts to serum. Science 283(5398):83-87.

Jacobs, P.A., and T.J. Hassold. 1995. The origin of numerical chromosome abnormalities. Adv. Genet. 33:101-133.

Jaffe, R. 1998. First trimester utero-placental circulation: Maternal-fetal interaction. J. Perinat. Med. 26(3):168-174.

Ji, Q.S., G.E. Winnier, K.D. Niswender, D. Horstman, R. Wisdom, M.A. Magnuson, and G. Carpenter. 1997. Essential role of the tyrosine kinase substrate phospholipase C-gamma1 in mammalian growth and development. Proc. Natl. Acad. Sci. U.S.A. 94(7):2999-3003.

Johnson, E.M., L.M. Newman, B.E.G. Gabel, T.F. Boerner, and L.A. Dansky. 1988. An analysis of the Hydra assay's applicability and reliability as a developmental toxicity prescreen. J. Am. Coll. Toxicol. 7(2):111-126.

Johnson, L., D. Greenbaum, K. Cichowski, K. Mercer, E. Murphy, E. Schmitt, R.T. Bronson, H. Umanoff, W. Edelmann, R. Kucherlapati, and T. Jacks. 1997. K-ras is an essential gene in the mouse with partial functional overlap with N-ras. Genes Dev. 11(19):2468-2481.

Johnson, R.L., and C.J. Tabin. 1997. Molecular models for vertebrate limb development. Cell 90(6):979-990.

Johnson, R.S., B.M. Spiegelman, and V. Papaioannou. 1992. Pleiotropic effects of a null mutation in the c-fos proto-oncogene. Cell 71(4):577-586.

Johnson, R.S., B. van Lingen, V.E. Papaioannou, and B.M. Spiegelman. 1993. A null mutation at the c-jun locus causes embryonic lethality and retarded cell growth in culture. Genes Dev. 7(7B):1309-1317.

Johnson, V.P., V.W. Swayze II, Y. Sato, and N.C. Andreasen. 1996. Fetal alcohol syndrome: Craniofacial and central nervous system manifestations. Am. J. Med. Genet. 61(4):329-339.

Jones, K.L., and D.W. Smith. 1973. Recognition of the fetal alcohol syndrome in early infancy. Lancet 2(7836):999-1001.

Jones, K.L., D.W. Smith, C.N. Ulleland, and A.P. Streissguth. 1973. Pattern of malformation in offspring of chronic alcoholic mothers. Lancet 1(7815):1267-1271.

Joshi, R.L., B. Lamothe, N. Cordonnier, K. Mesbah, E. Monthioux, J. Jami, and D. Bucchini. 1996. Targeted disruption of the insulin receptor gene in the mouse results in neonatal lethality. EMBO J. 15(7):1542-1547.

Juchau, M.R. 1980. Drug biotransformation in the placenta. Pharmacol. Ther. 8(3):501-524.

Juchau, M.R., S.T. Chao, and C.J. Omiecinski. 1980. Drug metabolism by the human fetus. Clin. Pharmacokinet. 5(4):320-339.

Kao, J., N.A. Brown, B. Schmid, E.H. Goulding, and S. Fabro. 1981. Teratogenicity of valproic acid: In vivo and in vitro investigations. Teratog. Carcinog. Mutagen. 1:367-382.

Kassim, N.M., S.W. McDonald, O. Reid, N.K. Bennett, D.P. Gilmore, and A.P. Payne. 1997. The effects of pre- and postnatal exposure to the nonsteroidal antiandrogen flutamide on testis descent and morphology in the albino Swiss rat. J. Anat. 190(Pt. 4):577-588.

Kastner, P., J.M. Grondona, M. Mark, A. Gansmuller, M. LeMeur, D. Decimo, J.L. Vonesch, P. Dolle, and P. Chambon. 1994. Genetic analysis of RXR alpha developmental function: Convergence of RXR and RAR signaling pathways in heart and eye morphogenesis. Cell 78(6):987-1003.

Kastner, P., M. Mark, N. Ghyselinck, W. Krezel, V. Dupe, J.M. Grondona, and P. Chambon. 1997a. Genetic evidence that the retinoid signal is transduced by heterodimeric RXR/RAR functional units during mouse development. Development 124(2):313-326.

Kastner, P., N. Messaddeq, M. Mark, O. Wendling, J.M. Grondona, S. Ward, N. Ghyselinck, and P. Chambon. 1997b. Vitamin A deficiency and mutations of RXRalpha, RXRbeta and RARalpha lead to early differentiation of embryonic ventricular cardiomyocytes. Development 124(23):4749-4758.

Kaufman, R.H., E. Adam, G.L. Binder, and E. Gerthoffer. 1980. Upper genital tract changes and pregnancy outcome in offspring exposed in utero to diethylstilbestrol. Am. J. Obstet. Gynecol. 137(3):299-308.

Kauffmann, R.C., Y. Qian, A. Vogt, S.M. Sebti, A.D. Hamilton, and R.W. Carthew. 1995. Activated Drosophila Ras1 is selectively suppressed by isoprenyl transferase inhibitors. Proc. Natl. Acad. Sc.i. U.S.A. 92(24):10919-10923.

Kavlock, R.J., B.C. Allen, E.M. Faustman, and C.A. Kimmel. 1995. Dose-response assessments for developmental toxicity. IV. Benchmark doses for fetal weight changes. Fundam. Appl. Toxicol. 26(2):211-222.

Keeler, R.F., and W. Binns. 1968. Teratogenic compounds of Veratrum californicum (Durand). V. Comparison of cyclopian effects of steroidal alkaloids from the plant and structurally related compounds from other sources. Teratology 1(1):5-10.

Keeler, R.F., and A.T. Tu. 1991. Toxicology of Plant and Fungal Compounds. Handbook of Natural Toxins, Vol. 6. New York: M. Dekker. 644pp.

Keller, S.J., and M.K. Smith. 1982. Animal virus screens for potential teratogens. I. Poxvirus morphogenesis. Teratog. Carcinog. Mutagen. 2(3-4):361-374.

Kelley, R.I., E. Roessler, R.C.M. Hennekam, G.L. Feldman, K. Kosaki, M.C. Jones, J.C. Palumbos, and M. Muenke. 1996. Holoprosencephaly in RSH/Smith-Lemli-Opitz syndrome: Does abnormal cholesterol metabolism affect the function of Sonic hedgehog? Am. J. Med. Genet. 66:478-484.

Kelsh, R.N., M. Brand, Y.J. Jiang, C.P. Heisenberg, S. Lin, P. Haffter, J. Odenthal, M.C. Mullins, F.J. van Eeden, M. Furutani-Seiki, M. Granato, M. Hammerschmidt, D.A. Kane, R.M. Warga, D. Beuchle, L. Vogelsang, and C. Nüsslein-Volhard. 1996. Zebrafish pigmentation mutations and the processes of neural crest development. Development 123:369-389.

Kennerdell, J.R., and R.W. Carthew. 1998. Use of dsRNA-mediated genetic interference to demonstrate that frizzled and frizzled 2 act in the wingless pathway. Cell 95(7):1017-1026.

Kerper, L.E., N. Ballatori, and T.W. Clarkson. 1992. Methylmercury transport across the blood-brain barrier by an amino acid carrier. Am. J. Physiol. 262(5):R761-R765.

Kerper, L.E., E.M. Mokrzan, T.W. Clarkson, and N. Ballatori. 1996. Methylmercury efflux from brain capillary endothelial cells is modulated by intracellular glutathione but not ATP. Toxicol. Appl. Pharmacol. 141(2):526-531.

Khan, S.A., R.B. Ball, and W.J. Hendry, III. 1998. Effects of neonatal administration of diethylstilbestrol in male hamsters: Disruption of reproductive function in adults after apparently normal pubertal development. Biol. Reprod. 58(1):137-142.

Khoury, M.J., T.H. Beaty, and B.H. Cohen, eds. 1993a. Scope and strategies of genetic epidemiology. Pp. 3-25 in Fundamentals of Genetic Epidemiology. New York: Oxford University Press.

Khoury, M.J., T.H. Beaty, and B.H. Cohen, eds. 1993b. Applications of genetic epidemiology in medicine and pubic health. Pp. 312-327 in Fundamentals of Genetic Epidemiology. New York: Oxford University Press.

Khoury, M.J., M.G. Farias, and J. Mulinare. 1989. Does maternal cigarette smoking during pregnancy cause cleft lip and palate in the offspring? Am. J. Dis. Child. 143:333-337.

Khoury, M.J., L.M. James, W.D. Flanders, and J.D. Erickson. 1992. Interpretation of recurring weak associations obtained from epidemiologic studies of suspected humans teratogens. Teratology 46:69-77.

Kimmel, C.A. 1997. Introduction to the symposium. Reprod. Toxicol. 11(2-3):261-263.

Kimmel, C.A., J.F. Holson, C.J. Hogue, and G. Carlo. 1984. Reliability of Experimental Studies for Predicting Hazards to Human Development. NCTR Technical Report for Experiment No. 6015. National Center for Toxicological Research, Jefferson, AR.

Kimmel, C.A., et al. 1997. New Approaches for Assessing the Etiology and Risks of Developmental Abnormalities From Chemical Exposure. Symposium Proceedings. National Academy of Sciences Auditorium, December 11-12, 1995, Washington, DC.

Kimmel, G.L., K. Smith, D.M. Kochhar, and R.M. Pratt. 1982. Overview of in vitro teratogenicity testing: Aspects of validation and application to screening. Teratog. Carcinog. Mutagen. 2(3-4):221-229.

Kisaalita, W.S. 1997. Susceptibility of differentiating neuroblastoma cells to developmental toxicants' cytotoxic effects is culture age-dependent. In Vitro Toxicol. 10(3):359-363.

Kistler, A., and W.B. Howard. 1990. Testing of retinoids for teratogenicity in vitro: Use of micromass limb bud cell culture. Methods Enzymol. 190:427-433.

Klambt, C., L. Glazer, and B.-Z Shilo. 1992. Breathless, a Drosophila FGF receptor homolog, is essential for migration of tracheal and specific midline glial cells. Genes Dev. 6:1668-1678.

Knudsen, T.B. 1997. Cell death. Pp. 209-244 in Drug Toxicity in Embryonic Development, Vol. 1., R.J. Kavlock, and G.P. Daston, eds. Heidelberg: Springer.

Kochhar, D.M. 1982. Embryonic limb bud organ culture in assessment of teratogenicity of environmental agents. Teratog. Carcinog. Mutagen. 2(3-4):303-312.

Kochhar, D.M., H. Jiang, J.D. Penner, R.L. Beard, and A.S. Chandraratna. 1996. Differential teratogenic response of mouse embryos to receptor selective analogs of retinoic acid. Chem. Biol. Interact. 100(1):1-12.

Kochhar, D.M., H. Jiang, J.D. Penner, A.T. Johnson, and R.A. Chandraratna. 1998. The use of a retinoid receptor antagonist in a new model to study vitamin A-dependent developmental events. Int. J. Dev. Biol. 42(4):601-608.

Koera, K., K. Nakamura, K. Nakao, J. Miyoshi, K. Toyoshima, T. Hatta, H. Otani, A. Aiba, and M. Katsuki. 1997. K-ras is essential for the development of the mouse embryo. Oncogene 15(10):1151-1159.

Konopka, R.J., and S. Benzer. 1971. Clock mutants of Drosophila melanogaster. Proc. Natl. Acad. Sci. U.S.A. 68(9):2112-2116.

Kost, T.A., and J.P. Condreay. 1999. Recombinant baculoviruses as expression vectors for insect and mammalian cells. Curr. Opin. Biotechnol. 10(5):428-433.

Kraft, J.C., and M.R. Juchau. 1993. Conceptional biotransformation of microinjected retinoids. Correlations with dysmorphogenic activities. Ann. N.Y. Acad. Sci. 678:338-340.

Kraft, J.C., T. Shepard, and M.R. Juchau. 1993. Tissue levels of retinoids in human embryos/fetuses. Reprod. Toxicol. 7(1):11-15.

Kraus, P., and T. Lufkin. 1999. Mammalian Dlx homeobox gene control of craniofacial and inner ear morphogenesis. J. Cell Biochem. Suppl. 32-33:133-140.

Kroll, K.L., and E. Amaya. 1996. Transgenic Xenopus embryos from sperm nuclear transplantations reveal FGF signaling requirements during gastrulation. Development 122(10):3173-3183.

Krumlauf, R. 1994. Hox genes in vertebrate development. Cell 78(2):191-201.

LaBonne, C., and M. Bronner-Fraser. 1999. Molecular mechanisms of neural crest formation. Annu. Rev. Cell Dev. Biol. 15:81-112.

Lammer, E.J., D.T. Chen, R.M. Hoar, N.D. Agnish, P.J. Benke, J.T. Braun, C.J. Curry, P.M. Fernhoff, A.W. Grix, Jr., I.T. Lott, J.M. Richard, and S.C. Sun. 1985. Retinoic acid embryopathy. N. Engl. J. Med. 313(14):837-841.

Lampen, A., S. Siehler, U. Ellerbeck, M. Gottlicher, and H. Nau. 1999. New molecular bioassays for the estimation of the teratogenic potency of valproic acid derivatives in vitro: Activation of the peroxisomal proliferator-activated receptor (PPARdelta). Toxicol. Appl. Pharmacol. 160(3):238-249.

Lampron, C., C. Rochette-Egly, P. Gorry, P. Dolle, M. Mark, T. Lufkin, M. LeMeur, and P. Chambon. 1995. Mice deficient in cellular retinoic acid binding protein II (CRABPII) or in both CRABPI and CRABPII are essentially normal. Development 121(2):539-548.

Lander, E.S., and N.J. Schork. 1994. Genetic dissection of complex traits. Science 265(5181):2037-2048.

Langenbach, R., P.B. Smith, and C. Crespi. 1992. Recombinant DNA approaches for the development of metabolic systems used in in vitro toxicology. Mutat. Res. 277(3):251-275.

Lau, M.M., C.E. Stewart, Z. Liu, H. Bhatt, P. Rotwein, and C.L. Stewart. 1994. Loss of the imprinted IGF2/cation-independent mannose 6-phosphate receptor results in fetal overgrowth and perinatal lethality. Genes Dev. 8(24):2953-2963.

Lave, L.B., and G.S. Omenn. 1986. Cost-effectiveness of short-term tests for carcinogenicity. Nature 324(6092):29-34.

Le Couteur, A., A. Bailey, S. Goode, A. Pickles, S. Robertson, I. Gottesman, and M. Rutter. 1996. A broader phenotype of autism: The clinical spectrum in twins. J. Child. Psychol. Psychiatry 37(7):785-801.

Lee, C.H., and L.N. Wei. 1999. Characterization of an inverted repeat with a zero spacer (IR0)-type retinoic acid response element from the mouse nuclear orphan receptor TR2-11 gene. Biochemistry 38(27):8820-8825.

Lee, K.J., and T.M. Jessell. 1999. The specification of dorsal cell fates in the vertebrate central nervous system. Annu. Rev. Neurosci. 22:261-294.

Lee, K.F., E. Li, L.J. Huber, S.C. Landis, A.H. Sharpe, M.V. Chao, and R. Jaenisch. 1992. Targeted mutation of the gene encoding the low affinity NGF receptor p75 leads to deficits in the peripheral sensory nervous system. Cell 69(5):737-749.

Lemoine, P., H. Harousseau, J.P. Borteyru, and J.C. Menuet. 1968. Les enfants de parents alcooligues: Anomalies observees, a propos de 127 cas. [in French]. Arch. Franc. Pediatr. 25:830-832.

Lenz, W. 1961. Kindliche Missbildungen nach Medikament-Einnahme während der Gravidität? In Fragen aus der Praxis. [in German]. Dtsch Med. Wschr. 86(32):2555-2556.

Lenz, W., and K. Knapp. 1962. Die Thalidomid-embryopathie. [in German]. Dtsch. Med. Wschr. 87:1232-1242.

Leppig, K.A., M.M. Werler, C.I. Cann, C.A. Cook, and L.B. Holmes. 1987. Predictive value of minor anomalies. I. Association with major malformations. J. Pediatr. 110(4):530-537.

Leroux, B.G., W.M. Leisenring, S.H. Moolgavkar, and E.M. Faustman. 1996. A biologically-based dose-response model for developmental toxicology. Risk Anal. 16(4):449-458.

Li, E., H.M. Sucov, K.F. Lee, R.M. Evans, and R. Jaenisch. 1993. Normal development and growth of mice carrying a targeted disruption of the alpha 1 retinoic acid receptor gene. Proc. Natl. Acad. Sci. U.S.A. 90(4):1590-1594.

Limbird, L.E., and P. Taylor. 1998. Endocrine disruptors signal the need for receptor models and mechanisms to inform policy. Cell 93(2):157-163.

Lin, Y.J., L. Seroude, and S. Benzer. 1998. Extended life-span and stress resistance in the Drosophila mutant methuselah. Science 282(5390):943-946.

Liu, F., H.Y. Wu, R. Wesselschmidt, T. Kornaga, and D.C. Link. 1996. Impaired production and increased apoptosis of neutrophils in granulocyte colony-stimulating factor receptor-deficient mice. Immunity 5(5):491-501.

Liu, J.P., J. Baker, A.S. Perkins, E.J. Robertson, and A. Efstratiadis. 1993. Mice carrying null mutations of the genes encoding insulin-like growth factor I (Igf-1) and type 1 IGF receptor (Igf1r). Cell 75(1):59-72.

Lohnes, D., P. Kastner, A. Dierich, M. Mark, M. LeMeur, and P. Chambon. 1993. Function of retinoic acid receptor gamma in the mouse. Cell 73(4):643-658.

Lohnes, D., M. Mark, C. Mendelsohn, P. Dolle, A. Dierich, P. Gorry, A. Gansmuller, and P. Chambon. 1994. Function of the retinoic acid receptors (RARs) during development (I). Craniofacial and skeletal abnormalities in RAR double mutants. Development 120(10):2723-2748.

Löscher, W. 1999. Valproate: A reappraisal of its pharmacodynamic properties and mechanisms of action. Prog. Neurobiol. 58(1):31-59.

Lufkin, T., D. Lohnes, M. Mark, A. Dierich, P. Gorry, M.P. Gaub, M. LeMeur, and P. Chambon. 1993. High postnatal lethality and testis degeneration in retinoic acid receptor alpha mutant mice. Proc. Natl. Acad. Sci. U.S.A. 90(15):7225-7229.

Luo, J., P. Pasceri, R.A. Conlon, J. Rossant, and V. Giguere. 1995. Mice lacking all isoforms of retinoic acid receptor beta develop normally and are susceptible to the teratogenic effects of retinoic acid. Mech. Dev. 53(1):61-71.

Lynch, D.W., R.L. Schuler, R.D. Hood, and D.G. Davis. 1991. Evaluation of Drosophila for screening developmental toxicants: Test results with 18 chemicals and presentation of a new Drosophila bioassay. Teratog. Carcinog. Mutagen. 11(3):147-173.

Lydon, J.P., F.J. DeMayo, O.M. Conneely, and B.W. O'Malley. 1996. Reproductive phenotpes of the progesterone receptor null mutant mouse. J. Steroid Biochem. Mol. Biol. 56(1-6 Spec. No.):67-77.

Lydon, J.P., F.J. DeMayo, C.R. Funk, S.K. Mani, A.R. Hughes, C.A. Montgomery, Jr., G. Shyamala, O.M. Conneely, and B.W. O'Malley. 1995. Mice lacking progesterone receptor exhibit pleiotropic reproductive abnormalities. Genes Dev. 9(18):2266-2278.

Makrides, S.C. 1999. Components of vectors for gene transfer and expression in mammalian cells. Protein Expr. Purif. 17(2):183-202.

Makris, S., K. Raffaele, W. Sette, and J. Seed. 1998. A Retrospective Analysis of Twelve Developmental Neurotoxicity Studies. Draft. Submitted to the U.S. EPA Office of Prevention, Pesticides and Toxic Substances (OPPTS). Dated 11/12/98. [Online]. Available: http://www.epa.gov/scipoly/sap/1998/index.htm

Malakoff, D. 2000. The rise of the mouse, biomedicine's model mammal. Science 288(5464):248-253.

Manchester, D.K., N.B. Parker, and C.M. Bowman. 1984. Maternal smoking increases xenobiotic metabolism in placenta but not umbilical vein endothelium. Pediatr. Res. 18(11):1071-1075.

Mansour, S.L. 1994. Targeted disruption of int-2 (fgf-3) causes developmental defects in the tail and inner ear. Mol. Reprod. Dev. 39(1):62-67.

Mangelsdorf, D.J., and R.M. Evans. 1995. The RXR heterodimers and orphan receptors. Cell 83(6):841-850.

March of Dimes. 1999. Facts and Figures of Birth Defects. [Online]. Available: http://www.modimes.org/HealthLibrary2/FactsFigures/stats.htm#Birth Defects

Marden P.M., D.W. Smith, and M.J. McDonald. 1964. Congenital anomalies in the newborn infant, including minor variations. J. Pediatr. 64(3):357-371.

Marshall, H., A. Morrison, M. Studer, H. Popperl, and R. Krumlauf. 1996. Retinoids and Hox genes. FASEB J. 10(9):969-978.

Marshall, H., S. Nonchev, M.H. Sham, I. Muchamore, A. Lumsden, and R. Krumlauf. 1992. Retinoic acid alters hindbrain Hox code and induces transformation of rhombomeres 2/3 into a 4/5 identity. Nature 360(6406):737-741.

Martin, G.R. 1998. The roles of FGFs in the early development of vertebrate limbs. Genes Dev. 12(11):1571-1586.

Martz, F., C. Failinger, III, and D.A. Blake. 1977. Phenytoin teratogenesis: Correlation between embryopathic effect and covalent binding of putative arene oxide metabolite in gestational tissue. J. Pharmacol. Exp. Ther. 203(1):231-239.

Mastroianni, A.C., R. Faden, and D. Federman, eds. 1994. Risks to reproduction and offspring. Pp. 175-202 in Women and Health Research: Ethical and Legal Issues of Including Women in Clinical Studies. Washington, DC: National Academy Press.

Mattison, D.R. 1997. Introduction. Session IV: Implications of new approaches for improved risk assessment for developmental abnormalities. Reprod. Toxicol. 11(2/3):437-441.

McAvoy, J.W., C.G. Chamberlain, R.U. de Iongh, A.M. Hales, and F.J. Lovicu. 1999. Lens development. Eye 13(Pt. 3b):425-437.

McBride, W.G. 1961. Thalidomide and congenital abnormalities. [Letter]. Lancet 2:1358.

McCartney-Francis, N.L., D.E. Mizel, M. Frazier-Jessen, A.B. Kulkarni, J.B. McCarthy, and S.M. Wahl. 1997. Lacrimal gland inflammation is responsible for ocular pathology in TGF-beta 1 null mice. Am. J. Pathol. 151(5):1281-1288.

McClung, C., and J. Hirsh. 1999. The trace amine tyramine is essential for sensitization to cocaine in Drosophila. Curr. Biol. 9(16):853-860.

McGinnis, W., M.S. Levine, E. Hafen, A. Kuroiwa, and W.J. Gehring. 1984a. A conserved DNA sequence in homoeotic genes of the Drosophila Antennapedia and bithorax complexes. Nature 308:428-433.

McGinnis, W., R.L. Garber, J. Wirz, A. Kuroiwa, and W.J. Gehring. 1984b. A homologous protein-coding sequence in Drosophila homeotic genes and its conservation in other metazoans. Cell 37(2):403-408.

McLachlan, J.A., R.R. Newbold, and B.C. Bullock. 1980. Long-term effects on the female mouse genital tract associated with prenatal exposure to diethylstil-bestrol. Cancer Res. 40(11):3988-3999.

McMahon, A.P., A.L. Joyner, A. Bradley, and J.A. McMahon. 1992. The midbrain-hindbrain phenotype of Wnt-1-/Wnt-1-mice results from stepwise deletion of engrailed-expressing cells by 9.5 days postcoitum. Cell 69(4):581-595.

Mello, C.C., J.M. Kramer, D. Stinchcomb, and V. Ambros. 1991. Efficient gene transfer in C. elegans: Extrachromosomal maintenance and integration of transforming sequences. EMBO J. 10(12):3959-3970.

Mendel, G. 1865. Versuche uber Pflanzen-Hybriden. Verh. Naturforshung Ver. Brunn. 4:3-47. Translated by W.A. Bateson as " Experiments in Plant Hybridization", and reprinted in J.A. Peters, 1959, Classic Papers in Genetics. Englewood Cliffs, NJ: Prentice-Hall.

Mendelsohn, C., D. Lohnes, D. Decimo, T. Lufkin, M. LeMeur, P. Chambon, and M. Mark. 1994. Function of the retinoic acid receptors (RARs) during development (II). Multiple abnormalities at various stages of organogenesis in RAR double mutants. Development 120(10):2749-2771.

Meraz, M.A., J.M. White, K.C. Sheehan, E.A. Bach, S.J. Rodig, A.S. Dighe, D.H. Kaplan, J.K. Riley, A.C. Greenlund, D. Campbell, K. Carver-Moore, R.N. DuBois, R. Clark, M. Aguet, and R.D. Schreiber. 1996. Targeted disruption of the Stat1 gene in mice reveals unexpected physiologic specificity in the JAK-STAT signaling pathway. Cell 84(3):431-442.

Mercola, M. 1999. Embryological basis for cardiac left-right asymmetry. Semin. Cell Dev. Biol. 10(1):109-116.

Metzstein, M.M., G.M. Stanfield, and H.R. Horvitz. 1998. Genetics of programmed cell death in C. elegans: Past, present and future. Trends Genet. 14(10):410-416.

Meyers, E.N., M. Lewandoski, and G.R. Martin. 1998. An Fgf8 mutant allelic series generated by Cre- and Flp-mediated recombination. Nat. Genet. 18(2):136-141.

Miettinen, P.J., D. Warburton, D. Bu, J.S. Zhao, J.E. Berger, P. Minoo, T. Koivisto, L. Allen, L. Dobbs, Z. Werb, and R. Derynck. 1997. Impaired lung branching morphogenesis in the absence of functional EGF receptor. Dev. Biol. 186(2):224-236.

Mikic, B., M.C. Van der Meulen, D.M. Kingsley, and D.R. Carter. 1996. Mechanical and geometric changes in the growing femora of BMP-5 deficient mice. Bone 18(6):601-607.

Miller, C., and D.A. Sassoon. 1998. Wnt-7a maintains appropriate uterine patterning during the development of the mouse female reproductive tract. Development 125(16):3201-3211.

Miller, C., K. Degenhardt, and D.A. Sassoon. 1998. Fetal exposure to DES results in de-regulation of Wnt7a during uterine morphogenesis. Nat. Genet. 20(3):228-230.

Miller, M.S., M.R. Juchau, F.P. Guengerich, D.W. Nebert, and J.L. Raucy. 1996. Drug metabolic enzymes in developmental toxicology. Fundam. Appl. Toxicol. 34(2):165-175.

Miller, R.K. 1991. Fetal drug therapy: Principles and issues. Clin. Obs. Gyn. 34(2):241-250.

Miller, R.K., M.E. Heckmann, and R.C. McKenzie. 1982. Diethylstilbestrol: Placental transfer, metabolism, covalent binding and fetal distribution in the Wistar rat. J. Pharmacol. Exp. Ther. 220(2):358-365.

Miller, R.K., A.G. Hendrickx, J.L. Mills, H. Hummler, and U.W. Wiegand. 1998. Periconceptional vitamin A use: How much is teratogenic? Reprod. Toxicol. 12(1):75-88.

Milunsky, A., J.W. Graef, and M.F. Gaynor, Jr. 1968. Methotrexate-induced congenital malformations. J. Pediatr. 72(6):790-795.

Mimura, J., K. Yamashita, K. Nakamura, M. Morita, T.N. Takagi, K. Nakao, M. Ema, K. Sogawa, M. Yasuda, M. Katsuki, and Y. Fujii-Kuriyama. 1997. Loss of teratogenic response to 2,3,7,8-tetrachlorodibenzo-p-dioxin (TCDD) in mice lacking the Ah (dioxin) receptor. Genes Cells 2(10):645-654.

Mirkes, P.E. 1996. Prospects for the development of validated screening tests that measure developmental toxicity potential: View of one skeptic. Teratology 53(6):334-338.

Mirkes, P.E., and S.A. Little. 1998. Teratogen-induced cell death in postimplantation mouse embryos: Differential tissue sensitivity and hallmarks of apoptosis. Cell Death Differ. 5(7):592-600.

Missero, C., G. Cobellis, M. De Felice, and R. Di Lauro. 1998. Molecular events involved in differentiation of thyroid follicular cells. Mol. Cell Endocrinol. 140(1-2):37-43.

Mizell, M., and E.S. Romig. 1997. The aquatic vertebrate embryo as a sentinel for toxins: Zebrafish embryo dechorionation and perivitelline space microinjection. Int. J. Dev. Biol. 41(2):411-423.

Moerman, D.G., and A. Fire. 1997. Muscle: Structure, function and development. Pp. 417-470 in C. Elegans II, D.L. Riddle, T. Blumenthal, B.J. Meyer, and J.R. Priess, eds. Plainview, NY: Cold Spring Harbor Laboratory Press.

Mokrzan, E.M., L.E. Kerper, N. Ballatori, and T.W. Clarkson. 1995. Methylmercury-thiol uptake into cultured brain capillary endothelial cells on amino acid system L. J. Pharmacol. Exp. Ther. 272 (3):1277-1284.

Montgomery, J.M., J.S. Fleming, and R.G. Mills. 1999. Evidence for a neural inhibitory factor which downregulates fetal-type acetylcholine receptor expression in skeletal muscle cell lines. Brain Res. 818(2):346-354.

Monod, J., and F. Jacob. 1961. General conclusions: Telonomic mechanisms in cellular metabolism, growth, and differentiation. Cold Spring Harbor Sympos. Quant. Biol. 26:389-401.

Moore, J.A., G.P. Daston, E. Faustman, M.S. Golub, W.L. Hart, C. Hughes, Jr., C.A. Kimmel, J.C. Lamb, IV, B.A. Schwetz, and A.R. Scialli. 1995. An evaluative process for assessing human reproductive and developmental toxicity of agents. Reprod. Toxicol. 9(1):61-95.

Moore, J.A., and an Expert Scientific Committee. 1997. An assessment of boric acid and borax using the IEHR Evaluative Process for Assessing Human Developmental and Reproductive Toxicity of Agents. Reprod. Toxicol. 11(1):123-160.

Moore, M.W., R.D. Klein, I. Farinas, H. Sauer, M. Armanini, H. Phillips, L.F. Reichardt, A.M. Ryan, K. Carver-Moore, and A. Rosenthal. 1996. Renal and neuronal abnormalities in mice lacking GDNF. Nature 382(6586):76-79.

Morales-Alcelay, S., S.G. Copin, J.A. Martinez, P. Morales, S. Minguet, M.L. Gaspar, and M.A. Marcos. 1998. Developmental hematopoiesis. Crit. Rev. Immunol. 8(6):485-501.

Morgan, T.H. 1934. Pp. 9-10 in Embryology and Genetics. New York: Columbia University Press.

Musselman, A.C., G.D. Bennett, K.A. Greer, J.H. Eberwine, and R.H. Finnell. 1994. Preliminary evidence of phenytoin-induced alterations in embryonic gene expression in a mouse model. Reprod. Toxicol. 8(5):383-395.

Myers, R.M., N. Risch, D. Spyker, and L. Lotspeich. 1998. A Full Genome Screen for Susceptibility Genes to Autism by Linkage Analysis in Ninety Multiplex Families. 48th Annual Meeting of the American Society of Human Genetics. Denver, CO, 27-31 Oct. 1998.

Nanson, J.L. 1992. Autism in fetal alcohol syndrome: A report of six cases. Alcohol Clin. Exp. Res. 16(3):558-565.

NAREP (North American Registry of Epilepsy and Pregnancy). 1998. A North American Registry for Epilepsy and Pregnancy, a unique public/private partnership of health surveillance. Epilepsia 39(7):793-798.

Nau, H. 1986. Species differences in pharmacokinetics and drug teratogenesis. Environ. Health Perspect. 70:113-129.

Nau, H. 1994. Valproic acid induced neural tube defects. Pp.144-160 in Neural Tube Defects, G. Bock, and J. Marsh, eds. Chichester, Sussex, UK: John Wiley.

Nau, H., M. Trotz, and C. Wegner. 1985. Controlled-rate drug administration in testing for toxicity, in particular teratogenicity: Toward interspecies bioequivalence. Pp. 143-157 in Topics in Pharmaceutical Sciences 1985, D.D. Breimer, and P. Speiser, eds. New York: Elsevier.

Nau, H., I. Chaoud, L. Dencker, E.J. Lammer, and W.J. Scott. 1994. Teratogenicity of vitamin A and retinoids. Pp. 615-663 in Vitamin A in Health and Diseases, R. Blomhoff, ed. New York: Marcel Dekker.

Nau, H., G. Tzimas, M. Mondry, C. Plum, and H.L. Spohr. 1995. Antiepileptic drugs alter endogenous retinoid concentrations: A possible mechanism of teratogenesis of anticonvulsant therapy. Life Sci. 57(1):53-60.

Nau, H., R. Zierer, H. Spielmann, D. Neubert, and C. Gansau. 1981. A new model for embryotoxicity testing: Teratogenicity and pharmacokinetics of valproic acid following constant-rate administration in the mouse using human therapeutic drug and metabolite concentrations. Life Sci. 29(26):2803-2814.

NBDPN (National Birth Defects Prevention Network). 2000. [Online]. Available: http://www.nbdpn.org/NBDPN.

NCHS. (National Center for Health Statistics). 1998. Report of Final Natality Statistics, 1996. In: Latest Birth Statistics for the Nation Released. Vol. 46(11)(suppl.). PHS 98-1120. National Center for Health Statistics. 100pp. [Online]. Available: http://www.cdc.gov/nchswww/releases/98news/98news/natal96.htm.

Nebert, D.W. 1989. The Ah locus: Genetic differences in toxicity, cancer, mutation and birth defects. CRC Crit. Rev. Toxicol. 20(3):153-174.

Nebert, D.W. 1994. Drug-metabolizing enzymes in ligand-modulated transcription. Biochem. Pharmacol. 47(1):25-37.

Nebert, D.W. 1999. Pharmacogenetics and pharmacogenomics: Why is this relevant to the clinical geneticist? Clin. Genet. 56(4):247-258.

Nebert, D.W., and J.J. Duffy. 1997. How knockout mouse lines will be used to study the role of drug-metabolizing enzymes and their receptors during reproduction, development, and environmental toxicity, cancer and oxidative stress. Biochem. Pharmacol. 53(3):249-254.

Nebert, D.W., J. Winker, and H.V. Gelboin. 1969. Aryl hydrocarbon hydroxylase activity in human placenta from cigarette smoking and nonsmoking women. Cancer Res. 29(10):1763-1769.

Nebert, D.W., A.L. Roe, M.Z. Dieter, W.A. Solis, Y. Yang, and T.P. Dalton. 2000. Role of the aromatic hydrocarbon receptor and [Ah] gene battery in the oxidative stress response, cell cycle control, and apoptosis. Biochem. Pharmacol. 59(1):65-85.

Nelson, K., and L.B. Holmes. 1989. Malformations due to presumed spontaneous mutations in newborn infants. N. Engl. J. Med. 320(1):19-23.

Neubert, R., and D. Neubert. 1997. Peculiarities and possible mode of actions of thalidomide. Pp. 41-119 in Drug Toxicity in Embryonic Development II, Handbook of Experimental Pharmacology, Vol. 124/II, R.J. Kavlock, and G.P. Daston, eds. Heidelberg: Springer.

Neul, J.L., and E.L. Ferguson. 1998. Spatially restricted activation of the SAX receptor by SCW modulates DPP/TKV signaling in Drosophila dorsal-ventral patterning. Cell 95(4):483-494.

Neumann, D.A., and C.A. Kimmel, eds. 1998. Human Variability in Response to Chemical Exposure Measures, Modeling and Risk Assessment. Boca Raton, FL: CRC Press. 257pp.

Newall, D.R., and K.E. Beedles. 1996. The stem-cell test: An in-vitro assay for teratogenic potential. Results of a blind trial with 25 compounds. Toxicol. in Vitro 10(2):229-240.

Newbold, R.R., R.B. Hanson, and W.N. Jefferson. 1997. Ontogeny of lactoferrin in the developing mouse uterus: A marker of early hormone response. Biol. Reprod. 56(5):1147-1157.

Newman, L.M., E.M. Johnson, and R.E. Staples. 1993. Assessment of the effectiveness of animal developmental toxicity testing for human safety. Reprod. Toxicol. 7(4):359-390.

Ng, J.K., K. Tamura, D. Buscher, and J.C. Izpisua-Belmonte. 1999. Molecular and cellular basis of pattern formation during vertebrate limb development. Curr. Top. Dev. Biol. 41:37-66.

Niederreither, K., V. Subbarayan, P. Dolle, and P. Chambon. 1999. Embryonic retinoic acid synthesis is essential for early mouse post-implantation development. Nat. Genet. 21(4):444-448.

NIEHS (National Institute of Environmental Health Sciences). 1997. Validation and Regulatory Acceptance of Toxicological Test Methods: A Report of the Ad Hoc Interagency Coordinating Committee on the Validation of Alternative Methods, NIH Pub. No. 97-3981, NIEHS, Research Triangle Park, NC. March.

NIEHS (National Institute of Environmental Health Sciences). 1998. NIEHS News: New NTP Centers Meet the Need to Know. Environ Health Perspect. 106(10). [Online]. Available: http://ehpnet1.niehs.nih.gov/docs/1998/106-10/niehsnews.html.

Nisbet, I.C.T., and N.J. Karch. 1983. Chemical Hazards to Human Reproduction. Noyes Data Corp., Park Ridge, IL.

Nishina, H., K.D. Fisher, L. Radvanyi, A. Shahinian, R. Hakem, E.A. Rubie, A. Bernstein, T.W. Mak, J.R. Woodgett, and J.M. Penninger. 1997. Stress-signalling kinase Sek1 protects thymocytes from apoptosis mediated by CD95 and CD3. Nature 385(6614):350-353.

Nolen, G.A. 1986. The effects of prenatal retinoic acid on the viability and behavior of the offspring. Neurobehav. Toxicol. Teratol. 8(6):643-654.

NRC (National Research Council). 1983. Risk Assessment in the Federal Government: Managing the Process. Washington DC: National Academy Press.

NRC (National Research Council). 1989. Biologic Markers in Reproductive Toxicology. Washington, DC: National Academy Press.

NRC (National Research Council). 1994. Science and Judgment in Risk Assessment. Washington, DC: National Academy Press.

NRC (National Research Council). 1996. Carcinogens and Anticarcinogens in the Human Diet. Washington, DC: National Academy Press.

Nuclear Receptors Committee. 1999. A unified nomenclature system for the nuclear receptor superfamily. [Letter]. Cell 97(2):161-163.

Nübler-Jung, K., and D. Arendt. 1996. Enteropneusts and chordate evolution. Curr. Biol. 6(4): 352-353.

Nüsslein-Volhard, C. 1991. Determination of the embryonic axes of Drosophila. Development (Suppl.) 1:1-10.

Nüsslein-Volhard C., and E. Wieschaus. 1980. Mutations affecting segment number and polarity in Drosophila. Nature 287:795-801.

Nuwaysir, E.F., M. Bittner, J. Trent, J.C. Barrett, and C.A. Afshari. 1999. Microarrays and toxicology: The advent of toxicogenomics. Mol. Carcinog. 24(3):153-159.

Oakley, G.P. 1986. Frequency of human congenital malformations. Clin. Perinatol. 13(3):545-554.

OECD (Organization for Economic Cooperation and Development). 1987. OECD Guidelines for the Testing of Chemicals.

OECD (Organization for Economic Cooperation and Development). 1998. OECD Guideline for the Testing of Chemicals. Proposal for Updating Guideline 414. Prenatal Development Toxicity Study. Draft Document. (March 1998).

Oeda, K., T. Sakaki, and H. Ohkawa. 1985. Expression of rat liver cytochrome P-450MC cDNA in Saccharomyces cerevisiae. DNA 4(3):203-210.

O'Flaherty, E. 1997. Pharmacokinetics, pharmacodynamics, and prediction of developmental abnormalities. Reprod. Toxicol. 11(2/3):413-416.

Omenn, G.S., and E.M. Faustman. 1997. Risk assessment, risk communication and risk management. Pp. 969-986 in The Scope of Public Health, Vol.2., Oxford Textbook of Public Health, 3rd Ed., R. Detels, W.W. Holland, J. McEwen, and G.S. Omenn, eds. New York: Oxford University Press.

Ortega, S., M. Ittmann, S.H. Tsang, M. Ehrlich, and C. Basilico. 1998. Neuronal defects and delayed wound healing in mice lacking fibroblast growth factor 2. Proc. Natl. Acad. Sci. U.S.A. 95(10):5672-5677.

Ou, Y.C., S.A. Thompson, S.C. Kirchner, T.J. Kavanagh, and E.M. Faustman. 1997. Induction of growth arrest and DNA damage-inducible genes Gadd45 and Gadd153 in primary rodent embryonic cells following exposure to methylmercury. Toxicol. Appl. Pharmacol. 147(1):31-38.

Ou, Y.C., S.A. Thompson, R.A. Ponce, J. Schroeder, T.J. Kavanagh, and E.M. Faustman. 1999. Induction of the cell cycle regulatory gene p21 (Waf1, Cip1) following methylmercury exposure in vitro and in vivo. Toxicol. Appl. Pharmacol. 157(3):203-212.

Oudenampsen, E., E.M. Kupsch, T. Wissel, F. Spener, and A. Lezius. 1990. Expression of fatty acid-binding protein from bovine heart in Escherichia coli. Mol. Cell Biochem. 98(1-2):75-79.

Pace, N.R. 1997. A molecular view of microbial diversity and the biosphere. Science 276(5313): 734-740.

Padgett, R.W., P. Das, and S. Krishna. 1998. TGF-beta signaling, Smads, and tumor suppressors. BioEssays 20:382-391.

Park, S., J. Frisen, and M. Barbacid. 1997. Aberrant axonal projections in mice lacking EphA8 (Eek) tyrosine protein kinase receptors. EMBO J. 16(11):3106-3114.

Palmiter, R.D., E.P. Sandgren, M.R. Avarbock, D.D. Allen, and R.L. Brinster. 1991. Heterologous introns can enhance expression of transgenes in mice. Proc. Natl. Acad. Sci. U.S.A. 88(2): 478-482.

Parker, K.L., A. Schedl, and B.P. Schimmer. 1999. Gene interactions in gonadal development. Annu. Rev. Physiol. 61:417-433.

Parsons, J.F., J. Rockley, and M. Richold. 1990. In-vitro micromass teratogen test- interpretation of results from a blind trial of 25 compounds using 3 separate criteria. Toxicol. in Vitro 4(4-5):609-611.

Payne, J., F. Shibasaki, and M. Mercola. 1997. Spina bifida occulta in homozygous Patch mouse embryos. Dev. Dyn. 209(1):105-116.

Pennisi, E. 1997. A catalog of cancer genes at the click of a mouse. Science 276(5315):1023-1024.

Perera, F.P., R.M. Whyatt, W. Jedrychowski, V. Rauh, D. Manchester, R.M. Santella, and R. Ottman. 1998. Recent developments in molecular epidemiology. A study of the effects of environmental polycyclic aromatic hydrocarbons on birth outcomes in Poland. Am. J. Epidem. 147(3): 309-314.

Perrimon, N., and A.P. McMahon. 1999. Negative feedback mechanisms and their roles during pattern formation. Cell 97(1):13-16.

Perrimon, N., L. Engstrom, and A.P. Mahowald. 1989. Zygotic lethals with specific maternal effect phenotypes in Drosophila melanogaster. I. Loci on the X chromosome. Genetics 121:333-352.

Peterka, M., and T. Pexieder. 1994. Embryotoxicity and teratogenicity of all-trans retinoic acid in chick embryo. Teratology 50(5):39A.

Peters, H., and R. Balling. 1999. Teeth. Where and how to make them. Trends Genet. 15(2):59-65.

Peterson, L.A., M.R. Brown, A.J. Carlisle, E.C. Kohn, L.A. Liotta, M.R. Emmert-Buck, and D.B. Krizman. 1998. An improved method for construction of directionally cloned cDNA libraries from microdissected cells. Cancer Res. 58(23):5326-5328.

Petrini, J., K. Damus, and R.B. Johnston, Jr. 1997. An overview of infant mortality and birth defects in the United States. Teratology 56(1/2):8-9.

Philippe, A., M. Martinez, M. Guilloud-Bataille, C. Gillberg, M. Rastam, E. Sponheim, M. Coleman, M. Zappella, H. Aschauer, L. van Malldergerme, C. Penet, J. Feingold, A. Brice, and M. Leboyer. 1999. Genome-wide scan for autism susceptibility genes. Paris Autism Research International Sibpair Study. Hum. Mol. Genet. 8(5):805-812.

Pichel, J.G., L. Shen, H.Z. Sheng, A.C. Granholm, J. Drago, A. Grinberg, E.J. Lee, S.P. Huang, M. Saarma, B.J. Hoffer, H. Sariola, and H. Westphal. 1996. Defects in enteric innervation and kidney development in mice lacking GDNF. Nature 382(6586):73-76.

Pignatello, M.A., F.C. Kauffman, and A.A. Levin. 1999. Multiple factors contribute to the toxicity of the aromatic retinoid TTNPB (Ro 13-7410): Interactions with the retinoic acid receptors. Toxicol. Appl. Pharmacol. 159(2):109-116.

Plasterk, R.H.A. 1995. Reverse genetics: From gene sequence to mutant worm. Pp. 59-80 in Methods in Cell Biology, Vol. 48, Caenorhabditis elegans: Modern Biological Analysis of an Organism, H.F. Epstein, and D.C. Shakes, eds. San Diego, CA: Academic Press.

Poland, A., D. Palen, and E. Glover. 1994. Analysis of the four alleles of the murine aryl hydrocarbon receptor. Mol. Pharmacol. 46(5):915-921.

Ponce, R.A., T.J. Kavanagh, N.K. Mottet, S.G. Whittaker, and E.M. Faustman. 1994. Effects of methyl mercury on the cell cycle of primary rat CNS cells in vitro. Toxicol. Appl. Pharmacol. 127(1):83-90.

Porter, J.A., K.E. Young, and P.A. Beachy. 1996. Cholesterol modification of hedgehog signaling proteins in animal development. Science 274(5285):255-259.

Powell-Braxton, L., P. Hollingshead, C. Warburton, M. Dowd, S. Pitts-Meek, D. Dalton, N. Gillett, and T.A. Stewart. 1993. IGF-I is required for normal embryonic growth in mice. Genes Dev. 7(12B):2609-2617.

Powell-Coffman, J.A., C.A. Bradfield, and W.B. Wood. 1998. Caenorhabditis elegans orthologs of the aryl hydrocarbon receptor and its heterodimerization partner the aryl hydrocarbon receptor nuclear translocator. Proc. Natl. Acad. Sci. U.S.A. 95:2844-2849.

Quig, D. 1998. Cysteine metabolism and metal toxicity. Altern. Med. Rev. 3(4):262-270.

Radatz, M., K. Ehlers, B. Yagen, M. Bialer, and H. Nau. 1998. Valnoctamide, valpromide and valnoctic acid are much less teratogenic in mice than valproic acid. Epilepsy Res. 30(1):41-48.

Raftery, L.A., V. Twombly, K. Wharton, and W.M. Gelbart. 1995. Genetic screens to identify elements of the decapentaplegic signaling pathway in Drosophila. Genetics 139:241-254.

Rawls, A., J. Wilson-Rawls, and E.N. Olson. 2000. Genetic regulation of somite formation. Curr. Top. Dev. Biol. 47:131-154.

Reiners, J., W. Wittfoht, H. Nau, R. Vogel, B. Tenschert, and H. Spielmann 1987. Teratogenesis and pharmacokinetics of cyclophosphamide after drug infusion as compared to injection in the mouse during day 10 of gestation. Pp. 41-48 in Pharmacokinetics in Teratogenesis, Vol. II, H. Nau, and W. Scott, eds. Boca Raton, FL: CRC Press.

Reinhardt, C.A. 1993. Neurodevelopmental toxicity in vitro: Primary cell culture models for screening and risk assessment [Review]. Reprod. Toxicol. 7(Suppl. 1):165-170.

Reichhardt, T. 1999. It's sink or swim as a tidal wave of data approaches. Nature 399(6736):517-520.

Relaix, F., and M. Buckingham. 1999. From insect eye to vertebrate muscle: Redeployment of a regulatory network. Genes Dev. 13(24):3171-3178.

Renwick, A.G. 1998. Toxicokinetics in infants and children in relation to the ADI and TDI. Food Addit. Contam. 15(Suppl.):17-35.

Riddle, D.L., and P.S. Albert. 1997. Genetic and environmental regulation of dauer larva development. Pp. 739-768 in C. elegans II, D.L. Riddle, T. Blumenthal, B.J. Meyer, and J.R. Priess, eds. Plainview, NY: Cold Spring Harbor Laboratory Press.

Riddle, D.L., T. Blumenthal, B.J. Meyer, and J.R. Priess, eds. 1997. C. elegans II. Plainview, NY: Cold Spring Harbor Laboratory Press. 1222 pp.

Robert, E. 1992. Detecting new teratogens: Valproic acid. Teratology 45(3):331.

Rodier, P.M. 1998. Neuroteratology of autism. Pp. 661-672 in Handbook of Developmental Neurotoxicology, W. Slikker, and L. Chang, eds. New York: Academic Press.

Rodier, P.M. 2000. The early origins of autism. Sci. Am. 282:56-63.

Rodier, P.M., M. Aschner, and P.R. Sager. 1984. Mitotic arrest in the developing CNS after prenatal exposure to methylmercury. Neurobehav. Toxicol. Teratol. 6(5):379-385.

Rodier, P.M., S.E. Bryson, and J.P. Welch. 1997. Minor malformations and physical measurements in autism: Data from Nova Scotia. Teratology 55(5):319-325.

Rodier, P.M., J.L. Ingram, B. Tisdale, S. Nelson, and J. Romano. 1996. Embryological origin for autism: Developmental anomalies of the cranial nerve motor nuclei. J. Comp. Neurol. 370(2):247-261.

Roelink, H., J.A. Porter, C. Chiang, Y. Tanabe, D.T. Chang, P.A. Beachy, and T.M. Jessell. 1995. Floor plate and motor neuron induction by different concentrations of the amino-terminal cleavage product of sonic hedgehog autoproteolysis. Cell 81(3):445-455.

Rosa, F.W. 1992. Epidemiology of drugs in pregnancy. Pp. 97-106 in Pediatric Pharmacology, 2nd Ed., S.J. Yaffe, and J.V. Aranda, eds. Philadelphia, PA: Saunders.

Rothman, K.J. 1986. Modern Epidemiology. Boston: Little, Brown.

Roux, C., and M. Aubry. 1966. Teratogenic action in the rat of an inhibitor of cholesterol synthesis, AY 9944 [in French]. C. R. Seances Soc. Biol. Fil. 160(7):1353-1357.

Rudner, A.D., and A.W. Murray. 1996. The spindle assembly checkpoint. Curr. Opin. Cell Biol. 8(6):773-780.

Rutherford, S.L., and S. Lindquist. 1998. Hsp90 as a capacitor for morphological evolution. Nature 396(6709):336-342.

Rutledge, J.C. 1997. Developmental toxicity induced during early stages of mammalian embryogenesis. Mutat. Res. 396(1-2):113-127.

Rutledge, J.C., A.G. Shourbaji, L.A. Hughes, J.E. Polifka, Y.P. Cruz, J.B. Bishop, and W.M. Generoso. 1994. Limb and lower-body duplications induced by retinoic acid in mice. Proc. Natl. Acad. Sci. U.S.A. 91(12):5436-5440.

Ruvkun, G. 1997. Patterning the nervous system. Pp. 543-581 in C. elegans II, D.L. Riddle, T. Blumenthal, B.J. Meyer, and J.R. Priess, eds. Plainview, NY: Cold Spring Harbor Laboratory Press.

Ruvkun, G., and O. Hobert. 1998. The taxonomy of developmental control in Caenorhabditis elegans. Science 282(5396):2033-2041.

Saint-Amant, L., and P. Drapeau. 1998. Time course of the development of motor behaviors in the zebrafish embryo. J. Neurobiol. 37(4):622-632.

Salvesen, G.S. 1999. Programmed cell death and the caspases. APMIS 107(1):73-79.

Sanchez, M.P., I. Silos-Santiago, J. Frisen, B. He, S.A. Lira, and M. Barbacid. 1996. Renal agenesis and the absence of enteric neurons in mice lacking GDNF. Nature 382(6586):70-73.

Sanford, L.P., I. Ormsby, A.C. Gittenberger-de Groot, H. Sariola, R. Friedman, G.P. Boivin, E.L. Cardell, and T. Doetschman. 1997. TGFbeta2 knockout mice have multiple developmental defects that are non-overlapping with other TGFbeta knockout phenotypes. Development 124(13):2659-2670.

Sariola, H., and K. Sainio. 1998. Cell lineages in the embryonic kidney: Their inductive interactions and signalling molecules. Biochem. Cell Biol. 76(6):1009-1016.

Saxton, T.M., M. Henkemeyer, S. Gasca, R. Shen, D.J. Rossi, F. Shalaby, G.S. Feng, and T. Pawson. 1997. Abnormal mesoderm patterning in mouse embryos mutant for the SH2 tyrosine phosphatase Shp-2. EMBO J. 16(9):2352-2364.

Schardein, J.L. 2000. Chemically Induced Birth Defects, 3th Ed. New York: Marcel Dekker.

Schilling, T.F., T. Piotrowski, H. Grandel, M. Brand, C.P. Heisenberg, Y.J. Jiang, D. Beuchle, M. Hammerschmidt, D.A. Kane, M.C. Mullins, F.J. van Eeden, R.N. Kelsh, M. Furutani-Seiki, M. Granato, P. Haffter, J. Odenthal, R.M. Warga, T. Trowe, and C. Nüsslein-Volhard. 1996. Jaw and branchial arch mutants in zebrafish I: branchial arches. Development 123:329-344.

Schmidt, C., F. Bladt, S. Goedecke, V. Brinkmann, W. Zschiesche, M. Sharpe, E. Gherardi, and C. Birshmeier. 1995. Scatter factor/hepatocyte growth factor is essential for liver development. Nature 373(6516):699-702.

Schneider, R.A., D. Hu, and J.A. Helms. 1999. From head to toe: Conservation of molecular signals regulating limb and craniofacial morphogenesis. Cell Tissue Res. 296(1):103-109.

Schuchardt, A., V. D'Agati, L. Larsson-Blomberg, F. Costantini, and V. Pachnis. 1994. Defects in the kidney and enteric nervous system of mice lacking the tyrosine kinase receptor Ret. Nature 367(6461):380-383.

Schuchardt, A., V. D'Agati, V. Pachnis, and F. Costantini. 1996. Renal agenesis and hypodysplasia in ret-k- mutant mice result from defects in ureteric bud development. Development 122(6): 1919-1929.

Schwartz, E.L., S. Hallam, R.E. Gallagher, and P.H. Wiernik. 1995. Inhibition of all-trans-retinoic acid metabolism by fluconazole in vitro and in patients with acute promyelocytic leukemia. Biochem. Pharmacol. 50(7):923-928.

Schwetz, B.A. 1993. In vitro approaches in developmental toxicology. [Review]. Reprod. Toxicol. 7(Suppl. 1):125-127.

Schwetz, B.A., and M.W. Harris. 1993. Developmental toxicology: Status of the field and contribution of the National Toxicology Program. [Review]. Environ. Health Perspect. 100:269-282.

Scott, W.J. 1977. Cell death and reduced proliferation rate. Pp. 81-98 in Handbook of Teratology, Vol. 2., J.G. Wilson, and F.C. Fraser, eds. New York: Plenum.

Selevan, S.G. 1985. Design of pregnancy outcome studies of industrial exposures. Pp. 219-229 in Occupational Hazards and Reproduction, K. Hemminki, M. Sorsa, and H. Vainio, eds. Washington, DC: Hemishere Pub.

Sever, L., M.C. Lynberg, and L.D. Edmonds. 1993. The impact of congenital malformations on public health. Teratology 48(6):547-549.

Shaw, G.M., C.R. Wasserman, E.J. Lammer, C.D. O'Malley, J.C. Murray, A.M. Basart, and M.M. Tolarova. 1996. Orofacial clefts, parental cigarette smoking and transforming growth factor-alpha gene variants. Am. J. Hum. Genet. 58:551-561.

Shepard, T.H. 1998. Catalog of Teratogenic Agents. 9th Ed. Baltimore: Johns Hopkins University Press. 593 pp.

Shepard, T.H., A.G. Fantel, P.E. Mirkes, J.C. Greenaway, E. Faustman-Watts, M. Campbell, and M.R. Juchau. 1983. Teratology testing: I. Development and status of short-term prescreens. II. Biotransformation of teratogens as studied in whole embryo culture. Prog. Clin. Biol. Res. 135:147-164.

Shuey, D.L., R.W. Setzer, C. Lau, R.M. Zucker, K.H. Elstein, M.G. Narotsky, R.J. Kavlock, and J.M. Rogers. 1995. Biological modeling of 5-fluorouracil developmental toxicity. Toxicology 102(1-2):207-213.

Shum, S., N.M. Jensen, and D.W. Nebert. 1979. The murine Ah locus: In utero toxicity and teratogenesis associated with genetic differences in benzo[a]pyrene metabolism. Teratology 20(3): 365-376.

Sibilia, M., and E.F. Wagner. 1995. Strain-dependent epithelial defects in mice lacking the EGF receptor. Science 269(5221):234-238.

Sidrauski, C., R. Chapman, and P. Walter. 1998. The unfolded protein response: An intracellular signalling pathway with many surprising features. Trends Cell. Biol. 8(6):245-249.

Silver, L.M. 1995. Mouse Genetics: Concepts and Applications. New York: Oxford University Press. 362pp.

Simone, N.L., R.F. Bonner, J.W. Gillespie, M.R. Emmert-Buck, and L.A. Liotta. 1998. Laser-capture microdissection: Opening the microscopic frontier to molecular analysis. Trends Genet. 14(7):272-276.

Simon, M.A., D.D. Bowtell, G.S. Dodson, T.R. Laverty, and G.M. Rubin. 1991. Ras1 and a putative guanine nucleotide exchange factor perform crucial steps in signaling by the sevenless protein tyrosine kinase. Cell 67(4):701-716.

Sleet, R.B., and K. Brendel. 1985. Homogeneous populations of Artemia nauplii and their potential use for in vitro testing in developmental toxicology. Teratog. Carcinog. Mutagen. 5(1):41-54.

Slikker, W., and R.K. Miller. 1994. Placental metabolism and transfer, role in developmental toxicology. Pp. 245-283 in Developmental Toxicology, 2nd Ed., C. Kimmel, and J. Buelke-Sam, eds. New York: Raven Press.

Smeyne, R.J., R. Klein, A. Schnapp, L.K. Long, S. Bryant, A. Lewin , S.A. Lira, and M. Barbacid. 1994. Severe sensory and sympathetic neuropathies in mice carrying a disrupted Trk/NGF receptor gene. Nature 368(6468):246-249.

Smith, J.L., and G.C. Schoenwolf. 1998. Getting organized: New insights into the organizer of higher vertebrates. Curr. Top. Dev. Biol. 40:79-110.

Smith, M.K., G.L. Kimmel, D.M. Kochhar, T.H. Shepard, S.P. Spielberg, and J.G. Wilson. 1983. A selection of candidate compounds for in vitro teratogenesis test validation. Teratog. Carcinog. Mutagen. 3(6):461-480.

Snow, E.T. 1997. The role of DNA repair in development. Reprod. Toxicol. 11(2/3):353-365.

Soriano, P. 1997. The PDGF alpha receptor is required for neural crest cell development and for normal patterning of the somites. Development 124(14):2691-2700.

Spellman, P.T., G. Sherlock, M.Q. Zhang, V.R. Iyer, K. Anders, M.B. Eisen, P.O. Brown, D. Botstein, and B. Futcher. 1998. Comprehensive identification of cell cycle-regulated genes of the yeast Saccharomyces cerevisiae by microarray hybridization. Mol. Biol. Cell 9(12):3273-3297.

Spemann, H., and H. Mangold. 1924. Über Induktion von Embryonalanlagen durch Implatation artfremder Organisatoren. [in German]. Arch. f. Mikr. Anat. u. Entw.-Mech. 100:599-638.

Spence, S., C. Anderson, M. Cukierski, and D. Patrick. 1999. Teratogenic effects of the endothelin receptor antagonist L-753,037 in the rat. Reprod. Toxicol. 13(1):15-29.

Spiegelstein, O., M. Bialer, M. Radatz, H. Nau, and B. Yagen. 1999. Enantioselective synthesis and teratogenicity of propylisopropyl acetamide, a CNS-active chiral amide analogue of valproic acid. Chirality 11(8):645-650.

Spielmann, H., I. Pohl, B. Doring, M. Liebsch, and F. Moldenhauer. 1997. The embryonic stem cell test, an in vitro embryotoxicity test using two permanent mouse cell lines: 3T3 fibroblasts and embryonic stem cells. In Vitro Toxicol. 10(1):119-127.

Spielmann, H., G. Scholtz, A. Seiler, I. Pohl, S. Bremer, N. Brown, and A. Piersma. 1998. Ergebnisse der ersten Phase des ECVAM-Projektes zur Prävalidierung und Validierung von drei in vitro Embryotoxizitätstests. [in German]. ALTEX 15:3-8.

Stainier, D.Y., and M.C. Fishman. 1992. Patterning the zebrafish heart tube: Acquisition of anteroposterior polarity. Dev. Biol. 153(1):91-101.

Stainier, D.Y., B. Fouquet, J.N. Chen, K.S. Warren, B.M. Weinstein, S.E. Meiler, M.A. Mohideen, S.C. Neuhauss, L. Solnica-Krezel, A.F. Schier, F. Zwartkruis, D.L. Stemple, J. Malicki, W. Driever, and M.C. Fishman. 1996. Mutations affecting the formation and function of the cardiovascular system in the zebrafish embryo. Development 123:285-292.

Stearnes, S.C. 1989. The evolutionary significance of phenotypic plasticity. Bioscience 39(7): 436-445.

Steele, V.E., R.E. Morrissey, E.L. Elmore, D. Gurganus-Rocha, B.P. Wilkinson, R.D. Curren, B.S. Schmetter, A.T. Louie, J.C. Lamb, IV, and L.L. Yang. 1988. Evaluation of two in vitro assays to screen for potential developmental toxicants. Fundam. Appl. Toxicol. 11(4):673-684.

Stephens, T.D. 1988. Proposed mechanisms of action in thalidomide embryopathy. Teratology 38(3):229-239.

Stephens, T.D., and B.J. Fillmore. 2000. Hypothesis: Thalidomide embryopathy—proposed mechanism of action. Teratology 61(3):189-195.

Sternberg, P.W., and H.R. Horvitz. 1991. Signal transduction during C. elegans vulval induction. Trends Genet. 7(11/12):366-371.

Stodgell, C.J., L.L. Knauber, J.L. Ingram, S.L. Hyman, D.A. Figlewicz, and P.M. Rodier. 1999. A structural variant of HOXA1 associated with autism. [Abstract]. Teratology 59:389.

Stohs, S.J., and D. Bagchi. 1995. Oxidative mechanisms in the toxicity of metal ions. Free Radic. Biol. Med. 18(2):321-336.

St-Onge, L., R. Wehr, and P. Gruss. 1999. Pancreas development and diabetes. Curr. Opin. Genet. Dev. 9(3):295-300.

Storm, E.E., and D.M. Kingsley. 1996. Joint patterning defects caused by single and double mutations in members of the bone morphogenetic protein (BMP) family. Development 122(12):3969-3979.

Streissguth, A.P., and P. Dehaene. 1993. Fetal alcohol syndrome in twins of alcoholic mothers: Concordance of diagnosis and IQ. Am. J. Med. Genet. 47(6):857-861.

Strom, B.L., ed. 1994. Systems available for pharmacoepidemiology studies. Pp 125-337 in Pharmacoepidemiology, 2nd Ed. New York: John Wiley & Sons.

Strömland, K., V. Nordin, M. Miller, B. Akerstrom, and C. Gillberg. 1994. Autism in thalidomide embryopathy: A population study. Dev. Med. Child Neurol. 36(4):351-356.

Studer, M., A. Lumsden, L. Ariza-McNaughton, A. Bradley, and R. Krumlauf. 1996. Altered segmental identity and abnormal migration of motor neurons in mice lacking Hoxb-1. Nature 384(6610):630-634.

Sucov, H.M., E. Dyson, C.L. Gumeringer, J. Price, K.R. Chien, and R.M. Evans. 1994. RXR alpha mutant mice establish a genetic basis for vitamin A signaling in heart morphogenesis. Genes Dev. 8(9):1007-1018.

Sucov, H.M., J.C. Izpisua-Belmonte, Y. Ganan, and R.M. Evans. 1995. Mouse embryos lacking RXR alpha are resistant to retinoic-acid-induced limb defects. Development 121(12):3997-4003.

Sugrue, S.P., and J.M. DeSesso. 1982. Altered glycosaminoglycan composition of rat forelimb-buds during hydroxyurea teratogenesis: An indication of repair. Teratology 26(1):71-83.

Sullivan, F.M., S.E. Smith, and P.R. McElhatton 1987. Interpretation of animal experiments as illustrated by studies on caffeine. Pp.123-130 in Pharmacokinetics in Teratogenesis, Vol. I. Interspecies Comparison and Maternal/Embryonic-Fetal Drug Transfer, H. Nau, and W. Scott, eds. Boca Raton, FL: CRC Press.

Sulston, J.E., and H.R. Horvitz. 1977. Post-embryonic cell lineages of the nematode, Caenorhabditis elegans. Dev. Biol. 56(1):110-156.

Sulston, J.E., E. Schierenberg, J.G. White, and J.N. Thomson. 1983. The embryonic cell lineage of the nematode Caenorhabditis elegans. Dev. Biol. 100:64-119.

Summerbell, D., and P.W. Rigby. 2000. Transcriptional regulation during somitogenesis. Curr. Top. Dev. Biol. 48:301-318.

Suzuki, Y., M.D. Yandell, P.J. Roy, S. Krishna, C. Savage-Dunn, R.M. Ross, R.W. Padgett, and W.B. Wood. 1999. A BMP homolog acts as a dose-dependent regulator of body size and male tail patterning in Caenorhabditis elegans. Development 126(2):241-250.

Swain, A., and R. Lovell-Badge. 1999. Mammalian sex determination: A molecular drama. Genes Dev. 13(7):755-767.

Swiatek, P.J., C.E. Lindsell, F.F. del Amo, G. Weinmaster, and T. Gridley. 1994. Notch1 is essential for postimplantation development in mice. Genes Dev. 8(6):707-719.

Tabin, C.J. 1998. A developmental model for thalidomide defects. Nature 396(6709):322-323.

Takahashi, J.S. 1995. Molecular neurobiology and genetics of circadian rhythms in mammals. Annu. Rev. Neurosci. 18:531-553.

Takeda, K., K. Noguchi, W. Shi, T. Tanaka, M. Matsumoto, N. Yoshida, T. Kishimoto, and S. Akira. 1997. Targeted disruption of the mouse Stat3 gene leads to early embryonic lethality. Proc. Natl. Acad. Sci. U.S.A. 94(8):3801-3804.

Tallquist, M.D., P. Soriano, and R.A. Klinghoffer. 1999. Growth factor signaling pathways in vascular development. Oncogene 18(55):7917-7932.

Tamayo, P., D. Slonim, J. Mesirov, Q. Zhu, S. Kitareewan, E. Dmitrovsky , E.S. Lander, and T.R. Golub. 1999. Interpreting patterns of gene expression with self-organizing maps: Methods and application to hematopoietic differentiation. Proc. Natl. Acad. Sci. U.S.A. 96(6):2907-2912.

Tanabe, Y., and T.M. Jessell. 1996. Diversity and pattern in the developing spinal cord. Science 274(5290):1115-1123.

Taubes, G. 1995. Epidemiology faces its limits. Science 269:164-169.

Terry, K.K., B.A. Elswick, D.B. Stedman, and F. Welsch. 1994. Developmental phase alters dosimetry-teratogenicity relationship for 2-methoxyethanol in CD-1 mice. Teratology 49:218-227.

Threadgill, D.W., A.A. Dlugosz, L.A. Hansen, T. Tennenbaum, U. Lichti, D. Yee, C. LaMantia, T. Mourton, K. Herrup, R.C. Harris, et al. 1995. Targeted disruption of mouse EGF receptor: effect of genetic background on mutant phenotype. Science 269(5221):230-234.

Tillner, J., T. Winckler, and T. Dingermann. 1996. Developmentally regulated promoters from Dictyostelium discoideum as molecular markers for testing potential teratogens. Pharmazie. 51(11):902-906.

Tint, G.S., M. Irons, E.R. Elias, A.K. Batta, R. Frieden, T.S. Chen, and G. Salen. 1994. Defective cholesterol biosynthesis associated with the Smith-Lemli-Opitz syndrome. N. Engl. J. Med. 330(2):107-113.

Toraason, M., J.S. Bohrman, E. Krieg, R.D. Combes, S.E. Willington, W. Zajac, and R. Langenbach. 1992. Evaluation of the V79 cell metabolic cooperation assay as a screen in vitro for developmental toxicants. Toxicol. in Vitro 6(2):165-174.

Treinen, K.A., C. Louden, M.J. Dennis, and P.J. Wier. 1999. Developmental toxicity and toxicokinetics of two endothelin receptor antagonists in rats and rabbits. Teratology 59(1):51-59.

Trofimova-Griffin, M.E., M.R. Brzezinski, and M.R. Juchau. 2000. Patterns of CYP26 expression in human prenatal cephalic and hepatic tissues indicate an important role during early brain development. Brain Res. Dev. Brain Res. 120(1):7-16.

Tucker, A.S., and P.T. Sharpe. 1999. Molecular genetics of tooth morphogenesis and patterning: The right shape in the right place. J. Dent. Res. 78(4):826-834.

Turner, M., P.J. Mee, A.E. Walters, M.E. Quinn, A.L. Mellor, R. Zamoyska, and V.L. Tybulewicz. 1997. A requirement for the Rho-family GTP exchange factor Vav in positive and negative selection of thymocytes. Immunity 7(4):451-460.

Tzimas, G., R. Thiel, I. Chahoud, and H. Nau. 1997. The area under the concentration-time curve of all-trans-retinoic acid is the most suitable pharmacokinetic correlate to the embryotoxicity of this retinoid in the rat. Toxicol. Appl. Pharmacol. 143:436-444.

Umanoff, H., W. Edelmann, A. Pellicer, and R. Kucherlapati. 1995. The murine N-ras gene is not essential for growth and development. Proc. Natl. Acad. Sci. U.S.A. 92(5):1709-1713.

Uphill, P.F., S.R. Wilkins, and J.A. Allen. 1990. In vitro Micromass Teratogen Test - results from a blind trial of 25 compounds. Toxicol. in Vitro 4(4-5):623-626.

Vaglia, J.L., and B.K. Hall. 1999. Regulation of neural crest cell populations: Occurrence, distribution and underlying mechanisms. Int. J. Dev. Biol. 43(2):95-110.

Van den Berg, M., L. Birnbaum, A.T.C. Bosveld, B. Brunstrom, P. Cook, M. Feeley, J.P. Giesy, A. Hanberg, R. Hasegawa, S.W. Kennedy, T. Kubiak, J.C. Larsen, F.X. van Leeuwen, A.K. Liem, C. Nolt, R.E. Peterson, L. Poellinger, S. Safe, D. Schrenk, D. Tillitt, M. Tysklind, M. Younes, F. Waern, and T. Zacharewski. 1998. Toxic equivalency factors (TEFs) for PCBs, PCDDs, PCDFs for humans and wildlife. Environ. Health Perspect. 106(12):775-792.

van Dyke, D.C., S.E. Hodge, F. Heide, and L.R. Hill. 1988. Family studies in fetal phenytoin exposure. J. Pediatr. 113(2):301-306.

Vanden Bossche, H., G. Willemsens, and P.A. Janssen. 1988. Cytochrome-P-450-dependent metabolism of retinoic acid in rat skinmicrosomes: Inhibition by ketoconazole. Skin Pharmacol. 1(3):176-185.

Vasicek, T.J., L. Zeng, X.J. Guan, T. Zhang, F. Costantini, and S.M. Tilghman. 1997. Two dominant mutations in the mouse fused gene are the result of transposon insertions. Genetics 147(2): 777-786.

Velie, E.M., and G.M. Shaw. 1996. Impact of prenatal diagnosis and elective termination on prevalence and risk estimates of neural tube defects in California, 1989-1991. Am. J. Epidemiol. 144(5):473-479.

Ventura, S.J., J.A. Martin, S.C. Curtin, and T.J. Matthews. 1997. Report of Final Natality Statistics. Hyattsville, MD: U.S. Department of Health and Human Services, CDC, National Center for Health Statistics, 45(11) Suppl. 2. PHS 97-1120.

Veraksa, A., M. Del Campo, and W. McGinnis. 2000. Developmental patterning genes and their conserved functions: From model organisms to humans. Mol. Genet. Metab. 69(2):85-100.

Vogel, D.G., P.S. Rabinovitch, and N.K. Mottet. 1986. Methylmercury effects on cell cycle kinetics. Cell Tissue Kinet. 19(2):227-242.

Vogt, T.F., and D. Duboule. 1999. Antagonists go out on a limb. Cell 99(6):563-566.

Waitzman, N.J., P.S. Romano, and R.M. Scheffler. 1994. Estimates of the economic costs of birth defects. Inquiry 31(2):188-205.

Wakayama, T., and R. Yanagimachi. 1999a. Cloning of male mice from adult tail-tip cells. Nat. Genet. 22(2):127-128.

Wakayama, T., and R. Yanagimachi. 1999b. Cloning the laboratory mouse. Semin. Cell Dev. Biol. 10(3):253-258.

Wakayama, T., A.C. Perry, M. Zuccotti, K.R. Johnson, and R. Yanagimachi. 1998. Full-term development of mice from enucleated oocytes injected with cumulus cell nuclei. Nature 394(6691): 369-374.

Wakayama, T., I. Rodriguez, A.C. Perry, R. Yanagimachi, and P. Mombaerts. 1999. Mice cloned from embryonic stem cells. Proc. Natl. Acad. Sci. U.S.A. 96(26):14984-14989.

Walker, B.E., and M.I. Haven. 1997. Intensity of multigenerational carcinogenesis from diethylstilbestrol in mice. Carcinogenesis 18(4):791-793.

Walmod, P.S., A. Foley, A. Berezin, U. Ellerbeck, H. Nau, E. Bock, and V. Berezin. 1998. Cell motility is inhibited by the antiepileptic compound, valproic acid and its teratogenic analogues. Cell Motil. Cytoskeleton 40(3):220-237.

Walmod, P.S., G. Skladchikova, A. Kawa, V. Berezin, and E. Bock. 1999. Antiepileptic teratogen valproic acid (VPA) modulates organisation and dynamics of the actin cytoskeleton. Cell Motil. Cytoskeleton. 42(3):241-255.

Walton, B.T. 1983. Use of the cricket embryo (Acheta domesticus) as an invertebrate teratology model. Fundam. Appl. Toxicol. 3(4):233-236.

Wang, Y.L., and N.A. Brown. 1994. Expression patterns of Hox-B1, -B2 and Krox 20 in triazole-treated rat embryos suggest rhombomere reductions and loss. Teratology 49:380.

Wang, Z.Q., M.R. Fung, D.P. Barlow, and E.F. Wagner. 1994. Regulation of embryonic growth and lysosomal targeting by the imprinted Igf2/Mpr gene. Nature 372(6505):464-467.

Wang, Z.Q., C. Ovitt, A.E. Grigoriadis, U. Mohle-Steinlein, U. Ruther, and E.F. Wagner. 1992. Bone and haematopoietic defects in mice lacking c-fos. Nature 360(6406):741-745.

Warburton, D., and M.K. Lee. 1999. Current concepts on lung development. Curr. Opin. Pediatr. 11(3):188-192.

Warkany, J. 1978. Teratogen update: Aminopterin and methotrexate. Folic acid deficiency. Teratology 17(Jan.):353-358.

Wassef, M., and A.L. Joyner. 1997. Early mesencephalon/metencephalon patterning and development of the cerebellum. Perspect. Dev. Neurobiol. 5(1):3-16.

Wasteneys, G.O., M. Cadrin, K.R. Reuhl, and D.L. Brown. 1988. The effects of methylmercury on the cytoskeleton of murine embryonal carcinoma cells. Cell Biol. Toxicol. 4(1):41-60.

Webster, W.S., P.D. Brown-Woodman, and H.E. Ritchie. 1997. A review of the contribution of whole embryo culture to the determination of hazard and risk in teratogenicity testing. Int. J. Dev. Biol. 41(2):329-335.

Webster, W.S., P.D. Brown-Woodman, M.D. Snow, and B.R. Danielsson. 1996. Teratogenic potential of almokalant, dofetilide, and d-sotalol: Drugs with potassium channel blocking activity. Teratology 53(3):168-175.

Weiner, A.J., M.P. Scott, and T.C. Kaufman. 1984. A molecular analysis of fushi tarazu, a gene in Drosophila melanogaster that encodes a product affecting embryonic segment number and cell fate. Cell 37(3):843-851.

Weinstein, D.C., and A. Hemmati-Brivanlou. 1999. Neural induction. Annu. Rev. Cell. Dev. Biol. 15:411-433.

Weinstein, M., X. Xu, K. Ohyama, and C.X. Deng. 1998. FGFR-3 and FGFR-4 function cooperatively to direct alveogenesis in the murine lung. Development 125(18):3615-3623.

Weiss, S.R., C.E. Cooke, L.R. Bradley, and J.M. Manson. 1999. Pharmacist's guide to pregnancy registry studies. J. Am. Pharm. Assoc. 39(6):830-834.

Welch, R.M., Y.E. Harrison, B.W. Gommi, P.J. Poppers, M. Finster, and A.H. Conney. 1969. Stimulatory effect of cigarette smoking on the hydroxylation of 3,4-benzo(a)pyrene and the N-demethylation of 3-methyl-4-monomethylamino-azobenzene by enzymes in human placenta. Clin. Pharm. Therap. 10(1):100-110.

Welch, W.J. 1991. The role of heat-shock proteins as molecular chaperones. Curr. Opin. Cell Biol. 3(6):1033-1038.

Weller, E., N. Long, A. Smith, P. Williams, S. Ravi, J. Gill, R. Henessey, W. Skornik, J. Brain, C. Kimmel, G. Kimmel, L. Holmes, and L. Ryan. 1999. Dose-rate effects of ethylene oxide exposure on developmental toxicity. Toxicol. Sci. 50(2):259-270.

Wellfelt, K., A.C. Skold, A. Wallin, and B.R. Danielsson. 1999. Teratogenicity of the class III antiarrhythmic drug almokalant. Role of hypoxia and reactive oxygen species. Reprod. Toxicol. 13(2):93-101.

Wells, P.G., and L.M. Winn. 1996. Biochemical toxicology of chemical teratogenesis. Crit. Rev. Biochem. Mol. Biol. 31(1):1-40.

Wells, P.G., P.M. Kim, C.J. Nicol, T. Parman, and L.M. Winn. 1997. Reactive intermediates. Pp. 451-516 in Drug Toxicity in Embryonic Development, Vol 1., Handbook of Experimental Pharmacology, Vol. 124/I, R.J. Kavlock, and G.P. Daston, eds. Heidelberg: Springer.

Wells, P.G, J.T. Zubovits, S.T. Wong, L.M. Molinari, and S. Ali. 1989. Modulation of phenytoin teratogenicity and embryonic covalent binding by acetylsalicylic acid, caffeic acid, and alpha-phenyl-N-t-butylnitrone: Implications for bioactivation by prostaglandin synthetase. Toxicol. Appl. Pharmacol. 97(2):192-202.

Welsch, F. 1992. In vitro approaches to the elucidation of mechanisms of chemical teratogenesis. [Review]. Teratology 46(1):3-14.

Welsch, F., G.M. Blumenthal, and R.B. Conolly. 1995. Physiologically based pharmacokinetic models applicable to organogenesis: Extrapolation between species and potential use in prenatal toxicity risk assessments. Toxicol. Lett. 82/83:539-547.

Welsch, F., K.K. Terry, D.B. Stedman, and B.A. Elswick. 1996. Linking embryo dosimetry and teratogenic responses to 2-methoxyethanol at different stages of gestation in mice. Occupat. Hyg. 2:121-130.

Wilcox, A.J., D.D. Baird, and C.R. Weinberg. 1999. Time of implantation of the conceptus and loss of pregnancy. N. Engl. J. Med. 340(23):1796-1799.

Willhite, C.C., P.J. Wier, and D.L. Berry. 1989. Dose response and structure-activity considerations in retinoid-induced dysmorphogenesis. Crit. Rev. Toxicol. 20(2):113-135.

Wilmut, I., A.E. Schnieke, J. McWhir, A.J. Kind, and K.H. Campbell. 1997. Viable offspring derived from fetal and adult mammalian cells. Nature 385(6619):810-813.

Wilson, C.L., and S. Safe. 1998. Mechanisms of ligand-induced aryl hydrocarbon receptor-mediated biochemical and toxic responses. Toxicol. Pathol. 26(5):657-671.

Wilson, J.G. 1973. Environment and Birth Defects. New York: Academic Press. 305 pp.

Wilson, J.G., and F.C. Fraser. 1977. Handbook of Teratology. New York: Plenum Press.

Wilson, K.L., L.M. Cornel, and P.E. Mirkes. 1999. Teratogen-induced activation of mitogen-activated protein kinases (MARKs) in postimplantation embryos. Teratology 59:385-386.

Winter, R.M. 1996. Analysing human developmental abnormalities. Bioessays 18(12):965-971.

Wise, L.D., R.L. Clark, J.O. Rundell, and R.T. Robertson. 1990. Examination of a rodent limb bud micromass assay as a prescreen for developmental toxicity. Teratology 41(3):341-351.

Wlodarczyk, B.C., J.C. Craig, G.D. Bennett, J.A. Calvin, and R.H. Finnell. 1996. Valproic acid-induced changes in gene expression during neurulation in a mouse model. Teratology 54(6):284-297.

Wolpert, L. 1969. Positional information and the spatial pattern of cellular differentiation. J. Theor. Biol. 25(1):1-47.

Wolpert, L., R. Beddington, J. Brockes, T. Jessell, P. Lawrence, and E. Mayerowitz. 1998. P. 484 in Principles of Development. New York: Current Biology.

Wood, W.B. 1998. Cell lineages in Caenorhabditis elegans development. Pp. 77-95 in Cell Lineage and Fate Determination, S.A. Moody, ed. San Diego, CA: Academic Press.

Wood, W.B. 1999. Caenorhabditis. Pp. 324 in Encyclopedia of Molecular Biology, T.E. Creighton, ed. New York: Wiley.

Wood, W. B. et al., eds. 1988. The Nematode Caenorhabditis elegans. Cold Spring Harbor, NY: Cold Spring Harbor Laboratory Press. 667 pp.

Wubah, J.A., M.M. Ibrahim, X. Gao, D. Nguyen, M.M. Pisano, and T.B. Knudsen. 1996. Teratogen-induced eye defects mediated by p53-dependent apoptosis. Curr. Biol. 6(1):60-69.

Wubah, J.A., C.A. Kimmel, G.L. Kimmel, R.J. Kavlock, C. Lau, R.W. Setzer, and T.B. Knudsen. 1997. Pax6 expression, a biomarker of teratogen-induced abnormalities of the eye. Teratology 55(1):33.

Xu, T., and S. Artavanis-Tsakonas. 1990. Deltex, a locus interacting with the neurogenic genes, Notch, Delta and mastermind in Drosophila melanogaster. Genetics 126:665-677.

Xu, T., and G.M. Rubin. 1993. Analysis of genetic mosaics in developing and adult Drosophila tissues. Development 117:1223-1237.

Yacobi, A., J.P. Skelly, V.P. Shah, and L.Z. Benet, eds. 1993. Integration of Pharmacokinetics, Pharmacodynamics and Toxicokinetics in Rational Drug Development. New York: Plenum Press. 270 pp.

Yamaguchi, T.P., K. Harpal, M. Henkemeyer, and J. Rossant. 1994. fgfr-1 is required for embryonic growth and mesodermal patterning during mouse gastrulation. Genes Dev. 8(24):3032-3044.

Yang, D.D., C.Y. Kuan, A.J. Whitemarsh, M. Rincon, T.S. Zheng, R.J. Davis, P. Rakic, and R.A. Flavell. 1997b. Absence of excitotoxicity-induced apoptosis in the hippocampus of mice lacking the Jnk3 gene. Nature 389(6653):865-870.

Yang, D., C. Tournier, M. Wysk, H.T. Lu, J. Xu, R.J. Davis, and R.A. Flavell. 1997a. Targeted disruption of the MKK4 gene causes embryonic death, inhibition of c-Jun NH2-terminal kinase activation, and defects in AP-1 transcriptional activity. Proc. Natl. Acad. Sci. U.S.A. 94(7):3004-3009.

Yates, J.R. III. 1998. Mass spectrometry and the age of the proteome. J. Mass Spectrom. 33:1-19.

Zakeri, Z., D. Quaglino, and H.S. Ahuja. 1994. Apoptotic cell death in the mouse limb and its suppression in the hammertoe mutant. Dev. Biol. 165(1):294-297.

Zeng, L., F. Fagotto, T. Zhang, W. Hsu, T.J. Vasicek, W.L. Perry, III, J.J. Lee, S.M. Tilghman, B.M. Gumbiner, and F. Costantini. 1997. The mouse fused locus encodes Axin, an inhibitor of the Wnt signaling pathway that regulates embryonic axis formation. Cell 90(1):181-192.

Zhao, F., K. Mayura, R.W. Hutchinson, R.P. Lewis, R.C. Burghardt, and T.D. Phillips. 1995. Developmental toxicity and structure-activity relationships of chlorophenols using human embryonic palatal mesenchymal cells. Toxicol. Lett. 78(1):35-42.

Zinaman, M.J., E.D. Clegg, C.C. Brown, J. O'Connor, and S.G. Selevan. 1996. Estimates of human fertility and pregnancy loss. Fertil. Steril. 65(3):503-509.

Zuber, M.X., E.R. Simpson, and M.R. Waterman. 1986. Expression of bovine 17 alpha-hydroxylase cytochrome P-450 cDNA in nonsteroidogenic (COS 1) cells. Science 234(4781):1258-1261.

Appendixes

Appendix A

Glossary

Acceptable daily intake (ADI)	An estimate of the daily exposure dose that is likely to be without deleterious effect even if continued exposure occurs over a lifetime.
A/D ratio	The ratio of the adult toxic dose to the developmentally toxic dose.
Allele	A gene's representation on one chromosome.
Benchmark dose (BMD)	An alternative approach to the NOAEL approach that uses all experimental data to fit one or more dose-response curves.
Bioinformatics	The study of how to most efficiently manage and utilize vast amounts of genomic sequence data.
Biological markers (biomarkers)	Indicators signaling events in biological systems or samples (NRC 1989). There are three classes of biomarkers—exposure, effect, and susceptibility. A marker of exposure is an exogenous substance or its metabolite(s) or the product of an interaction between a xenobiotic agent and some target molecule or cell that is measured in a compartment within an organism. A marker of effect is a measurable biochemical, physiological, or other alteration within an organism that, depending on magnitude, can be recognized as an established or potential health impairment or disease. A marker of susceptibility is an indicator of an

	inherent or acquired limitation of an organism's ability to respond to the challenge of exposure to a specific xenobiotic substance.
Checkpoint pathway	Type of pathway used in intracellular signaling. Checkpoint pathways are induced in response to a cell's own internal imbalance or errors in its synthetic activities. Induction of checkpoint pathways leads to a delay in certain synthetic processes until other processes are complete, thereby averting damage.
Chimera	An animal consisting of genetically different cells derived from two (or more) different zygotes.
Chromosomes	Structures that contain an organism's genes.
Cleavage	The first few divisions of an embryo following fertilization. There is little or no growth during these divisions and the cytoplasm is cleaved into smaller and smaller units.
Complementary DNA (cDNA)	DNA copy of mRNA from expressed genes made by the enzyme reverse transcriptase.
Cre/loxP	Bacterial recombinase system in which the Cre protein mediates DNA recombination between specific DNA sequences known as lox-P sites. This system is used in mammalian cells to delete (or invert) a stretch of DNA by flanking it with lox-P sites and then exposing the cell to Cre protein at some predetermined time.
Deoxyribonucleic acid (DNA)	A complex macromolecule that is composed of nucleic acids (adenine, guanine, cytosine, and thymine) and is found in cellular organisms. DNA carries all the genetic information necessary to determine the specific properties of an organism. In its native state, DNA exists as a double helix.
Developmental defect	A structural or functional anomaly that results from an alteration in normal development.
Developmental toxicant	A physical, chemical, or biological agent that is shown to affect development under specific conditions of exposure.

Developmental toxicology	The study of adverse effects on the developing organism that might result from exposure prior to conception (of either parent), during prenatal development, or from postnatal development to the time of sexual maturation.
Developmental biology	The study of the biology of normal development.
Developmental susceptibility gene	Any gene that encodes a gene product with which an environmental agent can interfere and cause a perturbation of normal development.
Drug metabolizing enzyme (DME)	Enzymes that metabolize endogenous substrates and also foreign chemicals. Among individuals, there are heritable differences in the catalytic activity of DMEs and in the ability to produce high levels of DMEs.
Ecogenetics	The study of the inherited basis of individuals' different responses to environmental agents.
Ectopic expression	Expression of a gene in a tissue in which it is not normally expressed.
Electroporation	A means of introducing molecules into cells by transiently permeabilizing their membranes with a brief electric shock.
Embryonic stem (ES) cells	Permanent *in vitro* stem-cell lines derived from the undifferentiated cells of a very early, preimplantation mammalian embryo. They have the potential to contribute to all cells of a developing embryo when they are placed into an early embryonic environment.
Environmental factor	Physical, chemical, and biological agents or conditions encountered by humans, such as infections, nutritional deficiencies and excesses, life-style factors (e.g., alcohol), hyperthermia, ultraviolet radiation, X-rays, and the myriad of manufactured chemicals (e.g., pharmaceuticals, synthetic chemicals, solvents, pesticides, fungicides, herbicides, cosmetics, and food additives) and natural materials (e.g., plant and animal toxins and products).
Epistasis	A situation in which the phenotypic expression of one gene obscures the phenotypic effects of another gene.
Eukaryotic	Cells in which the genetic material is separated from the cytoplasm by a nuclear membrane.

Exon	Nucleotide segment that codes for amino acids.
Expressed sequence tag (EST)	A DNA sequence complementary to a known piece of transcribed RNA.
Forward genetics	An approach to identify a particular gene starting with some knowledge about a mutant phenotype.
Functional genomics	The systematic and comprehensive analysis of gene products, including mRNAs and proteins.
Gene	The fundamental units of heredity carried by DNA.
Gene product	The mRNA resulting from transcription of a gene and the protein translated from the mRNA.
Gene targeting	The production of a mutation in a specific "target" gene using molecular biological techniques.
Genomics	The study of the genetic composition of organisms.
Genotype	An organism's genetic makeup (i.e., its DNA sequence).
Homeobox	A region of the homeotic genes that is highly conserved among several species, including *Drosophila*, frogs, and mammals. The homeobox sequence encodes a DNA-binding motif of the encoded protein that is a transcription factor.
Homologous recombination	Crossing over between two identical or homologous strands of DNA.
Human Genome Project (HGP)	A federally funded initiative to sequence and identify all human genes. HGP also involves studying the genome of a number of organisms other than humans, including insects, fish, plants, and other mammals.
Inbred strain	A strain of mice that has been inbred by brother-sister matings for more than 20 generations, and consequently all the individuals of the strain are more than 98% genetically identical.
Intron	DNA sequence between exons that is transcribed but not translated
Knockout mutation	A mutation that leads to the loss of function of a particular gene. Also called a null mutation.

Linkage disequilibrium	The condition in which two genes are close enough together on a chromosome that it is not likely that they will be separated by recombination during meiosis.
Linkage analysis	A means of statistically correlating phenotype with genotype using lod scores.
Lod score	The ratio of the likelihood of two or more loci remaining together when chromosomes recombine (true linkage) to the likelihood of chance alone.
Lowest-observed-adverse-effect level (LOAEL)	The lowest exposure level at which there are statistically or biologically significant increases in the frequency or severity of adverse effects in the exposed population and its appropriate control.
Margin of exposure (MOE)	The ratio of the NOAEL to the estimated exposure dose.
Maximum tolerated dose (MTD)	The maximum dose that an animal species can tolerate for a major portion of its lifetime without significant impairment or toxic effect other than carcinogenicity.
Mechanism of action	The detailed molecular knowledge of the events that result in an adverse developmental response in an organism. This knowledge includes the following types of mechanistic information: (1) The toxicant's kinetics and means of absorption, distribution, metabolism, and excretion within the mother and conceptus; (2) Its interaction (or those of a metabolite derived from it) with specific molecular components of cellular or developmental processes in the conceptus or with maternal or extraembryonic components of processes supporting development; (3) The consequences of the interactions on the function of the components in a cellular or developmental process; (4) The consequences of the altered process on a developmental outcome, namely, the generation of a defect. See also 'mode of action.'
Mendelian trait	Phenotype that shows simple pattern of inheritance. Mendelian traits are usually governed by a single genetic locus.
Mendelian gene	A gene located in a chromosome that obeys the laws of Mendelian inheritance.

Messenger RNA (mRNA)	RNA that is derived from transcription of expressed genes.
Microsatellites	A subclass of minisatellites in which the repeat unit consists of two base pairs (dinucleotide repeats).
Minisatellites	A class of restriction fragment length polymorphisms in which the restriction fragment length is caused by a variable number of tandem repeats (VNTRs).
Mode of action	The cascade of events that occurs during the development of a major adverse event, following maternal exposure to a toxicant. See also 'mechanism of action.' The description of mode of action contains less molecular information on components, interactions, and processes than does the mechanism of action of a toxicant.
Molecular epidemiology	The study of correlating genotype with phenotype in human population studies or family studies.
Multifactorial inheritance	The interaction between allelic variants of genes and environmental conditions in the production of disease.
Mutation	One or more altered bases in a nucleic acid sequence.
No-observed-adverse-effect level (NOAEL)	An exposure level at which there are no statistically or biologically significant increases in the frequency or severity of adverse effects between the exposed population and its appropriate control. Some effects may be observed at this level, but they are not considered adverse, nor precursors to adverse effects.
Open reading frame (ORF)	A sequence that codes for amino acids without any termination codons and is potentially translatable into protein.
Organogenesis	The organ-forming phase of embryonic development.
Phage display	A method that enables the presentation of large peptide and protein libraries on the surface of phage particles from which molecules of desired functional property(ies) can be rapidly selected.
Pharmacogenetics	The study of the heredity basis of differences in response among individuals to pharmacologic drugs.
Phenotype	A biological trait (e.g., eye color).

Pleiotropic	A gene having multiple phenotypic traits
Polymorphism	The presence of two or more alleles of a particular gene within a population of organisms.
Prokaryotic	Cells in which the nuclear region is not separated from the cytoplasm by a membrane.
Pronucleus	A haploid nucleus formed by either the sperm head or the egg nucleus after fertilization but before the nuclei fuse to form the diploid zygotic nucleus.
Proteomics	Analysis of an organism's protein composition and function.
Reference dose (RfD)	An estimate (with uncertainty spanning perhaps an order of magnitude) of a daily exposure to the human population (including sensitive subgroups) that is likely to be without an appreciable risk of deleterious effects during a lifetime.
Reference concentration (RfC)	An estimate (with uncertainty spanning perhaps an order of magnitude) of a continuous inhalation exposure to the human population (including sensitive subgroups) that is likely to be without an appreciable risk of deleterious noncancer effects during a lifetime.
Relative risk	The ratio of incidence or risk among exposed individuals to incidence or risk among nonexposed individuals.
Restriction fragment length polymorphism (RFLP)	Variations in DNA sequence that are indicated by the presence or absence of particular restriction sites in the DNA.
Restriction sites	Specific, short DNA sequences that are recognized by restriction enzymes that cleave the duplex DNA.
Reverse genetics	An approach to identify the function of a particular gene starting with a cloned copy of that gene.
Ribonucleic acid (RNA)	A complex macromolecule composed of adenine, cytosine, guanine, and uracil that is found in cellular organisms. RNA can serve several functions, including encoding the genetic information copied from DNA in the form of a sequence of bases that specifies a sequence of amino acids.

Risk assessment	The evaluation of scientific information on the hazardous properties of environmental agents and on the extent of human exposure to those agents. The product of the evaluation is a statement regarding the probability that populations (expressed qualitatively or quantitatively) so exposed will be harmed and to what degree.
Risk characterization	The integration of both qualitative and quantitative information from hazard identification and assessment of exposure and dose-response relationships. It is usually the last step in the risk assessment process and it includes an evaluation of uncertainty and variability in the assessment that would significantly influence the analysis.
RNAi	Inactivation of a gene by the introduction into cells of double-stranded mRNA for that gene.
Saturation mutagenesis	Mutagenesis screen in which enough mutants are produced to obtain all the different kinds of mutants affecting a particular process.
Selector genes	Genes that encode transcription factors. Transcription factors "select" which other genes will be expressed.
Sensitized strain	A model organism that contains a mutated pathway (e.g., signal transduction pathway), but the mutation causes no visible phenotype. The mutated pathway is close to a threshold of function and sensitive to an otherwise asymptomatic change in activity of a second element in the pathway.
Signal transduction pathway	Type of pathway used for cell-to-cell (intercellular) signaling. Signal transduction pathways can have from 1 to 10 intermediates. These pathways are used extensively during development.
Single nucleotide repeat (SNP)	A variation between individuals in one base pair at a specific location of a DNA sequence.
Spemann organizer	A group of cells that releases inducer proteins that are important in the placement, orientation, and scaling of later development by surrounding cells.

Stress pathway	Type of pathway used in intracellular signaling. Stress pathways are induced when a disruption of normal cell function and development occurs due to physical or chemical agents of the environment. Induction of stress pathways leads to cellular repair and counteraction.
Toxicity	The process by which a toxicant enters an organism, how it reacts with target molecules, how it exerts its deleterious effects, and how the organism reacts to the insult.
Toxicodynamics	The study of how a toxic chemical (or metabolites derived from it) interacts with specific molecular components of cellular and, in the context of this report, developmental processes in the body.
Toxicokinetics	The study of how a toxic chemical is absorbed, distributed, metabolized, and excreted into and from the body.
Transcription factor	A DNA binding protein that is involved in the formation of mRNA.
Transgenic	Any animal whose genome has been altered by addition of genetic material or by alteration of existing genes by gene targeting.
Uncertainty factor (UF)	One of several factors used in calculating an exposure level that will not cause toxicity from experimental data. UFs are used to account for the variation in susceptibility among humans, the uncertainty in extrapolating from experimental animal data to humans, the uncertainty in extrapolating from data from studies in which agents are given for less than a lifetime, and the uncertainty in using LOAEL data instead of NOAEL data.
Wild type	The normal or usual allele or phenotype. In cases in which multiple allelic variants have a normal phenotype, the wild-type allele is usually considered the most common one.
Zygote	An embryo newly formed as a result of fertilization.

Appendix B

Database Descriptions

Examples of Genome, Protein, Genetic Variant, and Toxicology Databases, as of the Beginning of 2000:

Protein, Nucleotide, 3D Structures, Genomes, Taxonomy, and PubMed Literature. The Entrez Browser at National Center for Biotechnology Information of the National Library of Medicine and National Institutes of Health
http://www.ncbi.nlm.nih.gov/Entrez/

Established in 1988 as a national resource for molecular biology information, NCBI creates public databases, conducts research in computational biology, develops software tools for analyzing genome data, and disseminates biomedical information - all for the better understanding of molecular processes affecting humans and disease. Linked databases for protein sequences, nucleotide sequences, 3D structures of proteins and nucleic acids, genomes, taxonomy, and the PubMed literature.

Biochemical Nomenclature
http://alpha.qmw.ac.uk/~ugca000/iupac/jcbn

The Biochemical Nomenclature Committee Web site has links to the International Union of Biochemistry and Molecular Biology and the International Union of Pure and Applied Chemistry.

Enzyme Nomenclature
http://www.expasy.ch/enzyme/

The Enzyme Nomenclature Database. This site is a repository of information relative to the nomenclature of enzymes. It is primarily based on the recommendations of the Nomenclature Committee of the International Union of Biochemistry and Molecular Biology (IUBMB) and it describes each type of characterized enzyme for which an EC (Enzyme Commission) number has been provided.

Protein Sequences
http://www.expasy.ch/sprot/sprot-top.html
Home of the SWISS-PROT Annotated Protein Sequence Database. SWISS-PROT is a curated protein sequence database that strives to provide a high level of annotations (such as the description of the function of a protein, its domains structure, post-translational modifications, variants, etc), a minimal level of redundancy, and a high level of integration with other databases.

Mendelian Inheritance in Man
http://www.ncbi.nlm.nih.gov/omim/
This database, developed by the National Center for Biotechnology Information (NCBI), is a catalogue of human genes and genetic disorders authored and edited by Dr. Victor A. McKusick and his colleagues at Johns Hopkins and elsewhere. It contains textual information, pictures, and reference information as well as links to NCBI's Entrez database of MEDLINE articles and sequence information.

Nomenclature for Human Gene Mutations
http://www.interscience.wiley.com/jpages/1059-7794/nomenclature.html
This Internet site contains recommendations for a Nomenclature System for Human Gene Mutations.

Human DNA Polymorphisms
http://research.marshfieldclinic.org/genetics
This site contains a significant amount of information on human DNA polymorphisms and their analysis.

Phosphodiesterase (PDE) Gene Family
http://depts.washington.edu/pde/
Recent phosphodiesterase (PDE) gene family nomenclature recommendations. Includes updates for nearly all of the gene families and many of the subfamilies.

Human Gene Nomenclature
http://www.gene.ucl.ac.uk/nomenclature/
Web site for the Human Gene Nomenclature Committee. It includes a nomenclature database and guidelines. In addition, the site contains information on gene families and access to other relevant links.

The Mouse Genome

http://www.informatics.jax.org/support/nomen

The Mouse Genome Database (MGD) contains information on mouse genetic markers, molecular segments, phenotypes, comparative mapping data, experimental mapping data, and graphical displays for genetic, physical, and cytogenetic maps. MGD is updated daily.

Mouse Knockout Mutants

http://www.bioscience.org/knockout/knochome.htm

The Knockout Mouse Database at this Web address presents information on the phenotypes rendered by the knockout of various molecules. Gene knockouts are classified according to the viability of the mice: (1) gene knockouts that are compatible with viability; (2) gene knockouts that result in prenatal mortality; (3) gene knockouts that result in postnatal mortality; and (4) gene knockouts that result in perinatal mortality.

Drosophila Genome

http://flybase.bio.indiana.edu/

FlyBase is a comprehensive database of information on the genetics and molecular biology of *Drosophila.* It includes data from the Drosophila Genome Projects and data curated from the literature. FlyBase is a joint project with the Berkeley and European Drosophila Genome Projects.

Rat Genetics and Genes

http://ratmap.gen.gu.se/ratmap/wwwnomen/nomen.html

The International Rat Genetic Committee (RGNC) is dedicated to developing an internationally accepted standard genetic nomenclature for rats and to bring this nomenclature to the attention of scientists working in the field of rat genetics. On its web site there is information on rat gene symbols, DNA symbols, chromosome nomenclature, and a brief summary of rat locus symbol nomenclature rules.

Chicken Genetics and Genes

http://www.ri.bbsrc.ac.uk/chickmap/nomenclature.html

This Web site is the home of Chickmap, a chicken gene mapping project, maintained by the Roslin Institute. It includes information on nomenclature for naming loci, alleles, linkage groups, and chromosomes to be used in poultry genome publications and databases.

Zebrafish Development

http://zfishstix.cs.uoregon.edu/

Home of Fish Net, a gateway to Zebrafish Research Databases, provided by the Institute of Nueroscience at the University of Oregon. Its links include information on: embryonic and larval anatomy, genomics, genetic staging, molecular probes, and ZFIN, the on-line database of zebrafish information.

Zebrafish Gene Nomenclature
http://zfish.uoregon.edu/zf_info/zfbook/chapt7/7.1.html
An exerpt from Chapter 7 (Genetic Methods) of the Zebrafish Book. It address conventions for naming zebrafish genes, including genes identified by mutation as well as the use of abbreviated names and alleles. It also discusses priority in naming.

Nematode (C. elegans) Genes, Transcripts, and Proteins
http://www.sanger.ac.uk/Projects/C_elegans/blast_servers.html
Home of the Sanger Centre's C. elegans BLAST server. This site allows the searching of a DNA database containing sequence data from both the Cambridge and St. Louis sequencing groups. You can also search the current database of C. elegans ESTs and the C. elegans protein database wormpep.

American Type Culture Collection (ATCC)
http://www.atcc.org/
ATCC is a global nonprofit bioscience organization that provides biological products, technical services, and educational programs to private industry, government, and academic organizations around the world. The mission of the ATCC is to acquire, authenticate, preserve, develop, and distribute biological materials, information, technology, intellectual property, and standards for the advancement, validation, and application of scientific knowledge.

Plant Genome Nomenclature
http://jiio6.jic.bbsrc.ac.uk/
The Mendel Bioinformatics Group at the John Innes Center in the UK maintains this database of plant genome nomenclature.

*Yeast (*S. cerevisiae*) Genome, Genes, and Proteins*
http://genome-www.stanford.edu/saccharomyces/
Home of the Saccharomyces Genome Database (SGD), covering the molecular biology and genetics of the yeast *Saccharomyces cerevisiae*, commonly known as baker's or budding yeast.

Human Cytochrome P450 Polymorphisms
http://www.imm.ki.se/CYPalleles/
Recommended nomenclature for the polymorphisms of human cytochrome P450 enzymes.

Signal Transduction
http://www.stke.org/
Sponsored by the journal *Science* and Stanford University, this site provides information on all signaling pathways, the variety of ligands, the protein intermediates, the variety of protein kinases, and the cross-talk of pathways.

Proteomics
http://expasy.nhri.org.tw/ch2d/
The Swiss database for identity of proteins by 2D polyacrylamide gel electrophoresis, as well as techniques and links.

Mouse Transgenic and Targeted Mutation Database
http://tbase.jax.org
The TBASE database attempts to organize information on transgenic animals and targeted mutations generated and analyzed worldwide.

Expressed Sequence Tag Web Site
http://www.ncbi.nlm.nih.gov/dbEST/index.html
The expressed sequence tag database contains sequence data and other information on expressed sequence tags from a number of organisms.

TOXNET (Toxicology Data Network)
http://sis.nlm.nih.gov/sis1/
TOXNET is a collection of databases on toxicology, hazardous chemicals, and related areas. The databases include TOXLINE, which contains citations from 1965 to the present on the pharmacological, biochemical, physiological, and toxicological effects of drugs and other chemicals; the Developmental and Reproductive Toxicology (DART) Database; and the Hazardous Substance Data Bank (HSDB).

Cancer Gene Anatomy Project
http://www.ncbi.nlm.nih.gov/ncicgap/
The Cancer Gene Anatomy Project (CGAP), administered by the National Cancer Institute, was established to generate information and technical tools needed to study the molecular anatomy of the cancer cell. Much of the information generated is presented in several databases maintained by the National Center for Biotechnology Information. These databases can be accessed through the CGAP Web site.

Environmental Genome Project
http://www.niehs.nih.gov/envgenom/
The goal of the Environmental Genome Project (EGP), administered by the National Institute of Environmental Health Sciences, is to understand the im-

pact and interaction of environmental exposures on human disease. Specifically, polymorphisms of environmental disease susceptibility genes are being identified and a central database of polymorphisms for these genes is being developed. This database, developed in conjunction with the University of Utah Genome Center, integrates gene sequence and polymorphism data; see www.genome.utah.edu/genesnps/. The EGP Web site also offers links to databases maintained by other organizations, such as Gene Map 99 (a map of more than 30,000 human genes) and the Oak Ridge National Laboratory Human Chromosome Map.

Appendix C

Signaling Pathways

Seventeen intercellular signal transduction pathways are currently known with regard to the identity of their ligands, transduction intermediates, kinases, and targets. These are illustrated in Panels 1-17. Several were discovered in the course of the analysis of developmental mutants of *Drosophila*. Since no new mutant has been discovered recently in the continuing analysis of mutants, it is expected that most are already known. Pathways 1-6 are used extensively in early development, for example, axis specification and germ layer formation, as well as later. Pathways 7-10 are used in later development, including organogenesis and tissue renewal. Pathways 11-17 are used extensively in the physiological function of differentiated cells of the fetus, juvenile, and adult.

Approximately twelve intracellular pathways of checkpoint controls and molecular stress responses are currently known. Three are illustrated in Panels 18 and 19, and several are listed in Table 6-7. Panel 18 shows the ER-Golgi unfolded protein response, a molecular stress response of the cell to various chemical (e.g., ethanol, dithiothreitol) or physical (e.g. heat) conditions that lead to loss of function by protein unfolding ("denaturation"). The response leads to increased chaperone protein levels in the ER and Golgi, which increase refolding and restore function. The pathway also acts as a G1/S checkpoint control when naturally unstable proteins are produced. Panel 19 shows the p53 related stress response to DNA damage (genotoxic stress) leading to G1/S or G2/M arrest until repair is completed (upper half of panel). The pathway also acts as a checkpoint control monitoring DNA synthesis, imposing G2/M arrest until it is finished, via phosphorylation of the cdc25 phosphatase (lower half of panel).

1A. Wnt pathway via β–catenin

Nematodes, Drosophila, vertebrates
Large family of ligands (Wnt 1-11) in vertebrates. Wg in Drosophila.
Several receptors of the Frizzled family and several GSK3 kinases in vertebrates.
Signaling affects transcription, spindle orientation, cell polarity, and maybe cadherin-mediated adhesion and gap junctions.
The pathway is used extensively in early development.

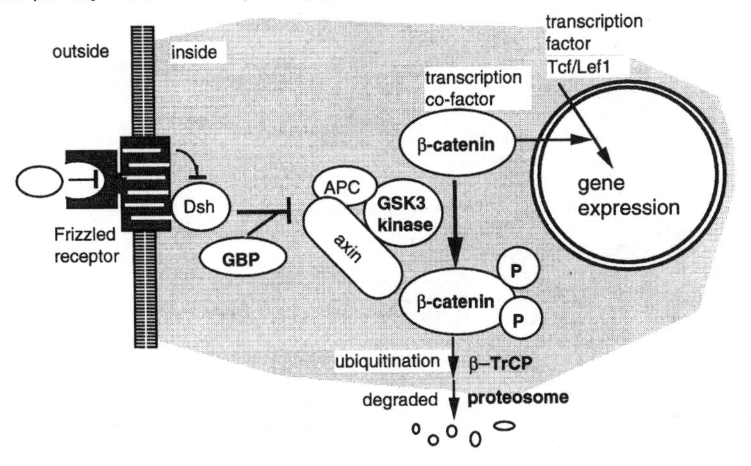

1B. Wnt pathway via JNK (planar cell polarity pathway)

Drosophila and vertebrates. Used in morphogenesis and establishment of cell polarity.

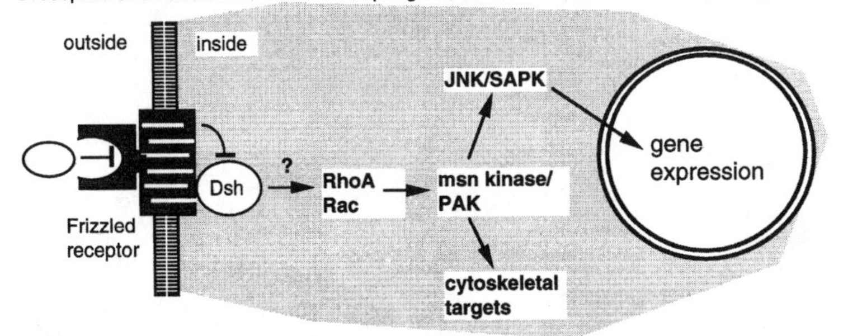

Abbreviations: APC=adenopolyposis coli protein; betaTrCP= beta-Transducin repeat containing protein, a ubiquitin ligase; Dsh=dishevelled; GBP=GSK3 binding protein; GSK3=glycogen synthase kinase 3; JNK=Jun N-terminal kinase; Lef1=lymphoid enhancer binding factor 1; msn=misshapened; PAK=p21 activated kinase; Rho, Rac=small GTP binding proteins; SAPK=serum activated protein kinase/S6 kinase; Tcf=T-cell transcription factor; Wnt=contraction of Wingless (Drosophila mutant) and Int (MMTV integration site in mouse).

2. Receptor serine/threonine kinase (TGF-β receptor) pathway

Nematodes, Drosophila, vertebrates. In Drosophila, there are seven ligands, all of the BMP family. In vertebrates >30 ligands: TGF-betas, BMPs, GDFs, activin, inhibin, nodals, MIS, dorsalin, "growth factors". Many receptors (>4), each containing a type I and type II chain. At least 9 Smads. Signaling affects transcription via Smad transcription factors. The pathway is used extensively in develoment.

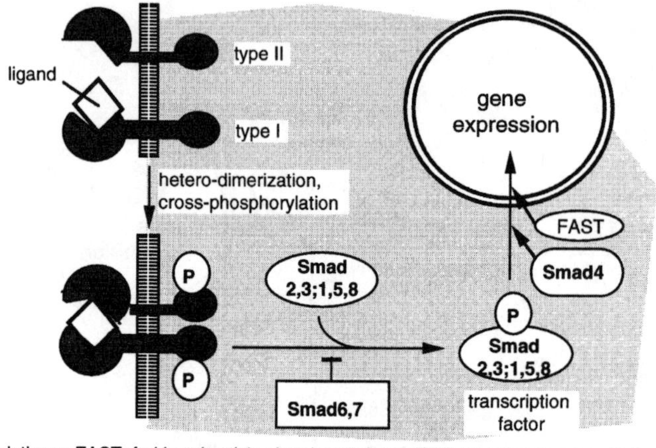

Abbreviations: FAST=forkhead activin signal transducer; TGF=transforming growth factor; Smad=transcription factors, "Sma" (a C. elegans gene) and "Mad" (a Drosophila gene), combined.

3. Hedgehog pathway (Patched receptor protein)

Drosophila, vertebrates, absent from nematodes. Several receptors and ligands in vertebrates (Sonic, Desert, Indian Hedgehog). Ligands require cholesterol addition and proteolytic cleavage for activity. Signaling affects mostly transcription via the Cubitus interuptus zinc-finger transcription factors (Gli family). When signaling occurs, the transcription factor is not cleaved and remains in its activating form. Without signaling, the factor is cleaved to a repressive form. The pathway is frequently used in early development.

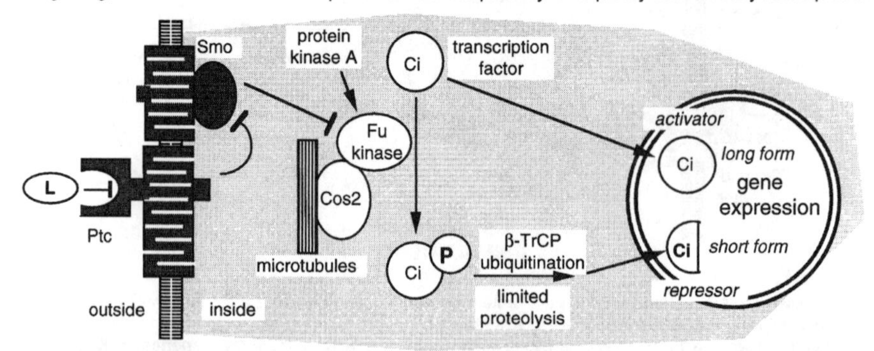

Abbreviations: beta-TrCP=beta-Transducin repeat containin protein; Ci=cubitus interruptus transcription factor in *Drosophila*; called Gli in vertebrates; Cos2=costal 2 kinesin like protein, with likely microtubule affinity; Fu=fused kinase; Ptc=patched receptor; Smo=smoothened receptor associated protein;

4. Receptor tyrosine kinase pathway (small G-protein [Ras] linked)

All metazoa
In vertebrates, many dimeric protein ligands: FGF 1-11; PDGF; EGF; VEGF; IGF; Ephrins (mitogens; "growth factors"). Many also in Drosophila, EGF-like, FGF-like, torso, Boss.
Many receptors in vertebrates: FGFR 1-4; PDGFR; EGFR; IGFR, Eph receptors; many small G proteins, and many (>40?) kinases.
Signaling affects transcription (c-Fos, Jun, Ets, AP-1 transcription factors), cell cycle proteins, cytoskeletal and motility proteins. Extensive cross-talk with other pathways. Mediates some stress responses (via p38K?). Often used in early development.

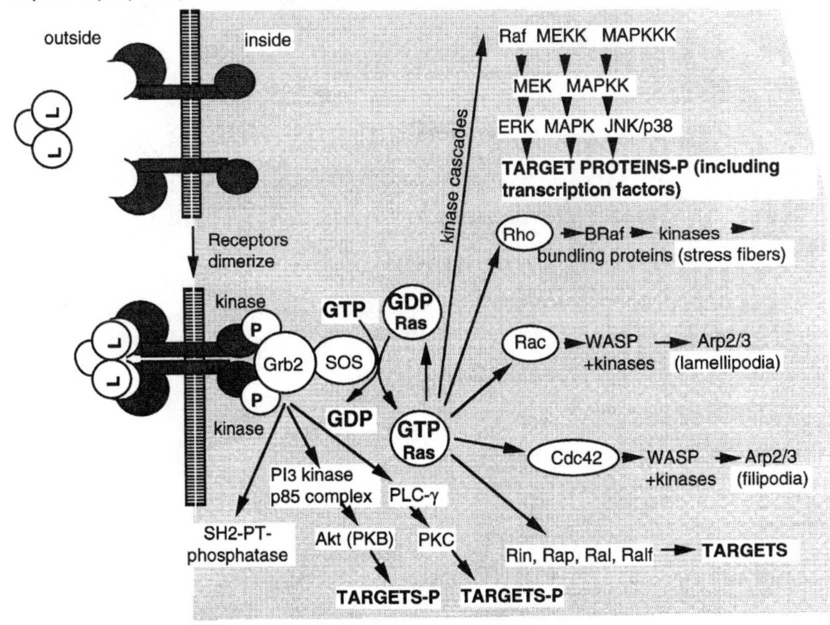

Abbreviations: ABP=actin binding proteins; Akt=a serine/threonine kinase, also called PKB; Arp2/3=actin related proteins 2 and 3; ERK=extracellular signal-regulated kinase; Grb2=growth factor receptor-bound protein 2, a docking protein; JNK=Jun N-terminal kinase; MAPK, MAPKK=mitogen activated kinase and the kinase kinase; p38=a kinase involved in stress pathways; PI3 kinase=a kinase adding phosphates to phosphoinositide triphosphate; PKB=protein kinase B; Rac/Ras/Rho=small G proteins; Raf=a serine/threonine kinase; SH2-PT-phosphatase=a phosphotyrosine phosphatase containing an SH2 domain; SOS=Son of Sevenless, a GTP/GDP exchange factor protein; WASP=Wiskott-Aldrich syndrome protein.

5. Notch-Delta pathway

Nematodes, Drosophila, vertebrates. Ligands, which include Delta, Serrate, LAG-2, and APX-1 family members, are transmembrane ligands (no diffusion from signaling cell). The receptors require two proteolytic steps for activity. A receptor is used only once. Signaling affects transcription. The pathway is used frequently in early development, for example, in lateral inhibition patterning.

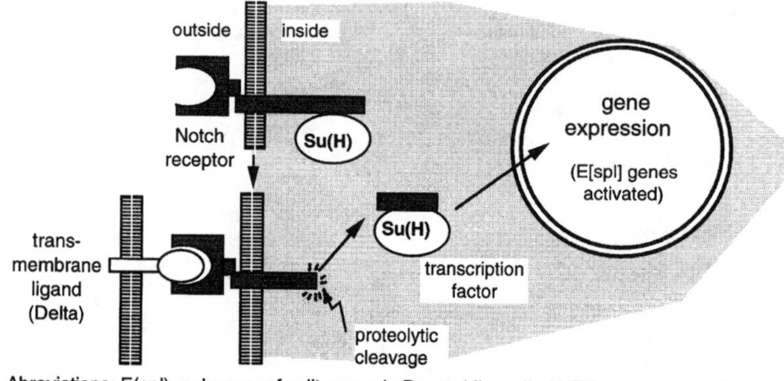

Abreviations: E(spl)=enhancer of split genes in Drosophila, called HER genes in vertebrates; Su(H)=suppressor of Hairless, a Drosophila transcription factor, called CSL in vertebrates.

6. Receptor-linked cytoplasmic tyrosine kinase (cytokine) pathway

Nematodes, Drosophila, and vertebrates. Some components are present in slime molds (non-metazoan eukaryotes). Many ligands in mammals: erythropoietin, GMCF, prolactin, thrombopoietin, growth hormone, LIF, CNTF, oncostatin M, cardiotrophin 1, IL-2, -3, -4, -6, -11, -12, -13, Interferons. At least three classes of receptors in mammals. Several kinases in mammals: e.g. JAK 1-3, Tyk2. Signaling affects transcription via activation of Stats 1-7.

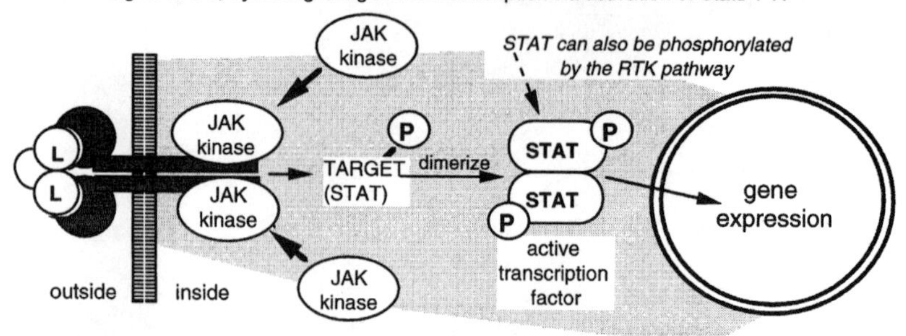

Abbreviations: JAK=Janus kinase, a class of tyrosine kinases; L=extracellular ligand; STAT=signal transducer and activator of transcription protein.

7. IL1/Toll receptor; NF-kappaB pathway

Drosophila, vertebrates, (absent in nematodes). The pathway is used in anti-bacterial innate immunity of most metazoa and plants. Receptor ligands in vertebrates include Interleukin-1 (of the IL-1 receptor) and in Drosophila include Spätzli (of the Toll receptor). Many other pathways can lead to NFκB activation as well, e.g., from TNF (see pathway 9), LPS, double stranded RNA, viruses, various stresses (via MAPKs, see pathway 4). Signaling effects are mostly on transcription via NFκB and rel family transcription factors. Used in Drosophila early development with Dorsal as the transcription factor and cactus as the IκB-like inhibitor.

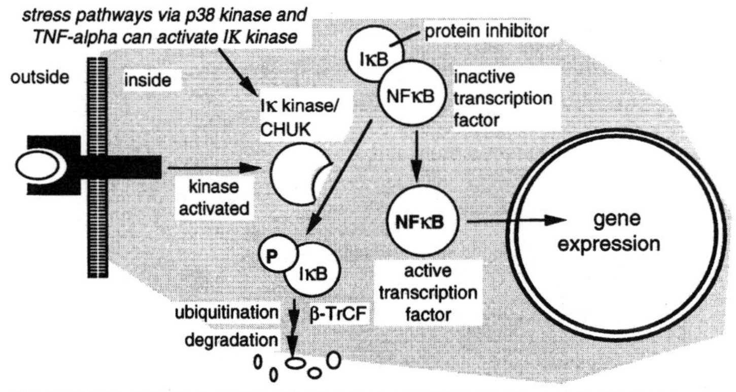

Abbreviations: β-TrCF=beta Transducin repeat containing protein, a ubiquitin ligase; IkB=inhibitor of NFkB; NFkB=nuclear factor kappaB.

8. Nuclear hormone receptor pathway

Drosophila and vertebrates have steroid and retinoic receptors. All metazoa probably have at least orphan receptors. Vertebrates have a great variety of ligands: steroid hormones (estrogen, glucocorticoids), retinoic acid, thyroxin, and some prostaglandins. All ligands pass the plasma membrane passively. Some receptors are cytoplasmic and some nuclear in the non-liganded state. None is a transmembrane protein. Receptors are zinc-finger DNA binding transcription factors (PPAR, steroid, thyroxin, vitamin D3, retinoic acid). Signaling affects transcription, both activation and repression.

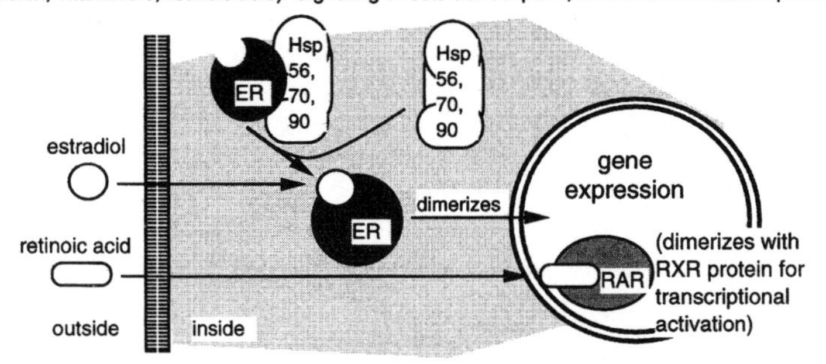

Abbreviations: ER=estrogen receptor; Hsp=heat shock protein, a protein folding chaperone; PPAR=peroxisome proliferator activated receptor; RAR=retinoic acid receptor; RXR=orphan receptor binding an unknown ligand or no ligand.

9. Apoptosis pathway (cell death pathway)

All metazoa. There are many paths of activation of the caspase proteolytic enzymes that ultimately impose cell death, e.g., fromextracellular signals such as TNF alpha or beta, and Fas ligand (from killer T cells), and intracellular signals such as DNA damage due to cellular stresses or checkpoint failures.

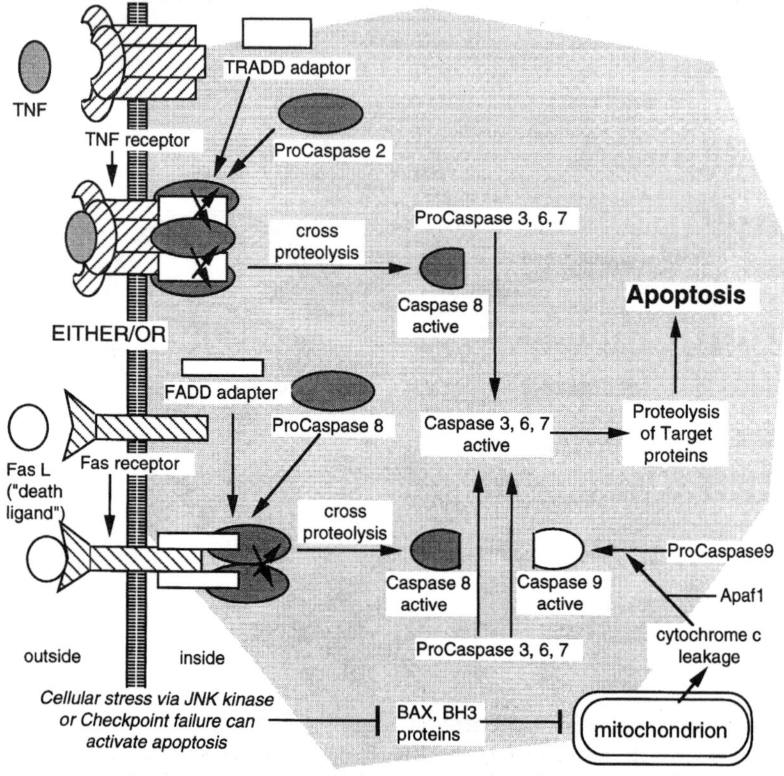

Abbreviations: Apaf=apoptosis activating factor; BAX= protein inhibitor of Bcl2 proteins; BH3 proteins=those with a domain for binding to Bcl2 proteins; Caspases=specific proteolytic enzymes; FADD=Fas receptor associated death domain protein ; FasL=FAS ligand; TNF=tumor necrosis factor; TRADD=TNF receptor associated death domain protein.

10. Receptor protein tyrosine phosphatase (RPTPs) pathway

Drosophila, vertebrates, nematodes. Ligands include ECM components (laminin and nidogen for the LAR receptor; chondroitin sulfate proteoglycan for RPTPzeta; ganglioside G(M3) for RPTPsigma). Vertebrates have many receptors (LAR, RPTPalpha, beta, mu/kappa, sigma, rho, zeta). These phosphatases may counteract other signaling pathways by dephosphorylating receptors and intermediates, such as p130Cas. The pathway is used in neuron pathfinding and in T-cell and B-cell receptor signaling (the CD45 protein). Stress agents may block the phosphatase, causing phosphorylated intermediates of other pathways to increase and signal.

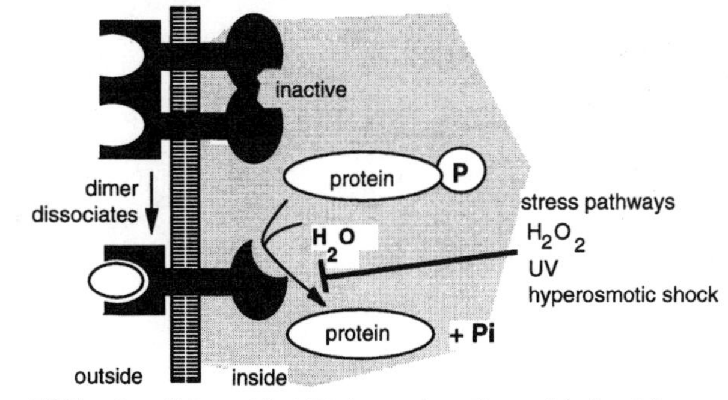

Abbreviations: ECM=extracellular matrix; LAR=leucocyte-antigen related protein; p130Cas=a docking protein of the Cas family.

11. Receptor guanylate cyclase pathway

Drosophila, vertebrates
In mammals, ligands include natriuretic peptides A,B,C and retinyl guanylate cyclase. In sea urchin sperm, there is a fertilization activated guanylate cyclase. Several cGMP-dependent kinases are present in vertebrates. Signaling affects transcription (via c-Fos, JunB, CREB, AP-1) and ion channels (retina).

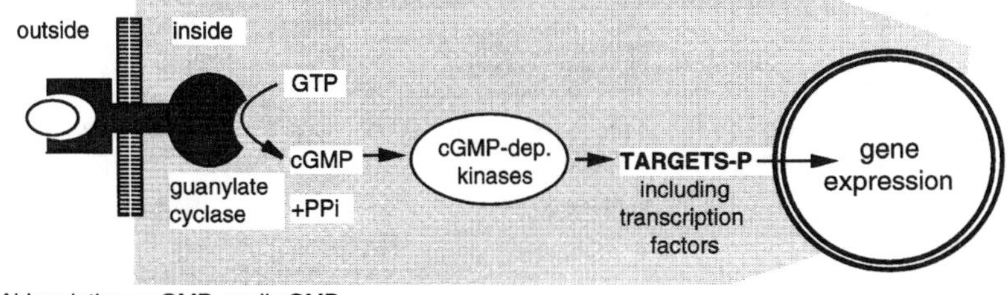

Abbreviations: cGMP=cyclic GMP.

12. Nitric oxide receptor pathway

Drosophila, vertebrates (nerve, smooth muscle, platelets), molluscs, probably not in nematodes (C. elegans).
The receptor is a cytoplasmic heme containing enzyme; NO binds at the heme group. There are several kinds of cGMP dependent kinases and many protein targets of the cGMP kinases. Signaling affects transcription (via c-Fos, CREB) and other cellular functions.

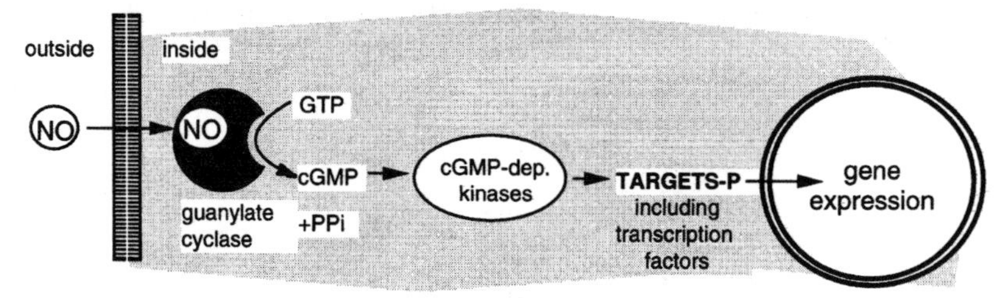

Abbreviations: NO=nitric oxide; cGMP=cyclic GMP..

13. G-protein coupled receptor (large G-protein) pathway

All metazoa. Also present in some single celled eukaryotes. In vertebrates, there are many ligands (proteins, peptides, small molecules) such as odorants, angiotensin, thrombin, epinephrin, glucagon, adenosine, seritonin, and various releasing factors. Many receptor proteins (1000 odorant receptors in vertebrates), rhodopsin (light), all of seven-pass transmembrane structure. Many G-proteins of the "large type", several adenyl cyclases, phospholipases, PKAs, and PKCs. Signaling affects transcription (e.g. via CREB), secretion, metabolism, motility, and other protein kinases (via MAPKKK).

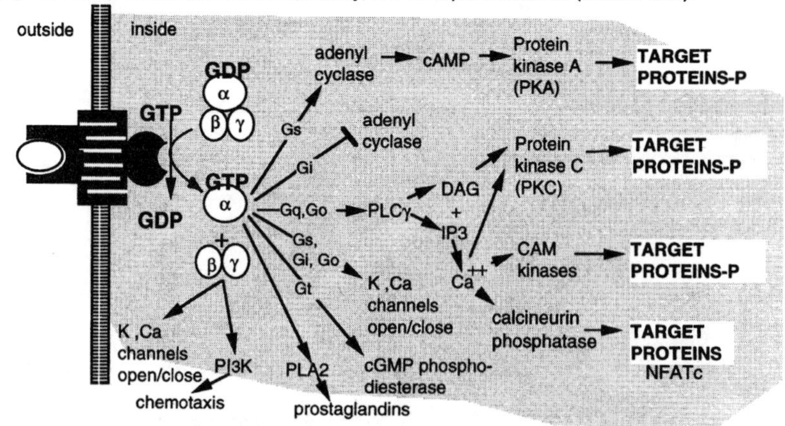

Abbreviations: α, β, γ=alpha, beta, gamma subunits of the large G-protein; DAG=diacyl glycerol; IP3= inositol triphosphate; NFAT=nuclear factor activating transcription; PI3K=3-phosphoinositde kinase; PLA2=phospholipase A2; PLCγ=phospholipase Cγ.

14. Integrin pathway

All metazoa. Ligands are mostly RGD-containing components of the extracellular matrix such as fibronectin, laminin, proteoglycans, and collagen. There are many integrin receptors in mammals consisting of combinations of different alpha (>8) and beta (>14) subunits.Several focal adhesion kinases (FAK) and Src family kinases are activated inside the cell.
Signaling affects motility, adhesion, and transcription.

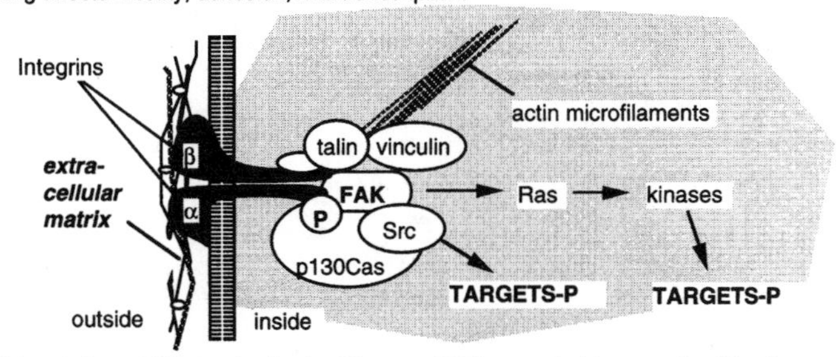

Abbreviations: FAK=focal adhesion kinase; p130Cas= a docking protein of the Cas family; Src=Sarcoma virus (Rous) kinase;

15. Cadherin pathway

All metazoa
In vertebrates, >10 kinds of cadherins (maybe many more) such as N, L, E. Cadherins bind calcium ion and mediate homophilic cell adhesion and microfilament assembly (in adherens junctions) using α- and β-catenin. Some transcription may be sensitive to the removal of beta-catenin by junction formation.

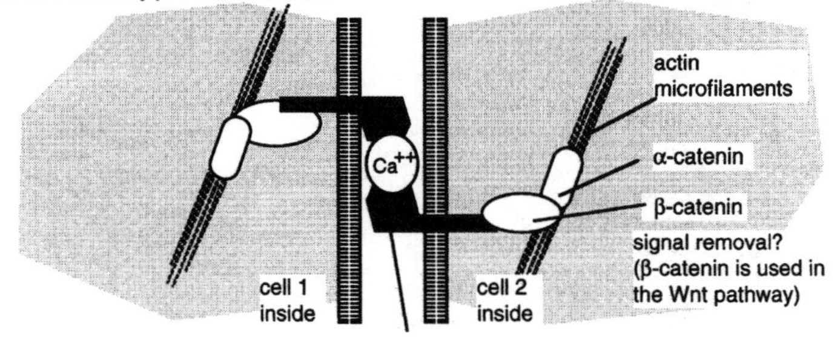

16. Gap junction pathway

Present in all metazoa; called gap junctions in vertebrates or septate junctions in invertebrates. In vertebrates, there are >8 kinds of junction proteins (connexins), some incompatable for a connection between two cells. Gap junctions have open and closed states. They allow ions, cAMP, and metabolites under 1200 daltons to pass. The membrane potential is equalized between cells when the junction is open. Some junctions are rectifying.

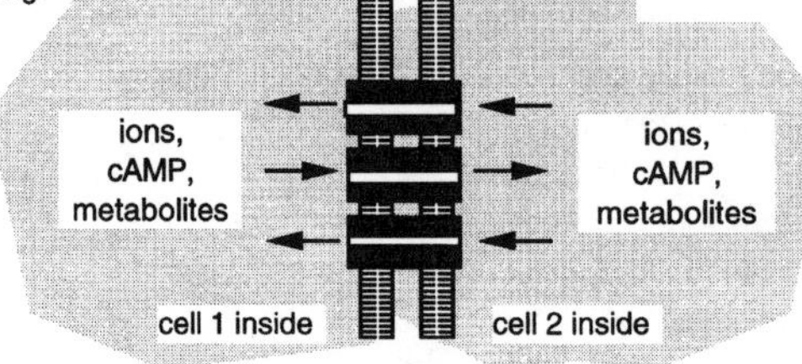

17. Ligand-gated cation channel pathway

All metazoa.

In vertebrates there are several receptors and several ligands, e.g. acetylcholine; glutamate, NMDA, GABA. Some channels are gated internally (e.g. by ATP, cGMP).

Signaling affects membrane potential leading to calcium dependent events (exocytosis, contraction).

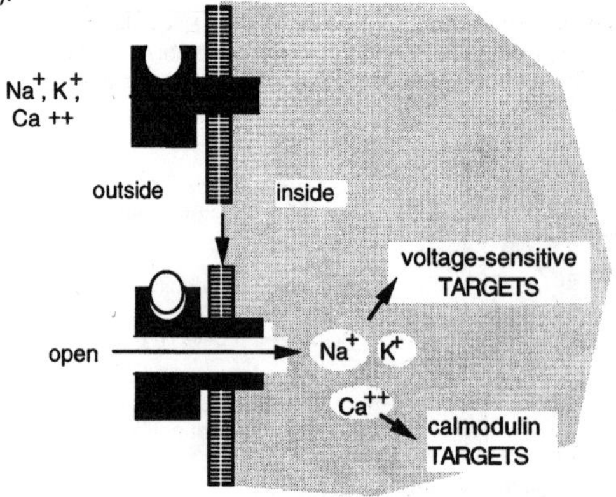

18. A stress response: The unfolded protein response (UPR)

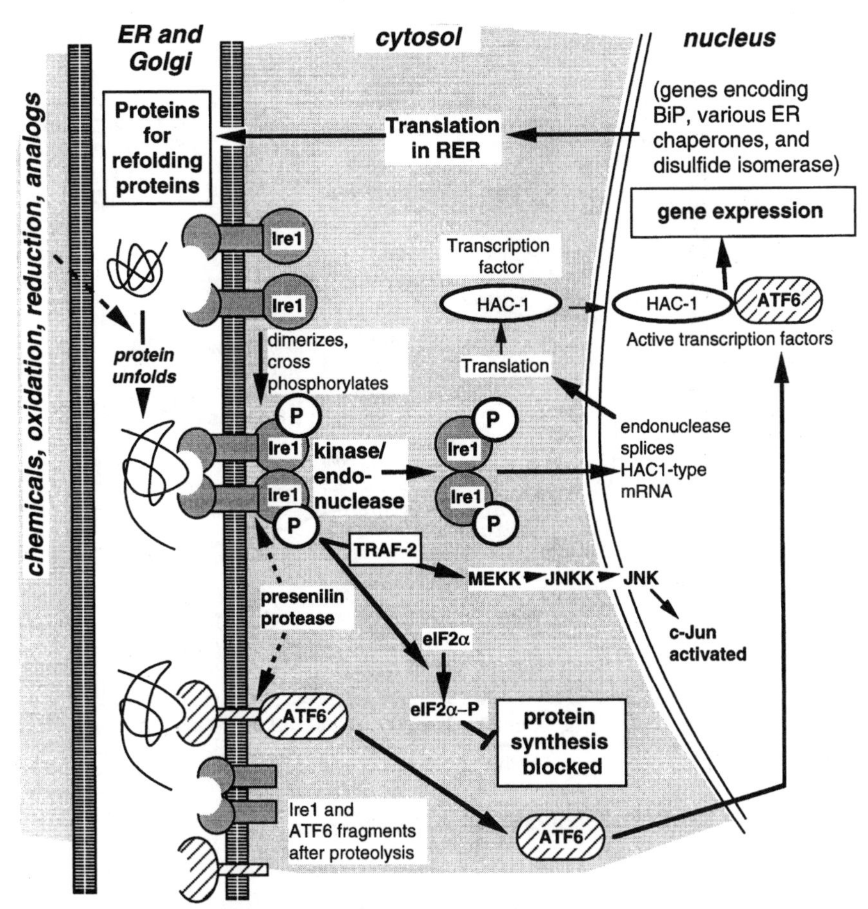

Abbreviations: ATF6=a basic leucine zipper transcription factor; BiP=a chaperone localized in the ER ; chaperone=a protein catalzying the folding of other proteins; eIF2a = a translation initiation factor; ER=endoplasmic reticulum; Ire1=a threonine-serine specific kinase and an endonuclease; HAC1=a yeast mRNA spliced by Ire1 nuclease, encoding a transcription factor; presenilin protease=a protease cleaving membrane associated proteins; RER=rough endoplasmic reticulum.

19. Stress responses and checkpoints for DNA damage and replication.

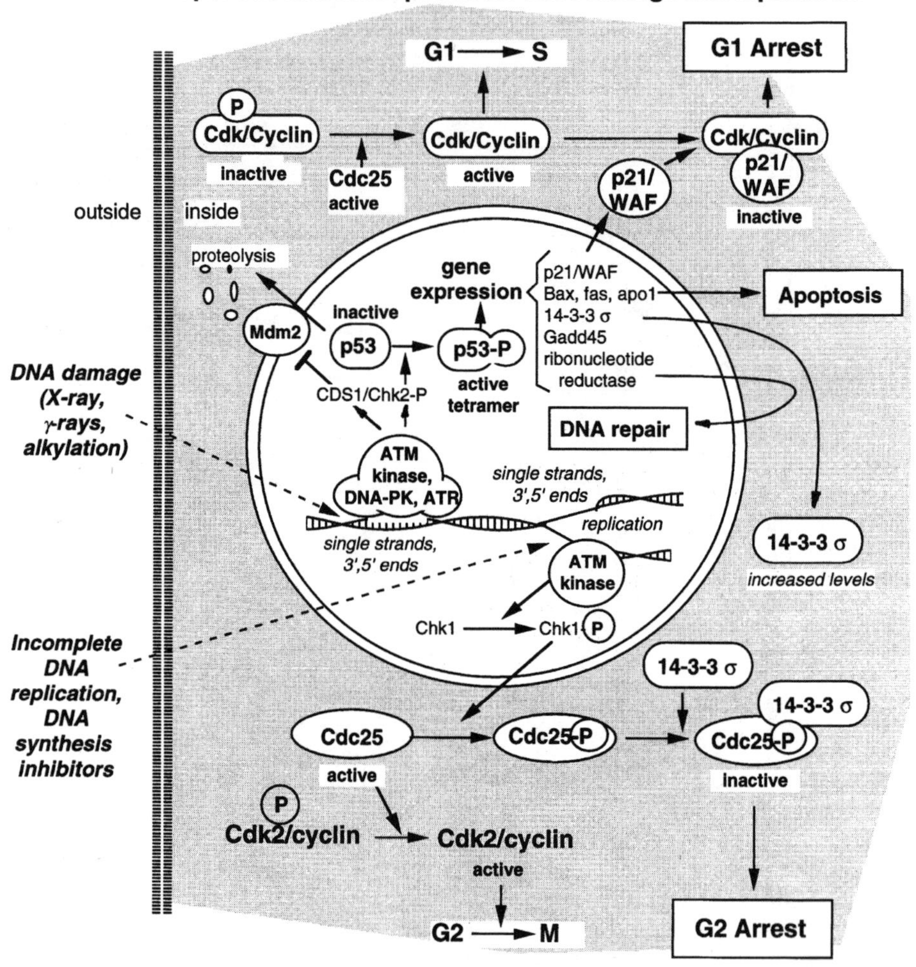

Abbreviations: 14-3-3=a sequestering protein; ATM=ataxia telangiectasia mutated kinase; ATR=ATM related kinase; Bax=protein activator of apoptosis; cdc25=a specific phosphatase; cdk=cyclin dependent kinase; chk1, chk2=checkpoint kinases 1 and 2; DNA-PK=DNA dependent protein kinase; p53=a 53kD transcription factor.

Appendix D

Biographical Information on the Committee on Developmental Toxicology

ELAINE FAUSTMAN (Chair), University of Washington, Seattle, WA. Dr. Faustman is a professor in the Department of Environmental Health and director of the Institute for Risk Analysis and Risk Communication at the University of Washington. She earned her Ph.D. in pharmacology and toxicology from Michigan State University. Her research interests include mechanistic investigations of the reproductive and developmental toxicity of metals and pesticides. She has developed quantitative risk assessment methods for non-cancer end points. Dr. Faustman previously served as a member of the National Research Council Committee on Toxicology. She is an elected fellow of the American Association for the Advancement of Science.

JOHN GERHART (Vice Chair), University of California, Berkeley, CA. Dr. Gerhart is a professor in the Department of Molecular and Cell Biology at the University of California, Berkeley. He earned his PhD in biochemistry from the same institution. His research interests include the development of *Xenopus* and the formation and function of Spemann's organizer. Dr. Gerhart is a member of the National Academy of Sciences.

NIGEL BROWN, St. George's Hospital Medical School, University of London, UK. Dr. Brown is professor of developmental biology in the Department of Anatomy and Developmental Biology at St. George's Hospital Medical School. He earned his Ph.D. in biochemistry from the University of Surrey, UK and did postdoctoral training at NIEHS. His research interests include mammalian heart development, mechanisms of teratogenesis, and left-right asymmetry. Dr. Brown is a former president of the European Teratology Society and chairman of the International Federation of Teratology Societies.

GEORGE DASTON, The Procter and Gamble Company, Cincinnati, OH. Dr. Daston is a toxicologist at Miami Valley Laboratories of the Procter and Gamble Company. He earned his Ph.D. in teratology from the University of Miami. His research interests include in vitro methodologies, teratogenic mechanisms, and risk assessment. Dr. Daston is a former member of the National Research Council Board on Environmental Studies and Toxicology.

MARK FISHMAN, Massachusetts General Hospital and Harvard Medical School, Boston, MA. Dr. Fishman is chief of cardiology, director of the Cardiovascular Research Center, and chief of the Developmental Biology Laboratory at the Massachusetts General Hospital. He also is a professor of medicine at Harvard Medical School. He earned his M.D. from Harvard Medical School. Dr. Fishman's research interests are in the development of the heart and other organ systems. He has helped pioneer the use of the zebrafish in large-scale genetic screens and has been instrumental in the discovery of many new genes critical to development of the early embryo.

JOSEPH HOLSON, WIL Research Laboratories, Inc., Ashland, OH. Dr. Holson is president and director of WIL Research Laboratories, Inc. He earned his Ph.D. in physiology from University of Cincinnati College of Medicine. Dr. Holson's scientific activities are primarily in product development and risk assessment where he specializes in developmental and reproductive toxicology.

HERMAN B.W.M. KOËTER, Organisation for Economic Cooperation and Development (OECD), Paris, France. Dr. Koëter is principal administrator of the OECD Environmental Health and Safety Division. He is responsible for and directs the OECD Test Guidelines Programme, the OECD Programme on Harmonization of Classification and Labeling, the OECD Special Activity on Endocrine Disrupters, and the OECD Special Activity on Animal Welfare Policies. He is also senior adviser for OECD on human health risk characterization and assessment issues. Dr. Koëter earned his masters degree in experimental pathology and his doctoral degree in biological toxicology from Utrecht State University, Utrecht, The Netherlands.

ANTHONY MAHOWALD, University of Chicago, Chicago, IL. Dr. Mahowald is Louis Block professor and chair of the Department of Molecular Genetics and Cell Biology and the Committee on Developmental Biology at the University of Chicago. He received his Ph.D. in biology from Johns Hopkins University. His research interests include the developmental genetics of oogenesis and germ-cell sex determination in *Drosophila*. Dr. Mahowald is a fellow of the American Academy of Arts and Sciences and a member of the National Academy of Sciences.

JEANNE MANSON, University of Pennsylvania, Philadelphia, PA. Dr. Manson is a fellow in the Center for Clinical Epidemiology and Biostatistics at the University of Pennsylvania, where she is obtaining her M.S. in clinical epidemiology. She earned her Ph.D. in developmental biology from Ohio State University. Dr. Manson previously worked in the pharmaceutical industry in the area of reproductive toxicology.

RICHARD MILLER, University of Rochester, Rochester, NY. Dr. Miller is professor and associate chair of obstetrics and gynecology, and professor of environmental medicine at the University of Rochester School of Medicine and Dentistry. He is also director of the Division of Research and Director of the PEDECS, which is a regional and national Teratogen Information Service. He earned his Ph.D. in pharmacology and toxicology from Dartmouth Medical School. Dr. Miller's research interests include the vertical transmission of the HIV-1 and the role of anti-HIV therapy, the toxicity of heavy metals, and the role of vitamins in normal and abnormal development.

PHILIP MIRKES, University of Washington, Seattle, WA. Dr. Mirkes is research professor in the Department of Pediatrics at the University of Washington. He earned his Ph.D. in zoology from the University of Michigan. Dr. Mirkes' research interests include teratology and developmental toxicology, heat shock response, and apoptosis.

DANIEL NEBERT, University of Cincinnati Medical Center, Cincinnati, OH. Dr. Nebert is a professor in the Department of Environmental Health at the University of Cincinnati Medical Center and in the Department of Pediatrics, Division of Human Genetics at Children's Hospital Medical Center. He earned his M.S. in biochemistry and M.D. from the University of Oregon Medical School. He is author or coauthor of more than 460 publications in the fields of pharmacogenetic disorders, toxicology, gene expression and signal transduction pathways, gene nomenclature and evolution, teratology, and developmental biology.

DREW NODEN, Cornell University, Ithaca, NY. Dr. Noden is professor of embryology in the Department of Biomedical Sciences of the College of Veterinary Medicine at Cornell University. He earned his Ph.D. in zoology from Washington University. Dr. Noden's research focuses on vertebrate craniofacial development, with particular emphasis on the migratory patterns of mesenchymal cells and the factors that influence their assembly into muscle, endothelial, and connective tissues.

VIRGINIA PAPAIOANNOU, Columbia University, NY. Dr. Papaioannou is professor of genetics and development at the College of Physicians and Surgeons of Columbia University. She earned her Ph.D. in genetics from the University of Cambridge, England. Her research interests include implantation, embryonic cell lineages and the genetics of early embryonic development, using mutations and transgenic technology.

GARY SCHOENWOLF, University of Utah, Salt Lake City, UT. Dr. Schoenwolf is professor of neurobiology and anatomy, and member of the Huntsman Cancer Institute at the University of Utah School of Medicine. He earned his M.S. and Ph.D. in zoology from the University of Illinois, Champaign-Urbana. His research interests include gastrulation, neurulation, and neuraxial patterning.

FRANK WELSCH, Chemical Industry Institute of Toxicology (CIIT), Research Triangle Park, NC. Dr. Welsch is senior scientist and head of the Teratology Laboratory at CIIT. He earned his qualification as a veterinarian and doctor medicinae veterinariae degree from the Veterinary School of Freie Universität, Berlin, Germany. His research interests include development of new testing methods to predict developmental toxicity hazards.

WILLIAM B. WOOD, University of Colorado, Boulder, CO. Dr. Wood is a professor in the Department of Molecular, Cellular and Developmental Biology at the University of Colorado, Boulder, and a member of the Cancer Institute, University of Colorado Health Sciences Center, Denver. He earned his Ph.D. in biochemistry from Stanford University. He studies the developmental genetics and molecular biology of embryonic pattern formation and sex determination in the nematode *Caenorhabditis elegans*. Dr. Wood is a fellow of the American Academy of Arts and Sciences and a member of the National Academy of Sciences.

Index

F

G

H

I

K

L

M

N

O

Q

R

T

U

V

W

X

Y

Z